SOLID STATE PHYSICS

VOLUME 59

Founding Editors

FREDERICK SEITZ
DAVID TURNBULL

SOLID STATE PHYSICS

Advances in
Research and Applications

Editors

HENRY EHRENREICH

FRANS SPAEPEN

Division of Engineering and Applied Sciences
Harvard University
Cambridge, Massachusetts

VOLUME 59

ELSEVIER
ACADEMIC
PRESS

Amsterdam Boston London New York Oxford Paris
San Diego San Francisco Singapore Sydney Tokyo

ELSEVIER B.V.	ELSEVIER Inc.	ELSEVIER Ltd	ELSEVIER Ltd
Sara Burgerhartstraat 25	525 B Street, Suite 1900	The Boulevard, Langford Lane	84 Theobalds Road
P.O. Box 211, 1000 AE Amsterdam	San Diego, CA 92101-4495	Kidlington, Oxford OX5 1GB London	WC1X 8RR
The Netherlands	USA	UK	UK

© 2004 Elsevier Inc. All rights reserved.

This work is protected under copyright by Elsevier Inc., and the following terms and conditions apply to its use:

Photocopying: Single photocopies of single chapters may be made for personal use as allowed by national copyright laws. Permission of the Publisher and payment of a fee is required for all other photocopying, including multiple or systematic copying, copying for advertising or promotional purposes, resale, and all forms of document delivery. Special rates are available for educational institutions that wish to make photocopies for non-profit educational classroom use.

Permissions may be sought directly from Elsevier's Rights Department in Oxford, UK: phone (+44) 1865 843830, fax (+44) 1865 853333, e-mail: permissions@elsevier.com. Requests may also be completed on-line via the Elsevier homepage (http://www.elsevier.com/locate/permissions).

In the USA, users may clear permissions and make payments through the Copyright Clearance Center, Inc., 222 Rosewood Drive, Danvers, MA 01923, USA; phone: (+1) (978) 7508400, fax: (+1) (978) 7504744, and in the UK through the Copyright Licensing Agency Rapid Clearance Service (CLARCS), 90 Tottenham Court Road, London W1P 0LP, UK; phone: (+44) 20 7631 5555; fax: (+44) 20 7631 5500. Other countries may have a local reprographic rights agency for payments.

Derivative Works: Tables of contents may be reproduced for internal circulation, but permission of the Publisher is required for external resale or distribution of such material. Permission of the Publisher is required for all other derivative works, including compilations and translations.

Electronic Storage or Usage: Permission of the Publisher is required to store or use electronically any material contained in this work, including any chapter or part of a chapter.

Except as outlined above, no part of this work may be reproduced, stored in a retrieval system or transmitted in any form or by any means, electronic, mechanical, photocopying, recording or otherwise, without prior written permission of the Publisher. Address permissions requests to: Elsevier's Rights Department, at the fax and e-mail addresses noted above.

Notice: No responsibility is assumed by the Publisher for any injury and/or damage to persons or property as a matter of products liability, negligence or otherwise, or from any use or operation of any methods, products, instructions or ideas contained in the material herein. Because of rapid advances in the medical sciences, in particular, independent verification of diagnoses and drug dosages should be made.

First edition 2004

ISBN: 0-12-607759-2
ISSN: 0081-1947

⊚ The paper used in this publication meets the requirements of ANSI/NISO Z39.48-1992 (Permanence of Paper). Printed in USA.

Contents

CONTRIBUTORS vii
PREFACE ix

The Thermodynamics of Elastically Stressed Crystals

P. W. VOORHEES AND WILLIAM C. JOHNSON

I.	Introduction	2
II.	Hydrostatically Stressed Crystals	12
III.	Deformation and Stress	34
IV.	Thermodynamics of a Single-Phase System	61
V.	Capillary and Interfacial Properties	94
VI.	Crystal-Fluid Equilibrium	109
VII.	Two-Phase Crystalline Systems	131
VIII.	Acknowledgements	198
IX.	Appendix A: Surface Area Change Owing to Accretion	199
X.	Appendix B: Continuity Condition at Two-Phase Crystalline Interface	200
XI.	Appendix C: Continuity Condition at Crystal-Fluid Interfaces	201

Solid Solutions of Hydrogen in Complex Materials

REINER KIRCHHEIM

I.	Introduction	203
II.	Fundamental Properties of Hydrogen in Metals	207
III.	Behavior of Hydrogen in Defective and Disordered Metals	214
IV.	Interaction of Hydrogen with Defects	228
V.	Hydrogen in Disordered and Amorphous Alloys	252
VI.	Other Interstitials in Amorphous Materials	262
VII.	Hydrogen in Systems with Reduced Dimensions	278

AUTHOR INDEX 293
SUBJECT INDEX 299

Contributors to Volume 59

Numbers in parentheses indicate the pages on which the authors' contributions begin.

WILLIAM C. JOHNSON (1) *Department of Materials Science and Engineering, University of Virginia, Charlottesville, VA 22904*

REINER KIRCHHEIM (?) *Institut für Materialphysik, Georg August Universität Göttigen Tammannstr, 1, D-37077, Göttingen, Germany*

P. W. VOORHEES (1) *Department of Materials Science and Engineering, Northwestern University, Evanston, IL 60208*

Preface

This volume contains two articles on topics in materials science of great importance: the thermodynamics of stressed solids, a fundamental problem that goes back to Gibbs, and hydrogen in materials, an area that is both scientifically rich and of great current technological importance.

The article on Thermodynamics of Elastically Stresses Solids by Peter Voorhees and William Johnson is a scholarly and pedagogical treatise on a deep, fundamental problem. Even though the thermodynamics of fluids applies to large, hydrostatically stressed solids in equilibrium, its extension to solids not in equilibrium, to non-hydrostatically stressed solids and to small particles where capillary effects become important, is far from straightforward.

Since Gibbs's initial work more than 100 year ago, our understanding of the solid has changed. We now know that diffusion occurs in the solid state and that solids can sustain non-dilatational stresses during diffusion. Over the years, this at times quite controversial problem has attracted some of the best minds in the field, including C. Herring, W. W. Mullins, J. W. Cahn, F. Larché, and M. E. Gurtin.

The present authors continue this tradition: they point out that much of the confusion in the field arises from a failure to distinguish two different thermodynamic approaches, in which either the vacancy concentration or the total number of lattice sites is treated as the independent variable. Using Gibbs's variational approach, they carefully track these descriptions while they develop the thermodynamics of small particles and non-hydrostatically stresses solids.

The subject is not just of academic importance. It is essential, for example, for a precise understanding of phase equilibria and surface morphologies in epitaxial alloy thin films. Much of the thermodynamics used in the current whirlwind of activity in nanoscience would be greatly improved by the rigor this article brings to the study of nanometer-size particles.

The article is deliberately pedagogical. It can be used to introduce graduate students to the subject, as the authors do in their own courses.

Hydrogen has a long history in materials science: as a reducing agent, as a highly mobile solute, as an embrittling agent, and as a structural probe. The resurgent interest in hydrogen as a fuel has prompted new research into the use of metals for its storage. In his article, Reiner Kirchheim focuses on solid solutions of hydrogen in complex materials, such as defective crystals and amorphous materials.

After a review of the basics of solubility of hydrogen in metals, the author starts with a fundamental treatment of how the hydrogen atoms are distributed according to Fermi-Dirac statistics over a complex energy landscape described by a density of site energies. He then lays out the implications for the diffusion of the hydrogen through such a landscape. The results are applied to the interaction with other solutes, vacancies, dislocations, grain boundaries and interfaces. The application to amorphous metals is particularly interesting, since hydrogen proves to be a remarkable tool to test the uniquely complex structure of these materials.

Even though the emphasis of the article is on metals, polymers, oxide glasses and metal-oxide interfaces are considered as well, as are systems with at least one small dimension, such as thin films and clusters.

HENRY EHRENREICH
FRANS SPAEPEN

SOLID STATE PHYSICS

VOLUME 59

The Thermodynamics of Elastically Stressed Crystals

P.W. Voorhees[*] and William C. Johnson[†]

[*]Department of Materials Science and Engineering,
Northwestern University, Evanston, IL 60208
Email: p-voorhees@northwestern.edu

[†]Department of Materials Science and Engineering, University of Virginia,
Charlottesville, VA 22904
Email: wcj2c@virginia.edu

I.	Introduction	2
II.	Hydrostatically Stressed Crystals	12
	1. Homogeneous Fluid	13
	2. Homogeneous Crystal–Binary Substitutional Alloy	14
	3. Equilibrium between a Crystal and Fluid	19
	4. Thermodynamic Relationships	21
III.	Deformation and Stress	34
	5. Deformation	34
	6. Stress	44
	7. Mechanical Equilibrium and Elastic Work	45
	8. Constitutive Equations for Small Strain	50
	9. Eigenstrains	52
IV.	Thermodynamics of a Single-Phase System	61
	10. Equilibrium	67
	11. Stress-dependence of Diffusion Potential	70
	12. Stress-induced Solute Redistribution	75
V.	Capillary and Interfacial Properties	94
	13. Introduction	94
	14. Surface Excess Quantities	95
	15. Thermodynamic Equilibrium Conditions	101
	16. Applications	106
VI.	Crystal–Fluid Equilibrium	109
	17. Introduction	109
	18. Energy Functional	111
	19. An Extremum in the Energy	112
	20. Small-strain Limit	120
	21. Applications	121
VII.	Two-Phase Crystalline Systems	131
	22. Introduction	131
	23. Equilibrium Conditions	132
	24. Applications of Equilibrium Conditions	141
VIII.	Acknowledgements	198
IX.	Appendix A: Surface Area Change Owing to Accretion	199
X.	Appendix B: Continuity Condition at Two-Phase Crystalline Interface	200
XI.	Appendix C: Continuity Condition at Crystal–Fluid Interfaces	201

I. Introduction

Elastic stresses arise naturally in crystals, even in the absence of external forces acting on the material. Solid-state phase transformations and the heteroepitaxial growth of thin films on a substrate are just two technologically important examples where self-stresses arise owing to a difference in the lattice parameters between the two crystals. Compositional heterogeneity, which develops during diffusional phase transformations, can engender stress when the crystal's lattice parameters are a function of the composition. Nonhydrostatic stresses are a natural byproduct of displacive transformations, where the lattices of the parent and product phase possess different lattice parameters and symmetries, and also occur during various second-order magnetic, ferroelectric, and order–disorder transitions. The deposition of a film with a different lattice parameter than the substrate often results in significant stresses, in some cases far in excess of the bulk yield stress of the film.

There is a significant body of experimental evidence that shows stresses can affect the evolution of a crystalline microstructure in several ways. One example, shown in Figure 1, illustrates the effect of a precipitate misfit strain resulting from the difference in lattice parameter between the precipitate and matrix, on the equilibrium precipitate shape as the precipitate size increases for a Ni-Al binary alloy. The precipitates evolve and grow by the diffusion of atoms through the matrix. In this alloy, the interfacial energy density is essentially isotropic, and the equilibrium precipitate shape in the absence of stress is given by the classical Wulff construction as a sphere. However, the elastic properties of the alloy reflect the anisotropic cubic symmetry of the the two crystals, with the elastically soft directions parallel to the $\langle 100 \rangle$ directions. Even though the misfit strain of the γ' precipitate is isotropic, the elastic energy of the precipitate is usually minimized when the precipitate assumes a plate-like shape.[1–3] The equilibrium shape of the precipitate is determined by minimizing the sum of the elastic and interfacial energies. Because the elastic energy scales with the volume of the precipitate, and the interfacial energy scales with the surface area of the precipitate, interfacial energy should be the dominant contribution to the system energy at sufficiently small particle sizes and, conversely, elastic energy should be the dominant energy contribution at large particle sizes. This behavior is similar to that observed in Figure 1. Equiaxed particles are observed at small particle sizes in Figure 1(a). With increasing particle radius, the precipitate shape begins to display some of the symmetry of the elastic field, developing regions of low curvature along the soft

---

[1] J. K. Lee, D. M. Barnett, and H. I. Aaronson, *Metall. Trans.* **8A**, 973 (1977).
[2] A. G. Khachaturyan, *Theory of Structural Phase Transformations in Solids*, John Wiley, New York (1983).
[3] T. Mura, *Micromechanics of Defects in Solids*, Kluwer Academic Publishers, Dordrecht, The Netherlands (1987).

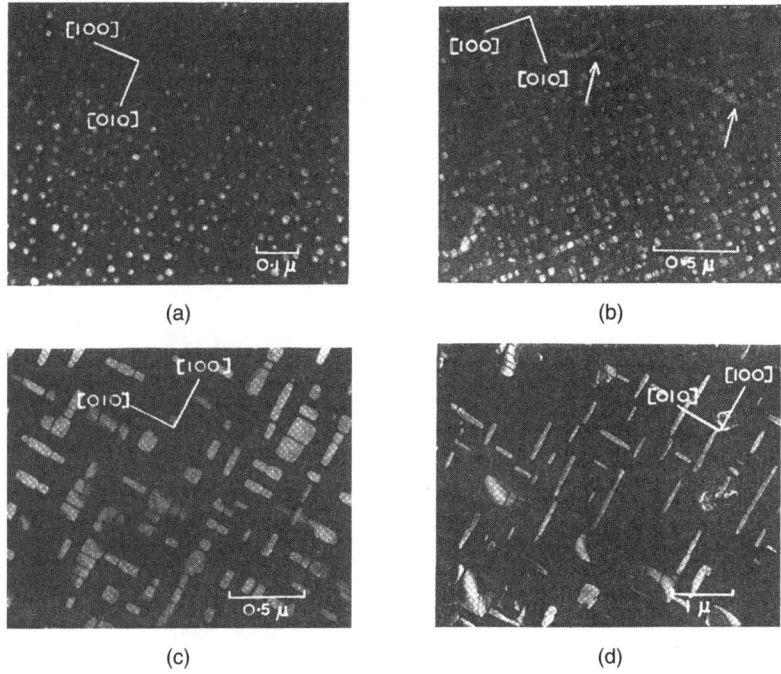

FIG. 1. Centered dark-field images show precipitate shape evolution in a Ni-6.71w%Al alloy aged for various times at 750°C: (a) 15 min, equiaxed γ' (Ni$_3$Al) precipitates; (b) 4h, cube-shaped precipitates; (c) 72h, cubical precipitates become plate-shaped and aligned along $\langle 100 \rangle$ elastically soft directions; and (d) 450 h. Precipitate shape transitions occur with increasing precipitate size in coherent systems owing to the different scaling of the elastic strain and interfacial energies with the precipitate radius: The interfacial energy dominates at small precipitate radii giving equiaxed precipitate shapes in this example, while the elastic energy dominates at large precipitate radii. Micrographs courtesy of A. J. Ardell.

$\langle 100 \rangle$ directions and assuming a shape that resembles cubes, as seen in Figure 1(b). At yet larger sizes the elastic energy contribution to the total energy becomes still more important, leading to a breaking of the four-fold symmetry of the precipitate in the (001) projection and development of plate-shaped or rod-shaped precipitates (Figure 1(c)); the aspect ratio increases with increasing particle size (Figure 1(d)).

Elastic stresses also influence the spatial correlation between particles as illustrated in the three-dimensional reconstruction of a $\gamma - \gamma'$ alloy.[4] These micrographs show that the particles are arranged in sheet-like arrays. In order to display this correlation, the microstructure has been separated into three regions: A, B, and C.

---

[4] A. C. Lund and P. W. Voorhees, *Acta mater.* **50**, 2585 (2002).

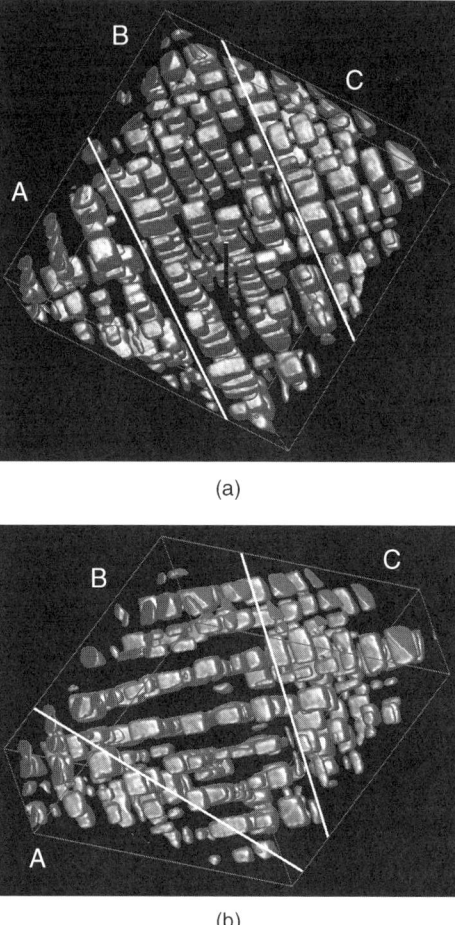

FIG. 2. A $10 \times 10 \times 3.6$ μm portion of a $\gamma - \gamma'$ microstructure showing two different views of the microstructure. In (b), the microstructure shown in (a) has been rotated, and the same regions A, B, and C are again labeled. There is a strong alignment of particles within each sheet along the two $\langle 100 \rangle$ directions, but not between sheets.

In all three regions, particles align along the $\langle 100 \rangle$ crystallographic directions. In region B in particular, most of this alignment occurs along two of the $\langle 100 \rangle$ directions while there is very little alignment along the third $\langle 100 \rangle$ direction. In Figure 2(b) the microstructure shown in Figure 2(a) has been rotated, and the same regions A, B, and C are again labeled. From this viewing angle it becomes obvious that the particles in region B are aligned in two-dimensional sheets, which are perpendicular to the $\langle 100 \rangle$ direction along which no alignment appears to be

TABLE 1. MATERIALS PARAMETERS FOR EXPERIMENTAL NI-AL-MO ALLOYS: FIGURES 2 AND 3

Alloy	Al [at%]	Mo [at%]	$f_{\gamma'}$ [%]	γ' solvus °C	ϵ^T (misfit)
A1	12.5	2.0	15 ± 5	≈ 1000	+0.65%
A2	9.9	5.0	19 ± 6	≈ 1000	+0.40%
A3	7.7	7.9	15 ± 5	≈ 950	<0.10%
A4	5.7	10.9	7 ± 2	?	−0.15%
A5	5.3	13.0	8 ± 2	≈ 900	−0.30%

occurring in Figure 2(a). The separation distances between the sheets are far larger than the interparticle separation distances within the sheets, and appear to be relatively constant. The strong alignment of particles within each sheet into lines of particles is also clear from Figure 2(a). This alignment is due to an attractive force between the corners of particles and a repulsive force between the faces of particles.[5]

The centered dark-field micrographs appearing in Figure 3 show the influence of the sign and magnitude of the misfit strain on the development of microstructure in a nickel-based alloy similar to that shown in Figure 2. The five microstructures correspond to five different ($\gamma + \gamma'$) Ni-Al-Mo ternary alloys after heat treating each alloy for 67h at 775°C. The compositions of each alloy, given by Table I, are chosen so that the volume fraction of the γ' precipitates in each alloy is roughly 10% but the misfit strain, ϵ^T, becomes progressively more negative with increasing Mo content.

The alloys of Figure 3 exhibit distinctive microstructures. The alloys with the largest magnitude of misfit, A1 (a) and A2 (b), display microstructures in which the precipitates are aligned along the elastically soft $\langle 100 \rangle$ directions of the matrix phase. The lowest misfit alloys, A3 (c) and A4 (d), show a more uniform dispersion of spherical precipitates while some of the precipitates of alloy A5 (e) are just beginning to assume a cube-like shape. After 430h of aging at 775°C, the precipitates of alloys A1 (a) and A2 (b) have lost their four-fold symmetric shapes and those of alloy A4 (d) have begun to assume a more cube-like morphology. The precipitates of alloy A3 (c), which have a vanishingly small misfit strain with respect to the matrix phase, still remain spherical and uniformly distributed. These results demonstrate that the long-range elastic stress engendered by the misfit strain strongly influences both the shape and the spatial distribution of precipitates of these alloys. This is particularly important as the Ni-Al system is the base alloy for the superalloys used in many high-temperature applications and the microstructure has a strong influence on the mechanical properties.

---

[5] C. H. Su and P. W. Voorhees, *Acta mater.* **44**, 1987 (1996).

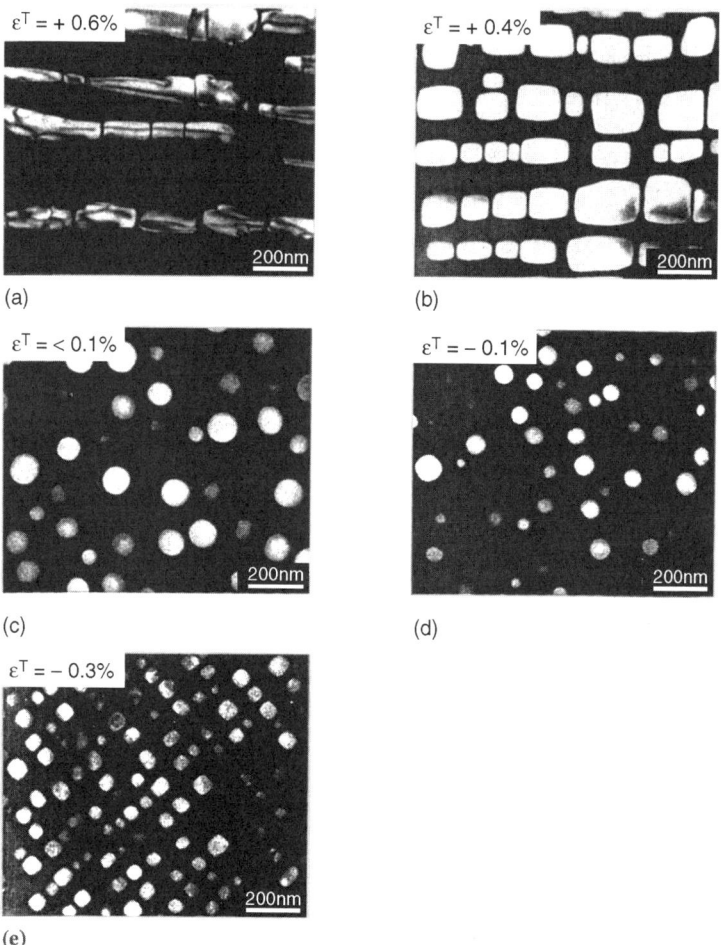

FIG. 3. Centered dark-field images show microstructural evolution in five Ni-Mo-Al alloys possessing roughly equal volume fractions of the γ' precipitates with misfit strains (a) $\epsilon^T = +0.6\%$; (b) $\epsilon^T = +0.4\%$; (c) $\epsilon^T \approx 0$; (d) $\epsilon^T = -0.15\%$; and (e) $\epsilon^T = -0.3\%$ after 67 h of aging at 775°C. Micrographs courtesy of M. Fährmann.

The morphological development of films deposited or grown on a substrate is also strongly influenced by stress.[6] Films deposited on a substrate are known to grow in many modes. If there is no difference in lattice parameter between the film and substrate, and the film wets the substrate, the film grows in a planar, layer-by-layer fashion. At the other extreme, in which the film does not wet the

---

[6] V. A. Shchukin and D. Bimberg, *Rev. Mod. Phys.* **71**, 1125 (1999).

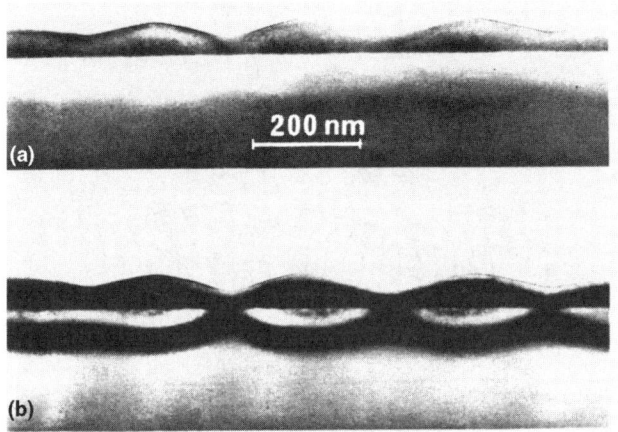

FIG. 4. Ripple formation in a $Si_{0.79}Ge_{0.21}$ alloy grown at 750°C on a (001) Si substrate. (a) bright field image, (b) dark field image showing strain.[7]

substrate, small islands of film form immediately on deposition. The intermediate case, Stranski-Kranstanow growth, occurs when the film wets the substrate and the film and substrate have different lattice parameters. In this case the film grows initially in a layer-by-layer mode and then forms islands at some later time in order to relieve the strain energy.

For many years it was thought that, during Stranski-Kranstanow growth, islands formed as a result of stress-relaxing dislocations forming at the interface of the substrate and film. However, more recently it has become clear that islands can form even in the absence of dislocations. As the film grows, the stress of this planar film can be reduced if the film undergoes a morphological instability. This process begins by the planar film becoming unstable to ripples on the surface, as seen in Figure 4.[7]

Figure 4(a) is the rippled surface of a nominally 40nm-thick Si-Ge film deposited on a Si substrate. There are no dislocations present in the film, as shown in Figure 4(b), yet there is clear evidence of strain. The undulations develop because the stress is relaxed at the peaks and concentrated at the troughs, similar to the stress concentration at the tip of a crack. The stress relaxation at the peaks and concentration at the troughs is illustrated in Figure 5. In this case the stress-free film has a larger lattice parameter than that of the substrate. The lack of lateral constraint near the peaks, as compared to a planar film, results in a splaying of the lattice planes near the peaks. This results in a film lattice parameter in this region that is closer to its natural lattice parameter. The converse is true at the

---

[7] A. G. Cullis, D. J. Robbins, A. J. Pidduck, and P. W. Smith, *J. Cryst. Grwth.* **123**, 333 (1992).

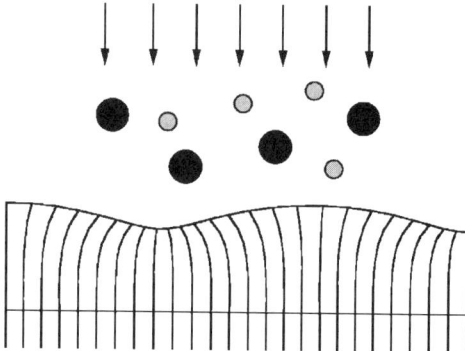

FIG. 5. A schematic depiction showing the elastic distortion of lattice planes for a film possessing a rippled surface. Atoms of two different sizes are being deposited from above onto the film, which is in a state of compression with respect to the substrate. Stresses are relaxed at the surface peaks and concentrated in the troughs.

troughs, where the distance between lattice planes is further removed from the stress-free lattice parameter of the film. This variation of stress along the surface induces atoms to diffuse from the troughs to the peaks, causing the amplitude of the corrugations to increase in time. The amplitude of these undulations continues to increase, forming islands when the amplitude is sufficiently large[8,9] (see Fig. 6). The Ge islands shown in this experiment are dislocation-free. There is a thin layer of film, a wetting layer, along the substrate between islands due to the tendency for the film to wet the substrate. Under different conditions, an array of islands of different morphologies can result (Fig. 7).[10] These islands are of varying size and morphology with the smaller islands or huts having ⟨105⟩ facets. The pathway by which islands form in this system is of intense interest as the islands are sufficiently small to be used as quantum dots. Such dots have unique, nearly delta-function-like, density of states and thus offer the potential for use in novel electronic devices.

The lattice parameter of these films is a function of composition; i.e. the constituent atomic species have different atomic radii. Thus, if an alloy film is being deposited, there can be an interaction between the local stress and the local composition of the film. To illustrate this interaction, Figure 5 shows large and small atoms being deposited while the morphological instability is developing. Due to the larger lattice parameter at the peak, there tends to be a higher concentration of the larger atoms than average in this region, while the smaller atoms have

---

[8] D. J. Eaglesham and M. Cerullo, *Phys. Rev. Lett.* **64**, 1943 (1990).
[9] S. Guha, A. Madhukar, and K. C. Rajkumar, *Appl. Phys. Lett.* **57**, 2110 (1990).
[10] G. Mederios-Ribeiro, A. M. Bratkovski, T. I. Kamins, D. A. A. Ohlberg, and R. S. Williams, *Science* **279**, 353 (1998).

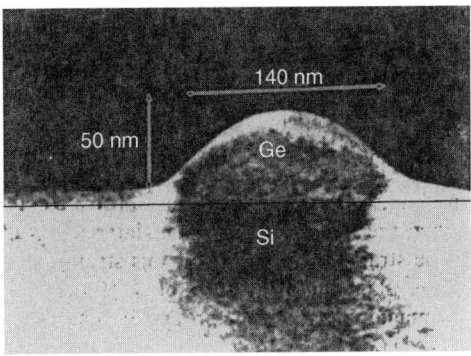

FIG. 6. Coherent Ge island on a Si substrate. The line denotes the film–substrate interface.[8]

a higher concentration than average at the troughs. This will give rise to composition modulations in the film. Because the nonplanar front is propagating in the vertical direction, these composition modulations will be along the vertical direction. An example of such modulations is shown in Figure 8.[11,12] Thin layers of InAs and AlAs were deposited on an InP substrate. The resulting stress induces a small-amplitude morphological instability of the surface, as shown in the cross-sectional micrograph that is taken parallel to the film growth direction (Figure 8(a)). The smaller In atoms tend of segregate near the troughs and are shown as the light regions of the film in Figure 8(a). A section through the film perpendicular to the growth direction is shown in Figure 8(b). There is clearly a very complex In-distribution pattern. Interestingly, the periodicity of the pattern corresponds quite well to the nonplanar morphology of the film surface as measured by atomic force microscopy (AFM). The magnitude of the In segregation can be significant, indicating that there is a strong interaction between composition and stress in this system. Given the large degree of segregation, such a growth process leads naturally to the formation of quantum wires along the growth direction.

A thermodynamic foundation clarifying the interaction between stress and composition is essential to describe the morphological evolution of the processes just discussed. A self-consistent thermodynamic framework should yield conditions for chemical, thermal, and mechanical equilibrium, both within the crystal and at any interphase interfaces. In addition, the thermodynamics should provide a means to formulate flux conditions and field equations for systems that are not in global equilibrium.

---

[11] A. G. Norman, S. P. Ahrenkiel, H. Moutinho, M. M. Al-Jassim, A. Mascrenhas, J. Mirecki-Millunchick, S. R. Lee, R. D. Twesten, S. M. Follstaedt, J. L. Reno, *et al.*, *Appl. Phys. Lett.* **73**, 1844 (1998).

[12] J. Mirecki-Millunchick, R. D. Twesten, D. M. Follstaedt, S. R. Lee, E. D. Jones, Y. Zhang, S. P. Ahrenkiel, and A. Mascarenhas, *Appl. Phys. Lett.* **70**, 1402 (1997).

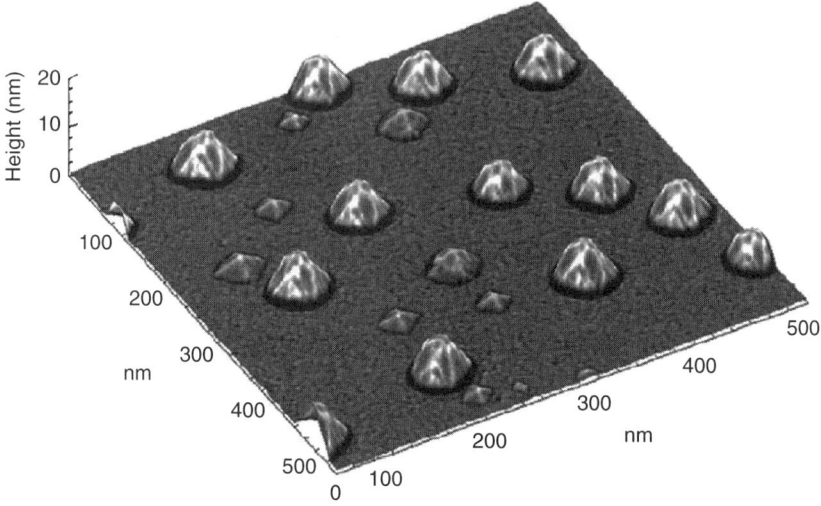

FIG. 7. STM micrograph of coherent Ge islands on a Si substrate. The smaller islands form pyramidal-shaped huts, whereas the larger islands are more equiaxed.[10]

There are at least a couple of different starting points for obtaining the governing equations for microstructural evolution in stressed crystals. A variational approach, developed by Gibbs and used effectively by Larché and Cahn and others to study more complicated crystals, requires first obtaining the conditions for thermodynamic equilibrium. This powerful approach allows all the various relationships between the physical properties of the crystal, such as the dependence of chemical potential on state of stress, to be developed in a thermodynamically self-consistent manner. A more general approach to dynamical problems one in which it is not necessary to invoke the equilibrium conditions in formulating the boundary value problem, has also been developed.[13] Both approaches yield a self-consistent thermodynamic description of non-hydrostatically stressed crystals. Here, we employ Gibbs' variational approach to formulate the equilibrium conditions and the relationships between the various thermodynamic parameters and then, where appropriate, compare these results to the more general dynamical theory.

The conditions governing thermodynamic equilibrium in multi-phase fluids have been known for more than 100 years. While Gibbs determined the equilibrium conditions for solids as well, he was limited by the understanding of the structure of solids at that time. Solids were assumed to be compounds of definite proportion wherein there was always one component that did not diffuse or

---

[13] M. E. Gurtin, *Arch. Rational. Mech. Anal.* **131**, 67 (1995).

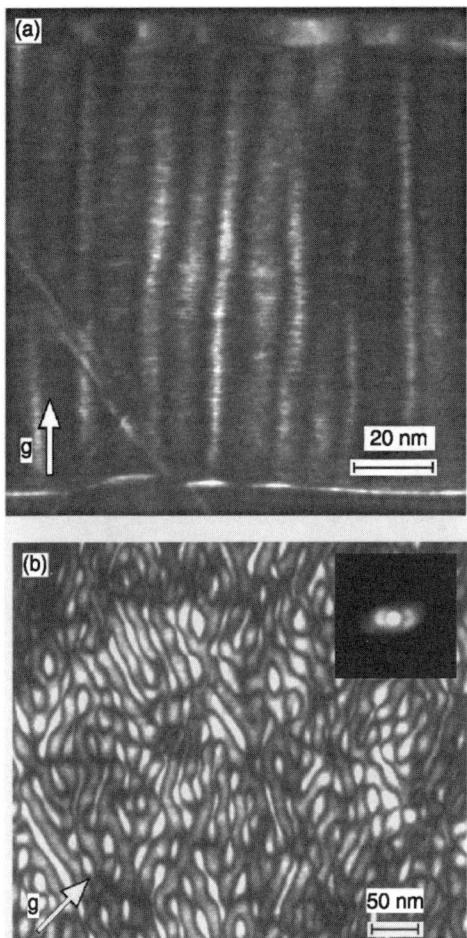

FIG. 8. (a) A dark-field image of a cross section showing narrow In-rich columns extending vertically through the film. Note the slightly rippled surface. (b) Plan-view dark-field image showing wire and dot structures. The inset to (b) is an in-plane 200 spot showing strong satellite intensity broadly distributed about [110].

dissolve in a surrounding fluid. Examples of such solids are polymer–solvent mixtures, interstitial solid solutions, and a simple sponge immersed in water. In the time since Gibbs' work, it has become clear that solid-state diffusion is possible in crystalline solids, and that such crystals can withstand nondilatational stresses during the diffusion process. As these processes were not considered by Gibbs, it is necessary to develop the equilibrium conditions in such systems. There has been much work in this area, some of which has been quite controversial, and it is still an area of active research. Nevertheless, the field has progressed to the point where the equilibrium conditions for a wide range of single- and multi-phase systems under stress are well accepted. It thus seems appropriate to collect the results of this work on the thermodynamics of crystals in a single article. We do not intend simply to restate the work of the many scientists who have contributed to the development of this field, however, but to present a consistent pedagogical derivation of the conditions for equilibrium. As our primary objective is to present the material in a form that is accessible to a wide audience, we focus only on the key assumptions and central results of the field and direct the reader to papers in literature for exploration of more advanced topics.

II. Hydrostatically Stressed Crystals

In this section, we present two different thermodynamic descriptions of a homogeneous, hydrostatically stressed crystal. One development assumes that vacancies can be treated as an independent species, while the other assumes that the number of lattice sites in the crystal is the independent variable. The importance of this difference and the resulting equilibrium conditions for hydrostatic crystals has been discussed by Herring[14,15] and Mullins.[16] A generalization of this treatment to nonequilibrium systems using entropy production arguments is given by Gurtin and Fried.[17] The two definitions of the chemical potentials are shown to be equivalent and to reduce to the familiar thermodynamic description for crystals at equilibrium when the crystal is hydrostatically stressed and large enough that capillarity effects can be neglected. The distinction between the two definitions and the corresponding equilibrium conditions becomes critical for hydrostatically stressed crystals not in thermodynamic equilibrium, for non-hydrostatically stressed crystals, and for various types of nanoparticles. A failure to account for these differences is a source of some confusion in this field, and is the basis for competing claims as to whether the chemical potential is or is not equal between contiguous phases in systems containing nanoparticles or stressed crystals. We therefore attempt to present the thermodynamic description of each definition as carefully as possible and will refer back to these definitions when developing the thermodynamics of small particles and nonhydrostatically stressed crystals.

We begin by considering a homogeneous binary substitutional alloy under hydrostatic pressure. The extension to a multicomponent system is straightforward

---

[14] C. Herring, in *The Physics of Powder Metallurgy*, ed., McGraw Hill (1950), 143–179.

[15] C. Herring, in *Structure and Properties of Solid Surfaces*, ed. R. Gomer and C. Smith, University of Chicago Press, Chicago (1952), 5–81.

[16] W. W. Mullins, in *Int. Conf. Solid-solid Phase Transformations*, eds. H. I. Aaronson, D. E. Laughlin, R. F. Sekerka, and C. M. Wayman (Warrendale, PA, TMS-AIME, 1981), 49–66.

[17] M. E. Gurtin and E. Fried, *Rev. Appl. Mech.* (in press).

and is given elsewhere.[18] By establishing equilibrium between the crystal and a fluid comprised of the same components, the substitutional binary system serves to illustrate the similarities and differences between the two thermodynamic descriptions. Expressions for the free energy densities are then obtained. The energy densities form the basis for the treatment of evolving systems under hydrostatic pressure and for systems experiencing nonhydrostatic stresses. This section ends with a treatment of interstitial solutions and solutions for which a specific atom species can occupy both substitutional and interstitial sites.

1. HOMOGENEOUS FLUID

If the internal energy of a homogeneous, multicomponent fluid ($\mathcal{E}$) is assumed to be a function of its entropy ($\mathcal{S}$), volume ($\mathcal{V}$), and number of atoms of component i ($\mathcal{N}_i$), the combined form of the first two laws of thermodynamics gives

$$d\mathcal{E} = \theta d\mathcal{S} - P d\mathcal{V} + \sum_{i=1}^{n} \mu_i d\mathcal{N}_i \qquad (2.1)$$

where θ is the temperature,[19] P is the pressure, μ_i is the chemical potential of species i, and n is the number of independent components. We use the calligraphic notation to denote extensive quantities that refer to the entire system. Because $\mathcal{E} = \mathcal{E}(\mathcal{S}, \mathcal{V}, \mathcal{N}_i)$, the intensive variables are also defined as partial derivatives of the energy. In particular, the pressure and chemical potential of component i are given by

$$P = -\left.\frac{\partial \mathcal{E}}{\partial \mathcal{V}}\right|_{\mathcal{S},\mathcal{N}_i} \quad \text{and} \quad \mu_i = \left.\frac{\partial \mathcal{E}}{\partial \mathcal{N}_i}\right|_{\mathcal{S},\mathcal{V},\mathcal{N}_j \neq \mathcal{N}_i}. \qquad (2.2)$$

The partial derivative giving the pressure assumes that the entropy and number of atoms of each component are held constant as the volume is changed. As such, a variation in $\mathcal{V}$ is recognized to be a result of the application of mechanical forces and not the result of addition or removal of fluid mass. The partial derivative for the chemical potential is taken such that the fluid entropy, volume, and number of atoms of each mass component j, excluding mass component i, are held constant. Thus the chemical potential measures the change in the internal energy of a large fluid system accompanying a small increase in the number of atoms of species i in the fluid holding the entropy, volume, and number of atoms of other species fixed. For a fluid, the number of atoms of each chemical component can be varied

---

[18] F. C. Larché and J. W. Cahn, *Acta metall.* **33**, 331 (1985).
[19] We use θ for the temperature instead of T so as to avoid confusion with the Piola-Kirchhoff stress tensor.

independently of the other, and the addition of the atom can be imagined to take place in any region of the homogeneous fluid.

Finally, we note that the internal energy of a homogeneous fluid is a mathematically homogeneous function of order one in the extrinsic thermodynamic variables. That is, if the system entropy, volume, and number of components of each species are scaled by a factor λ, then $E(\lambda \mathcal{S}, \lambda \mathcal{V}, \lambda \mathcal{N}_A, \lambda \mathcal{N}_B) = \lambda E(\mathcal{S}, \mathcal{V}, \mathcal{N}_A, \mathcal{N}_B)$. This property permits the energy of the system to be given as[20]

$$\mathcal{E} = \theta \mathcal{S} - P\mathcal{V} + \sum_{i=1}^{n} \mu_i \mathcal{N}_i. \qquad (2.3)$$

Differentiating Eq. (2.3) and combining with Eq. (2.1) yields the Gibbs-Duhem equation for a fluid.

$$\mathcal{S}d\theta - \mathcal{V}dP + \sum_{i=1}^{n} \mathcal{N}_i d\mu_i = 0 \qquad (2.4)$$

2. HOMOGENEOUS CRYSTAL–BINARY SUBSTITUTIONAL ALLOY

The spatially periodic structure of the crystal is an important property when constructing a thermodynamic model of the crystal[18,21]. For a given crystal structure, the position each atom species can occupy on each of the crystal sublattices determines the number of different elements that must be included explicitly in the thermodynamic description of each sublattice of the crystal. For example, a binary substitutional alloy possessing a simple cubic structure can be viewed as consisting of two identical sublattices. One sublattice can be designated as the substitutional lattice. It is occupied by the two components A and B and, if a substitutional site is vacant, by a substitutional vacancy. The other sublattice can be designated as the interstitial lattice. If the A and B atoms do not occupy an interstitial site, the interstitial lattice is composed entirely of interstitial vacancies. The thermodynamic description of such a crystal thus depends only on the number of A and B atoms and the number of substitutional vacancies. The explicit inclusion of the interstitial lattice in the thermodynamic description is not necessary. If A or B atoms can also occupy the interstitial site, this possibility must be incorporated into the thermodynamic description of the crystal. The presence of a crystalline lattice requires that defects be included in the thermodynamic description. The number of such defects affects the system energy, and a certain number of defects are usually present at

---

[20] J. W. Gibbs, *The Scientific Papers of J. Willard Gibbs Volume 1: Thermodynamics*, Mineola, NY, Dover Publications (1961).
[21] F. C. Larché and J. W. Cahn, *Acta metall.* **21**, 1051 (1973).

equilibrium. Explicit consideration of the defects is often necessary to formulate a consistent thermodynamic description of the crystal.

For other crystal structures, there can exist several different sublattices. In a face-centered cubic alloy, there are two distinct interstitial lattices composed of the octahedral and tetrahedral sites, respectively. The ratio of the number of interstitial sites to each substitutional site is always a constant for a particular crystal structure. Later, we will show how nonhydrostatic stresses can break the degeneracy of interstitial sites that are otherwise equivalent under hydrostatic pressure.

The importance of the lattice and various crystalline defects to a self-consistent thermodynamic description of the crystal can be better understood by first considering a binary substitutional alloy for which the two atoms reside on only one lattice and for which substitutional vacancies can be present; interstitial atoms are temporarily neglected. This assumption is justifiable for many systems, in which equilibrium concentration is extremely small. Let $\mathcal{N}_o$ be the number of lattice sites in the crystal. The number of atoms of components A and B in the crystal, $\mathcal{N}_A$ and $\mathcal{N}_B$, and the number of vacancies, $\mathcal{N}_v$, are related by

$$\mathcal{N}_A + \mathcal{N}_B + \mathcal{N}_v = \mathcal{N}_o. \tag{2.5}$$

Equation (2.5) relates four thermodynamic variables, any three of which can be chosen to describe the thermodynamic state of the crystal. Each choice leads to a different physical interpretation of mass transfer and to a different conjugate variable, the chemical potential. For example, there are two ways in which an A atom can be imagined to be added to the crystal while holding the number of B atoms fixed. The first entails holding the number of vacancies fixed so that addition of the A atom results in the creation of a new lattice site. This new lattice site appears on the surface of the crystal provided there is no disruption of the lattice within the crystal owing, for example, to dislocations or grain boundaries. The second manner in which an A atom can be added to the crystal is by holding the total number of lattice sites fixed. Addition of an A atom thereby requires destruction of a vacancy. These two processes give rise to two possible definitions of the chemical potential for a homogeneous crystal under hydrostatic pressure.

As for the fluid, we first consider the internal energy of an isolated, homogeneous crystalline solid under hydrostatic stress or pressure. The assumption of hydrostatic stress allows us to highlight the differences between solids and fluids without considering the important ability of crystals to withstand non-dilatational, or shear stresses. This also implies, as is demonstrated later, that the crystal is uniform in its thermodynamic state at equilibrium. After writing the equations for an isolated crystal, we examine equilibrium between a crystal immersed in a fluid of the same species.

We consider a binary substitutional crystal of components A and B. Owing to the presence of the lattice, only three of the variables, $\mathcal{N}_A$, $\mathcal{N}_B$, $\mathcal{N}_v$, and $\mathcal{N}_o$, are independent (see Eq. (2.5)). We initially choose the internal energy of the crystal $\mathcal{E}$ to be a function of the variable set $\mathcal{N}_A$, $\mathcal{N}_B$, and $\mathcal{N}_o$ before exploring a second set ($\mathcal{N}_A$, $\mathcal{N}_B$, $\mathcal{N}_V$) later. For the first variable set, the number of vacancies is a dependent variable determined by $\mathcal{N}_A$, $\mathcal{N}_B$, and $\mathcal{N}_o$ according to Eq. (2.5). Allowing the energy to depend on $\mathcal{N}_o$ implies that lattice sites are created or destroyed on the crystal surface. Because the crystal is by design hydrostatically stressed and thus homogeneous at equilibrium, it is not necessary to commit to a specific surface location where these sites are to be created.

The internal energy of the crystal is a function of $\mathcal{S}$, $\mathcal{V}$, $\mathcal{N}_A$, $\mathcal{N}_B$, and $\mathcal{N}_o$: $\mathcal{E} = \mathcal{E}(\mathcal{S}, \mathcal{V}, \mathcal{N}_A, \mathcal{N}_B, \mathcal{N}_o)$. The total derivative of the energy is thus

$$d\mathcal{E} = \left.\frac{\partial \mathcal{E}}{\partial \mathcal{S}}\right|_{\mathcal{V},\mathcal{N}_A,\mathcal{N}_B,\mathcal{N}_o} d\mathcal{S} + \left.\frac{\partial \mathcal{E}}{\partial \mathcal{V}}\right|_{\mathcal{S},\mathcal{N}_A,\mathcal{N}_B,\mathcal{N}_o} d\mathcal{V} + \left.\frac{\partial \mathcal{E}}{\partial \mathcal{N}_A}\right|_{\mathcal{S},\mathcal{V},\mathcal{N}_B,\mathcal{N}_o} d\mathcal{N}_A$$
$$+ \left.\frac{\partial \mathcal{E}}{\partial \mathcal{N}_B}\right|_{\mathcal{S},\mathcal{V},\mathcal{N}_A,\mathcal{N}_o} d\mathcal{N}_B + \left.\frac{\partial \mathcal{E}}{\partial \mathcal{N}_o}\right|_{\mathcal{S},\mathcal{V},\mathcal{N}_A,\mathcal{N}_B} d\mathcal{N}_o. \quad (2.6)$$

The first two partial derivatives can be replaced by the temperature (θ) and pressure (P), respectively.

$$\theta = \left.\frac{\partial \mathcal{E}}{\partial \mathcal{S}}\right|_{\mathcal{V},\mathcal{N}_A,\mathcal{N}_B,\mathcal{N}_o} \quad (2.7)$$

and

$$-P = \left.\frac{\partial \mathcal{E}}{\partial \mathcal{V}}\right|_{\mathcal{S},\mathcal{N}_A,\mathcal{N}_B,\mathcal{N}_o}. \quad (2.8)$$

The change in volume associated with the second partial derivative in Eq. (2.6) is a consequence of mechanical forces, not a change associated with an increase or decrease in the number of lattice sites. The third and fourth terms of Eq. (2.6) involve changes in the energy of the crystal owing to the addition of an atom of the specified chemical component holding the number of atoms of the second chemical component and the total number of lattice sites fixed. We denote these derivatives as

$$\mu_A^c = \left.\frac{\partial \mathcal{E}}{\partial \mathcal{N}_A}\right|_{\mathcal{S},\mathcal{V},\mathcal{N}_B,\mathcal{N}_o}$$
$$\mu_B^c = \left.\frac{\partial \mathcal{E}}{\partial \mathcal{N}_B}\right|_{\mathcal{S},\mathcal{V},\mathcal{N}_A,\mathcal{N}_o}. \quad (2.9)$$

The last partial derivative of Eq. (2.6) gives the change in crystal energy associated with a change in the number of lattice sites under conditions of constant crystal volume. The addition of a lattice site is accomplished by moving an atom within the crystal to the surface and the simultaneous creation of a vacancy within the crystal. We denote this derivative as

$$\mu_o^c = \frac{\partial \mathcal{E}}{\partial \mathcal{N}_o}\bigg|_{S,V,\mathcal{N}_A,\mathcal{N}_B}. \tag{2.10}$$

This process leads to a change in the deformation state and an increase in pressure of the crystal when the partial molar volume of the vacancy is nonzero. For the hydrostatically stressed crystal, Eq. (2.6) can be rewritten in a simplified form as

$$d\mathcal{E} = \theta dS - PdV + \mu_A^c d\mathcal{N}_A + \mu_B^c d\mathcal{N}_B + \mu_o^c d\mathcal{N}_o. \tag{2.11}$$

The chemical potentials defined by Eq. (2.11) require the elimination of a vacancy in the crystal when an A or B atom is added to the crystal. Because the conditions for the addition of a component to the crystal do not have an equivalent in a fluid, the superscript c is used to emphasize that this chemical potential is defined for a crystal, and that it necessitates an exchange of a vacancy and an atom. Later the energy associated with this exchange process will be identified with the diffusion potential.

Although the chemical potentials defined here are appealing, as they have a form very similar to that for a fluid, they may not always be a convenient, or useful, choice. For sufficiently large systems, a large number of vacancies are present, which allows for the addition of an atom holding the number of lattice sites fixed. However, for sufficiently small systems at lower temperatures, it is conceivable that no vacancies are present at equilibrium, and the exchange of an atom with a vacancy could not occur. As we shall encounter later, additional considerations can arise when diffusion is considered in the interior regions of a heterogeneous crystal.

We thus explore an alternate definition of the chemical potentials, which follows from considering a different set of independent variables for the energy. In the previous analysis, we assumed $\mathcal{E} = \mathcal{E}(S, V, \mathcal{N}_A, \mathcal{N}_B, \mathcal{N}_o)$. Here, we can choose the variable set $\mathcal{E} = \mathcal{E}(S, V, \mathcal{N}_A, \mathcal{N}_B, \mathcal{N}_V)$. In this case, the differential $d\mathcal{E}$ is written

$$d\mathcal{E} = \frac{\partial \mathcal{E}}{\partial S}\bigg|_{V,\mathcal{N}_A,\mathcal{N}_B,\mathcal{N}_V} dS + \frac{\partial \mathcal{E}}{\partial V}\bigg|_{S,\mathcal{N}_A,\mathcal{N}_B,\mathcal{N}_V} dV + \frac{\partial \mathcal{E}}{\partial \mathcal{N}_A}\bigg|_{S,V,\mathcal{N}_B,\mathcal{N}_V} d\mathcal{N}_A$$

$$+ \frac{\partial \mathcal{E}}{\partial \mathcal{N}_B}\bigg|_{S,V,\mathcal{N}_A,\mathcal{N}_V} d\mathcal{N}_B + \frac{\partial \mathcal{E}}{\partial \mathcal{N}_v}\bigg|_{S,V,\mathcal{N}_A,\mathcal{N}_B} d\mathcal{N}_V. \tag{2.12}$$

As before, the first two partial derivatives can be associated with the temperature and negative of the pressure, respectively; they are identical to Eqs. (2.7) and (2.8) as fixing $\mathcal{N}_A$, $\mathcal{N}_B$, and $\mathcal{N}_V$ fixes $\mathcal{N}_o$ through the lattice constraint, Eq. (2.5). The remaining partial derivatives give energy changes that are, in general, different from those defined by Eqs. (2.9) and (2.10) as they include the implicit addition of a lattice site to the crystal. We denote the chemical potentials defined by atom addition holding the number of vacancies fixed with a superscript v. The fundamental equation becomes

$$d\mathcal{E} = \theta d\mathcal{S} - Pd\mathcal{V} + \mu_A^v d\mathcal{N}_A + \mu_B^v d\mathcal{N}_B + \mu_V^v d\mathcal{N}_V \tag{2.13}$$

where we have defined

$$\mu_A^v = \left.\frac{\partial \mathcal{E}}{\partial \mathcal{N}_A}\right|_{\mathcal{S},\mathcal{V},\mathcal{N}_B,\mathcal{N}_V} \quad \mu_B^v = \left.\frac{\partial \mathcal{E}}{\partial \mathcal{N}_B}\right|_{\mathcal{S},\mathcal{V},\mathcal{N}_A,\mathcal{N}_V} \quad \text{and} \quad \mu_V^v = \left.\frac{\partial \mathcal{E}}{\partial \mathcal{N}_V}\right|_{\mathcal{S},\mathcal{V},\mathcal{N}_A,\mathcal{N}_B}. \tag{2.14}$$

The difference between the two chemical potentials that arises from the variable set chosen for the thermodynamic description of the crystal can be understood by considering a thought experiment in which a crystal of a given number of lattice sites is surrounded by a rigid wall. The wall fixes the volume of the crystal at a particular value, which is different from the volume the crystal would have if it were free from all forces, i.e., under zero pressure. There is no change in the actual volume of the crystal when an A atom is added to, and a vacancy simultaneously removed from, the crystal when the number of lattice sites is held fixed on account of the constraint imposed by the rigid wall. However, there would be a change in the volume of the crystal if the crystal were observed in its stress-free condition, because the A atom and vacancy have different partial molar volumes. This change in volume results in a change in the pressure of the crystal when it is constrained by the wall, the magnitude of which depends on the bulk elastic properties (bulk modulus) of the crystal. Similarly, there would be no change in the actual volume of the constrained crystal when an A atom is added by creating a new lattice site. However, the resulting change in the volume of the crystal as observed in its stress-free state would differ from the volume change in the stress-free state of the crystal for which the number of lattice sites is held fixed. If many such additions of atoms are made in each case, the respective systems would be pushed away from equilibrium. However, each system would be homogeneous and have a well-defined, calculable energy state, but the energy of the system for which the number of vacancies is held constant would differ from the energy of the system for which the number of lattice sites is held constant. The difference in stress-free volumes results in slightly different pressure changes and, hence, energy changes when the A atom is added to the crystal. Consequently, the two processes define different chemical potentials.

Each thermodynamic representation describes the same crystal, and any predictions made by the representations must be in complete agreement. Indeed, when the crystal is in equilibrium under a prescribed set of conditions, there is no difference between the values of the variously defined chemical potentials, as we now demonstrate by considering equilibrium between a crystal and a fluid.

3. Equilibrium Between a Crystal and Fluid

In this section, we derive the conditions for thermodynamic equilibrium of a two-phase binary system consisting of a homogeneous fluid and a homogeneous substitutional crystal. The equilibrium conditions are obtained by performing a thought experiment in which the thermodynamic state of the system is perturbed. Possible perturbations include changing the phase compositions, allowing small amounts of heat to flow between the phases, changing the defect concentration of the crystal, and allowing the transformation of a small amount of one phase into the other phase. Gibbs showed that equilibrium is obtained for an *isolated* system undergoing such perturbations if the corresponding change in the energy of the system vanishes.[20] By an isolated system, Gibbs meant a system that is not influenced by any external forces. Thus, to avoid external mechanical forces, the volume is held fixed. To avoid changes in entropy, no heat flow into the system is permitted. And, to prevent changes in chemical energy, the total number of atoms of each chemical component of the system must remain constant. These conditions—constant entropy, mass, and volume of the system—constrain the possible variations. If δ is used to represent a perturbation or variation in the thermodynamic state of each homogeneous phase, the four constraints can be written as

$$\begin{aligned} \delta S^s + \delta S^f &= 0 \\ \delta \mathcal{V}^s + \delta \mathcal{V}^f &= 0 \\ \delta \mathcal{N}_A^s + \delta \mathcal{N}_A^f &= 0 \\ \delta \mathcal{N}_B^s + \delta \mathcal{N}_B^f &= 0 \end{aligned} \qquad (2.15)$$

where the superscripts s and f have been used to designate the solid, or crystal, and the fluid phases, respectively. Note that these constraints do not prohibit the exchange of mass or heat between phases.

The variation in the energy of the fluid, $\delta \mathcal{E}^f$, is obtained from Eq. (2.1) while we first choose Eq. (2.11) for the energy change in the crystal, $\delta \mathcal{E}^s$. The number of lattice sites in the crystal is thus treated as the independent variable. The system energy $\mathcal{E} = \mathcal{E}^s + \mathcal{E}^f$, so that the variation in the system energy becomes

$$\delta \mathcal{E} = \delta \mathcal{E}^s + \delta \mathcal{E}^f = \theta^f \delta S^f - P^f \delta \mathcal{V}^f + \mu_A^f \delta \mathcal{N}_A^f + \mu_B^f \delta \mathcal{N}_B^f + \theta^s \delta S^s \\ - P^s \delta \mathcal{V}^s + \mu_A^c \delta \mathcal{N}_A^s + \mu_B^c \delta \mathcal{N}_B^s + \mu_o^c \delta \mathcal{N}_o^s. \qquad (2.16)$$

Using the constraints, Eq. (2.15), the change in system energy owing to the perturbation in the thermodynamic state is

$$\delta \mathcal{E} = (\theta^c - \theta^f)\delta S^s - (P^c - P^f)\delta \mathcal{V}^s + \left(\mu_A^c - \mu_A^f\right)\delta \mathcal{N}_A^s \\ + \left(\mu_B^c - \mu_B^f\right)\delta \mathcal{N}_B^s + \mu_o^c \delta \mathcal{N}_o^s. \qquad (2.17)$$

The system is in thermodynamic equilibrium when the variation of the total energy, $\delta \mathcal{E}$, vanishes. As the variations still retained in Eq. (2.17) can be considered to be independent, $\delta \mathcal{E}$ can vanish identically only when each coefficient in parentheses vanishes. Consequently, the conditions for thermal equilibrium require uniform temperature.

$$\theta^s = \theta^f \qquad (2.18)$$

Mechanical equilibrium requires equality of the pressures,

$$P^s = P^f \qquad (2.19)$$

while chemical equilibrium requires

$$\mu_A^c = \mu_A^f \quad \text{and} \quad \mu_B^c = \mu_B^f. \qquad (2.20)$$

Equation (2.20) shows the equality of the chemical potential of the crystal, *defined by an exchange of an atom and a vacancy*, with the chemical potential of the fluid at equilibrium. In addition, there is the equilibrium condition

$$\mu_o^c = 0. \qquad (2.21)$$

This equation is unique to a crystal, and is a condition on the equilibrium number of lattice sites. Because the number of lattice sites is not conserved since the solid is in contact with a fluid, the condition for equilibrium in the system is simply that the change in the energy of the crystal with respect to a change in the number of lattice sites must vanish. This condition states that the energy is an extremum and can be used to calculate equilibrium vacancy concentrations.[16]

The preceding derivation of the equilibrium conditions can be repeated using $\mathcal{N}_V$ as the independent variable rather than $\mathcal{N}_o$. In this case, the conditions of chemical equilibrium are

$$\mu_A^v = \mu_A^f \quad \text{and} \quad \mu_B^v = \mu_B^f \qquad (2.22)$$

and the condition

$$\mu_V^v = 0. \qquad (2.23)$$

Comparison of Eq. (2.20) with Eq. (2.22) shows that, for a system *in equilibrium*, the two chemical potentials are equal in value, although their meanings are different. Furthermore, at equilibrium, $\mu_o^c = \mu_V^v = 0$, and the fundamental equation for the hydrostatically stressed crystal with vacancies assumes the familiar form

$$d\mathcal{E} = \theta d\mathcal{S} - Pd\mathcal{V} + \mu_A d\mathcal{N}_A + \mu_B d\mathcal{N}_B \qquad (2.24)$$

where $\mu_i^c = \mu_i^v \equiv \mu$. So long as the crystal is in equilibrium, differentiating between the two definitions of chemical potentials is unnecessary. If the system is not in equilibrium, for example, if the hydrostatically stressed crystal has an excess or dearth of vacancies, as would result from a sudden change in temperature, the chemical potentials are different, and the way in which atoms are added to the system becomes important.

4. Thermodynamic Relationships

In this section, we develop some thermodynamic relationships for crystalline systems under hydrostatic pressure that we will use later when treating non-hydrostatically stressed systems. We begin by presenting the Gibbs-Duhem equations and formal expressions for the system free energies. In doing so, we first examine the equations for the case in which the number of vacancies is treated as an independent variable and then show that the usual form of these equations is recovered for systems in which vacancies are in global equilibrium. We then show that the alternate choice leads to the same result. These equations are then used to write the free energy densities, which form the basis for describing the system free energy of non-hydrostatically stressed crystals. These equations also provide a definition of crystalline diffusion potentials. Finally, we extend the treatment to interstitial systems in which at least one species can occupy multiple sublattices.

a. Gibbs-Duhem Equation

Equation (2.13) is the combined form of the first two laws of thermodynamics for a substitutional binary crystal with vacancies subjected to hydrostatic pressure. Increasing the size of the system by a factor of two leads to a corresponding doubling of all the extensive quantities and, like the fluid system, the crystal energy is thus a homogeneous function of order one. Note that this relationship is true only for hydrostatically stressed systems because, in this case, the mechanical work done is unaffected by the manner in which the volume of the system is doubled. Because the energy of the crystal is a homogeneous function of degree one, Eq. (2.13) yields

$$\mathcal{E} = \theta\mathcal{S} - P\mathcal{V} + \mu_A^v \mathcal{N}_A + \mu_B^v \mathcal{N}_B + \mu_V^v \mathcal{N}_V. \qquad (2.25)$$

Taking the total differential of Eq. (2.25) and using Eq. (2.13) yields the Gibbs-Duhem equation for the crystal.

$$0 = Sd\theta - VdP + \mathcal{N}_A d\mu_A^v + \mathcal{N}_B d\mu_B^v + \mathcal{N}_V d\mu_V^v \qquad (2.26)$$

When vacancies are in global equilibrium, $\mu_i^v = \mu_i^c = \mu_i$ and $\mu_V^v = 0$. This allows Eq. (2.26) to be expressed in the more common form

$$0 = Sd\theta - VdP + \mathcal{N}_A d\mu_A + \mathcal{N}_B d\mu_B. \qquad (2.27)$$

b. Free Energies

Other energy functions follow from the internal energy in the usual way used for homogeneous systems. The Helmholtz free energy of the crystal, $\mathcal{F}$, is

$$\mathcal{F} = \mathcal{E} - \theta S. \qquad (2.28)$$

Taking the total differential of Eq. (2.28) and using Eq. (2.13) yields the differential form for $\mathcal{F}$.

$$d\mathcal{F} = -Sd\theta - PdV + \mu_A^v d\mathcal{N}_A + \mu_B^v d\mathcal{N}_B + \mu_V^v d\mathcal{N}_V \qquad (2.29)$$

If vacancies are in equilibrium, Eq. (2.29) reduces to

$$d\mathcal{F} = -Sd\theta - PdV + \mu_A d\mathcal{N}_A + \mu_B d\mathcal{N}_B \qquad (2.30)$$

where the global equilibrium relationship $\mu_i^c = \mu_i^v = \mu_i$ has been used. Similarly, the Gibbs free energy of the system $\mathcal{G}$ is defined as

$$\mathcal{G} = \mathcal{E} - \theta S + PV. \qquad (2.31)$$

Differentiating Eq. (2.31) and using Eq. (2.13) yields

$$d\mathcal{G} = -Sd\theta + VdP + \mu_A^v d\mathcal{N}_A + \mu_B^v d\mathcal{N}_B + \mu_V^v d\mathcal{N}_V. \qquad (2.32)$$

For a system in which vacancies are in global equilibrium, Eq. (2.32) becomes

$$d\mathcal{G} = -Sd\theta + VdP + \mu_A d\mathcal{N}_A + \mu_B d\mathcal{N}_B. \qquad (2.33)$$

The chemical potentials of components A and B in the crystal can also be defined by Eq. (2.33) as

$$\mu_A = \left.\frac{\partial \mathcal{G}}{\partial \mathcal{N}_A}\right|_{\theta, P, \mathcal{N}_B} \quad \text{and} \quad \mu_B = \left.\frac{\partial \mathcal{G}}{\partial \mathcal{N}_B}\right|_{\theta, P, \mathcal{N}_A}. \qquad (2.34)$$

However, this definition implicitly assumes that the vacancies are in equilibrium and that the addition of atoms must obey the lattice constraint. For example, the addition of an A atom holding $\mathcal{N}_B$ fixed implies the creation of a new lattice site (as did the definition of μ_A^c) or the removal of a vacancy in the interior (as did the definition of μ_A^v). Differentiating between these two processes is not necessary for a hydrostatic system with a global equilibrium of vacancies.

For a system held at constant pressure and temperature, $dP = d\theta = 0$, and Eq. (2.32) becomes homogeneous of order one and thus,

$$\mathcal{G} = \mu_A^v \mathcal{N}_A + \mu_B^v \mathcal{N}_B + \mu_V^v \mathcal{N}_V. \tag{2.35}$$

which, when the system is in equilibrium, becomes

$$\mathcal{G} = \mu_A \mathcal{N}_A + \mu_B \mathcal{N}_B. \tag{2.36}$$

Expressions similar to Eqs. (2.27), (2.29), and (2.36) are obtained when $\mathcal{N}_A$, $\mathcal{N}_B$, and $\mathcal{N}_o$ are chosen as the independent variable set.

c. Free Energy Densities

The thermodynamic fields associated with most evolving systems are functions of position and time, and the energy associated with such a system must be calculated by integrating the energy density over the volume of the system. The energy density can be expressed in various ways, including on a per-unit-volume basis or on a per-atom (or lattice site) basis. The representation chosen for the energy density depends on the problem considered. In elastically deformed crystals, stress and deformation tensors are usually referred either to the current state of the system or to some convenient reference state. Consequently, the thermodynamic fields must also be expressible as densities referred to either the current state of the crystal (Eulerian description) or to some other convenient reference state (Lagrangian description).

In this section, we identify the natural variables and their conjugates for the internal and Helmholtz free energy densities. The densities are defined on a per-unit-volume basis for the current state of the system, on a per-unit-volume basis for the reference state, and on a per-atom basis. Reference state quantities are denoted with a prime superscript. A binary fluid is treated first in order to illustrate the similarities and differences between a crystal and fluid. An explicit expression for the three representations of the free energy density is then obtained for a substitutional binary crystal. An extension to interstitial solutions then follows.

a. Energy Densities for a Fluid System. A fundamental equation for the internal energy and the Gibbs-Duhem equation for a fluid are given by Eqs. (2.1) and (2.4), respectively. An expression for the internal energy density, as measured per-unit-volume of the current state of a binary fluid, is obtained by dividing Eq. (2.3) by the system volume $\mathcal{V}$,

$$e_v = \theta s_v - P + \mu_A \rho_A + \mu_B \rho_B \tag{2.37}$$

and then differentiating to give

$$de_v = \theta ds_v + s_v d\theta - dP + \mu_A d\rho_A + \rho_A d\mu_A + \mu_B d\rho_B + \rho_B d\mu_B \tag{2.38}$$

where e_v, s_v, and ρ_i are the energy density, entropy density, and atom density of species i, respectively. Dividing the Gibbs-Duhem equation, Eq. (2.4), by the system volume and subtracting the result from Eq. (2.38) yields

$$de_v = \theta ds_v + \mu_A d\rho_A + \mu_B d\rho_B. \tag{2.39}$$

Equation (2.39) is a fundamental equation in the Gibbsian sense: Given the equation of state $e_v = e_v(s_v, \rho_A, \rho_B)$, all remaining unknown thermodynamic variables can be calculated. θ, μ_A, and μ_B are obtained by differentiation of e_v with respect to s_v, ρ_A, and ρ_B, respectively, while the pressure is determined from Eq. (2.37). s_v, ρ_A, and ρ_B are the independent thermodynamic variables, which are naturally associated with the energy density, e_v, and θ, μ_A, and μ_B are their conjugate variables.

The energy density measured per atom of fluid can be obtained similarly. First, divide Eq. (2.3) by the total number of fluid atoms, $\mathcal{N}_o = \mathcal{N}_A + \mathcal{N}_B$, and then differentiate the result to give

$$de_o = \theta ds_o + s_o d\theta - V_o dP - PdV_o + \mu_A dc_A + c_A d\mu_A + \mu_B dc_B + c_B d\mu_B \tag{2.40}$$

where c is the mole fraction of the indicated species ($c_i = N_i/N_o$) and the subscript "o" denotes a density measured on a per-atom basis (e.g., $V_o = V/N_o$ is the mean atomic volume). Dividing the Gibbs-Duhem equation, Eq. (2.4), by the total number of atoms, $\mathcal{N}_o$, and subtracting the result from Eq. (2.40) yields

$$de_o = \theta ds_o - PdV_o + \mu_A dc_A + \mu_B dc_B. \tag{2.41}$$

Because $c_A + c_B = 1$, Eq. (2.41) reduces to

$$de_o = \theta ds_o - PdV_o + (\mu_B - \mu_A)dc_B. \tag{2.42}$$

The energy density e_o is a function of the three independent variables, s_o, V_o, and c_B; $e_o = e_o(s_o, V_o, c_B)$. Given an expression for $e_o(s_o, V_o, c_B)$, θ, P, and the difference $\mu_B - \mu_A$ are obtained by differentiation with respect to s_o, V_o, and c_B, respectively. The individual chemical potentials are determined by using the Gibbs-Duhem equation. s_o, V_o, and c_B are the natural variables to associate with the energy density, e_o, and θ, P, and $\mu_B - \mu_A$ are their conjugate variables.

The conjugate variable associated with c_B in Eq. (2.42) is the difference in the chemical potentials. The difference arises because the mole fraction is the independent thermodynamic variable and a change in mole fraction occurs at constant number of atoms. Thus, the addition of a B atom (increase in energy of μ_B) necessitates the concurrent removal of an A atom (decrease in energy μ_A) and the corresponding energy change reflects the difference in chemical potentials $(\mu_B - \mu_A)$. This difference in chemical potentials is often referred to as the diffusion potential, M_{BA}, defined as:[21]

$$M_{BA} \equiv \mu_B - \mu_A = \left.\frac{\partial e_o}{\partial c_B}\right|_{s_o, V_o}. \qquad (2.43)$$

The Helmholtz free energy density can also be expressed on a per-atom or on a per-unit-volume basis. The derivation is identical to that used for the energy densities. Using the definition $\mathcal{F} = \mathcal{E} - \theta S$,

$$df_v = -s_v d\theta + \mu_A d\rho_A + \mu_B d\rho_B \qquad (2.44)$$

and

$$df_o = -s_o d\theta - P dV_o + (\mu_B - \mu_A) dc_B. \qquad (2.45)$$

The natural variables to associate with f_v are θ, ρ_A, and ρ_B and the natural variables to associate with f_o are θ, V_o, and c_B. An equivalent expression for the diffusion potential is obtained from Eq. (2.45) as

$$M_{BA} = \mu_B - \mu_A = \left.\frac{\partial f_o}{\partial c_B}\right|_{\theta, V_o}. \qquad (2.46)$$

The Gibbs free energy density can also be expressed on a per-atom basis. The derivation is identical to that used for the energy densities. Using the definition $\mathcal{G} = \mathcal{E} - \theta S + P\mathcal{V}$,

$$dg_o = -s_o d\theta + V_o dP + (\mu_B - \mu_A) dc_B. \qquad (2.47)$$

An equivalent expression for the diffusion potential is obtained from Eq. (2.45) as

$$M_{BA} = \mu_B - \mu_A = \left.\frac{\partial g_o}{\partial c_B}\right|_{\theta, P}. \qquad (2.48)$$

Differentiating Eq. (2.48) yields

$$\left.\frac{\partial^2 g_o}{\partial c_B^2}\right|_{\theta, P} = \left.\frac{\partial \mu_B}{\partial c_B}\right|_{\theta, P} - \left.\frac{\partial \mu_A}{\partial c_B}\right|_{\theta, P}. \qquad (2.49)$$

The Gibbs-Duhem equation on a per atom basis is

$$0 = -s_o d\theta + V_o dP + c_B d\mu_B + c_A d\mu_A. \tag{2.50}$$

The chemical potentials in a binary alloy are functions of the mole fraction of one component at constant θ and P so that we can write

$$d\mu_i = \left.\frac{\partial \mu_i}{\partial c_B}\right|_{\theta,P} dc_B \quad \text{for } i = A, B. \tag{2.51}$$

Using Eq. (2.51) in Eq. (2.50) and assuming the temperature and pressure are constant yields

$$0 = \left(c_A \left.\frac{\partial \mu_A}{\partial c_B}\right|_{\theta,P} + c_B \left.\frac{\partial \mu_B}{\partial c_B}\right|_{\theta,P}\right) dc_B. \tag{2.52}$$

For Eq. (2.52) to hold for all dc_B, the term in the parenthesis must be zero. Using this result in Eq. (2.49) yields

$$\left.\frac{\partial \mu_B}{\partial c_B}\right|_{\theta,P} = (1 - c_B) \left.\frac{\partial^2 g_o}{\partial c_B^2}\right|_{\theta,P} \tag{2.53}$$

and

$$\left.\frac{\partial \mu_A}{\partial c_B}\right|_{\theta,P} = -c_B \left.\frac{\partial^2 g_o}{\partial c_B^2}\right|_{\theta,P}. \tag{2.54}$$

b. Energy Densities for a Crystalline System. An expression for the energy density on a per-unit-volume of the actual state basis can be obtained for a binary substitutional crystal with vacancies in the same way as for the fluid system. In this case the volume is that of the system at the pressure P. Divide Eqs. (2.25) and (2.26) by the volume of the system. Differentiate the first equation and then combine the two resulting equations to obtain

$$de_v = \theta ds_v + \mu_A^v d\rho_A + \mu_B^v d\rho_B + \mu_V^v d\rho_V. \tag{2.55}$$

For the crystal, $e_v = e_v(s_v, \rho_A, \rho_B, \rho_V)$ is a fundamental equation. In this representation, s_v, ρ_A, ρ_B, and ρ_V are the independent thermodynamic variables to associate with e_v, and the temperature and the chemical potentials μ_i^v are the corresponding conjugate variables. All three concentrations can be changed independently, as might occur if the pressure were changed. The chemical potentials implicitly require the addition of a lattice site within the volume when the density of one component is increased holding the densities of the other components fixed; i.e., the density of the lattice sites will change with the thermodynamic state of the system. If vacancies are in equilibrium, $\mu_V^v = 0$, and the resulting expression for the energy density is identical in form to that of the fluid.

The internal energy density of a homogeneous crystal can also be expressed on a per-mole or per-atom site basis. Dividing Eq. (2.25) by the total number of atom sites, $\mathcal{N}_o = \mathcal{N}_A + \mathcal{N}_B + \mathcal{N}_V$, yields

$$e_o = \theta s_o - P V_o + \mu_A^v c_A + \mu_B^v c_B + \mu_V^v c_V. \tag{2.56}$$

Because $c_A + c_B + c_V = 1$, any one of the mole fractions appearing in Eq. (2.56) can be eliminated in favor of the other two mole fractions. If we eliminate c_V, Eq. (2.56) becomes

$$e_o = \theta s_o - P V_o + M_{AV} c_A + M_{BV} c_B + \mu_V^v \tag{2.57}$$

where the diffusion potentials have been defined as

$$M_{AV} = \mu_A^v - \mu_V^v \quad \text{and} \quad M_{BV} = \mu_B^v - \mu_V^v. \tag{2.58}$$

Similar to their definition for a fluid system, the diffusion potentials for the crystal represent the change in energy owing to an exchange of component species; for the case of M_{BV}, the exchange requires the addition of a B atom and the removal of a vacancy. The individual chemical potentials used to define the diffusion potential, μ_i^v, are well defined for the bulk crystal.

Differentiating Eq. (2.57) and using the Gibbs-Duhem equation, Eq. (2.26), yields

$$de_o = \theta ds_o - P dV_o + M_{AV} c_A + M_{BV} c_B. \tag{2.59}$$

The natural thermodynamic variables to associate with e_o are s_o, V_o, c_A, and c_B and the fundamental equation is of the form $e_o = e_o(s_o, V_o, c_A, c_B)$. Similar to the fluid system, when the energy density of the crystal is expressed on a per-unit-lattice basis, the resultant conjugate variable for the composition is the diffusion potential.

In dealing with non-hydrostatically stressed crystals, the unstressed state of the crystal is frequently chosen as the reference state for the measurement of strain. Such formulations require that the free energy densities be expressible on a per-unit-volume basis when referred to the reference state, as opposed to that done in Eq. (2.55), where the volume of the actual state was used. Consider the crystal occupying a domain R of volume $\mathcal{V}$. Imagine a uniform mapping of the crystal into the region R' of volume $\mathcal{V}'$. The prime is used to designate the reference (Lagrangian) state of the system. For all subsequent thermodynamic states of the crystal, its properties and densities are to be referred back to the domain R'. The energy density of the crystal, $e_{v'}$, when it occupies this state is obtained by dividing

Eq. (2.25) by the volume $\mathcal{V}'$ to obtain

$$e_{v'} = E/\mathcal{V}' = \theta s_{v'} - PJ + \mu_A^v \rho_A' + \mu_B^v \rho_B' + \mu_V^v \rho_V' \tag{2.60}$$

where $s_{v'} = \mathcal{S}/\mathcal{V}'$ is the entropy density of the crystal when referred to the reference state, $\rho_i' = \mathcal{N}_i/\mathcal{V}'$ is the reference state density of component i, and J is a measure of the deformation

$$J \equiv \mathcal{V}/\mathcal{V}'. \tag{2.61}$$

J is the ratio of the volume of a given region in the crystal to its corresponding volume in the reference state. The variables ρ_A', ρ_B', and ρ_V' are not independent and Eq. (2.61) cannot be differentiated directly. Because the domain R' is taken to be constant and does not change with the thermodynamic state of the crystal, the density of lattice sites in the reference state, ρ_o', is a constant.

$$\rho_o' = \rho_A' + \rho_B' + \rho_V' = \text{constant} \tag{2.62}$$

This condition is in contrast to the actual density of lattice sites of the crystal; ρ_o is not a constant but a function of the thermodynamic state. For example, ρ_o changes as a function of pressure, if the crystal is compressible, or as a function of composition, if components A and B have different partial molar volumes.

Using Eq. (2.62) to eliminate ρ_V' from Eq. (2.60) in terms of the constant ρ_o' yields

$$e_{v'} = \theta s_{v'} - PJ + M_{AV}\rho_A' + M_{BV}\rho_B' + \mu_V^v \rho_o' \tag{2.63}$$

where the diffusion potentials are defined by Eq. (2.58). Other representations are also possible. For example, if ρ_B' were to be eliminated, Eq. (2.60) would become

$$e_{v'} = \theta s_{v'} - PJ + M_{AB}\rho_A' + M_{VB}\rho_V' + \mu_B^v \rho_o' \tag{2.64}$$

where

$$M_{AB} = \mu_A^v - \mu_B^v. \tag{2.65}$$

Dividing the Gibbs-Duhem equation by $\mathcal{V}'$ and using Eq. (2.62) gives

$$s_{v'}d\theta - JdP + \rho_A'dM_{AV} + \rho_B'dM_{BV} + \rho_o'd\mu_V^v = 0. \tag{2.66}$$

Differentiating Eq. (2.63) and using Eq. (2.66) yields the following equation for $de_{v'}$:

$$de_{v'} = \theta ds_{v'} - PdJ + M_{AV}d\rho_A' + M_{BV}d\rho_B'. \tag{2.67}$$

Here $e_{v'} = e_{v'}(s_{v'}, J, \rho'_A, \rho'_B)$ with natural variables $s_{v'}, J, \rho'_A$, and ρ'_B. Alternate expressions for the energy density are also possible using a different set of concentration variables. If ρ'_B is eliminated from the Gibbs-Duhem equation and the result is combined with the differential form of Eq. (2.64), the expression for $de_{v'}$ is

$$de_{v'} = \theta ds_{v'} - P dJ + M_{AB} d\rho'_A + M_{VB} d\rho'_V \qquad (2.68)$$

with a fundamental equation of the form $e_{v'} = e_{v'}(s_{v'}, J, \rho'_A, \rho'_V)$.

The Helmholtz free energy density, $f_{v'}$, is

$$df_{v'} = -s_{v'} d\theta - P dJ + M_{AV} d\rho'_A + M_{BV} d\rho'_B \qquad (2.69)$$

with $f_{v'} = f_{v'}(\theta, J, \rho'_A, \rho'_B)$ being a fundamental equation and θ, J, ρ'_A, and ρ'_B being the natural variables to associate with $f_{v'}$.

Equation (2.69) can also be derived by using the variable set $\mathcal{N}_A, \mathcal{N}_B$, and $\mathcal{N}_V$. Beginning with Eq. (2.11), instead of Eq. (2.13), one obtains

$$df_{v'} = -s_{v'} d\theta - P dJ + \mu_A^c d\rho'_A + \mu_B^c d\rho'_B. \qquad (2.70)$$

Physically, μ_i^c corresponds to the addition of an atom of component i to a vacant lattice site. Because $\mu_i^c = M_{iV}$, Eq. (2.70) is equivalent to Eq. (1.69). Henceforth, we employ the diffusion potential notation to emphasize that the conjugate variables of the concentrations refer to an exchange process.

c. Free Energy Density of a Pressure-free Crystal. It is often convenient to take R' as the stress-free state of the crystal at a prescribed temperature and concentration, ρ'_A and ρ'_B. When the crystal itself is stress free at the same temperature and concentration as the reference state, $J = 1$ (by definition), and a simple relationship exists between the energy density referred to the reference state and the energy density expressed on a per-atom-site basis.

$$e_{v'} = \rho'_o e_o \quad \text{and} \quad f_{v'} = \rho'_o f_o. \qquad (2.71)$$

Because the energy or free energy densities of a solution are usually expressed in the stress-free state on a per-mole basis, their representation in terms of a volume density referred to a stress-free reference state is given simply by Eq. (2.71). For example, if A and B form an ideal solution of substitutional atoms and there are no vacancies, the free energy density measured per-atom-site at zero pressure $(J = 1)$ is

$$f_o(\theta, 1, c_B) = (1 - c_B)\left[f_A^o + k\theta \ln(1 - c_B)\right] + c_B\left[f_B^o + k\theta \ln(c_B)\right] \qquad (2.72)$$

where k is Boltzmann's constant and f_i^o is the free energy per atom of pure component i at the given temperature. Taking the reference state to be the unstressed crystal at zero pressure and composition c_B allows the free energy density to be written immediately as

$$f_{v'}(\theta, 1, \rho'_B) = \rho'_o f_o(\theta, 1, c_B) = \rho'_o \{(1 - c_B)[f_A^o + k\theta \ln(1 - c_B)] + c_B[f_B^o + k\theta \ln(c_B)]\}. \tag{2.73}$$

The diffusion potential, corresponding to the change in energy associated with removing an A atom and adding a B atom, is

$$M_{BA} = \left.\frac{\partial f_{v'}}{\partial \rho'_B}\right|_{\theta, J} = \frac{1}{\rho'_o} \left.\frac{\partial f_{v'}}{\partial c_B}\right|_{\theta, J} = (f_B^o - f_A^o)c_B + k\theta[\ln(c_B) - \ln(1 - c_B)]. \tag{2.74}$$

d. *Dependence of the Free Energy Density on Deformation.* Equation (2.73) gives the free energy density as a function of temperature and composition for the stress-free crystal ($J = 1$) without vacancies. If the reference state is the stress-free state of the crystal, the free energy density of the vacancy-free crystal at temperature θ, composition c_B, and deformation J, $f_{v'}(\theta, J, c_B)$ can be obtained from Eq. (2.69) as

$$\int_{f_{v'}(\theta, 1, c_B)}^{f_{v'}(\theta, J, c_B)} df_{v'} = \int_1^J P dJ \tag{2.75}$$

or

$$f_{v'}(\theta, J, c_B) = f_{v'}(\theta, 1, c_B) + \int_1^J P dJ \tag{2.76}$$

where the integration takes place at constant temperature and composition and the free energy density $f_{v'}(\theta, 1, c_B)$ is obtained from solution thermodynamics such as Eq. (2.73).

In order to complete the integration of Eq. (2.76), a constitutive equation connecting the pressure to the deformation J is necessary. For a simple isotropic system at reasonable pressures, the isothermal bulk modulus, K_θ, can be treated as a constant, independent of the volume (or J). The bulk modulus is defined as

$$K_\theta = -V \left(\frac{\partial P}{\partial V}\right)_\theta. \tag{2.77}$$

As K_θ is taken to be a constant, Eq. (2.77) yields

$$\int_0^P dP = -\int_{\mathcal{V}'}^{\mathcal{V}} K_\theta \frac{d\mathcal{V}}{\mathcal{V}}. \qquad (2.78)$$

Integration gives

$$P = -K_\theta \ln(\mathcal{V}/\mathcal{V}') = -K_\theta \ln J. \qquad (2.79)$$

Equation (2.79) is a constitutive equation that connects the pressure and deformation J. Substituting Eq. (2.79) into Eq. (2.76) yields

$$f_{v'}(\theta, J, c_B) = f_{v'}(\theta, 1, c_B) - \int_1^J K_\theta \ln J \, dJ. \qquad (2.80)$$

For most crystals, $\mathcal{V} \approx \mathcal{V}'$ so that $J \approx 1$. With this assumption, $\ln J$ can be expanded in a series about $J = 1$ which gives to first order

$$P = -K_\theta(J - 1) = -K_\theta \left(\frac{\mathcal{V} - \mathcal{V}'}{\mathcal{V}'}\right) \qquad (2.81)$$

where $J - 1$ is the change in volume per unit reference volume of the crystal. Substituting Eq. (2.81) into Eq. (2.76) and integrating yields

$$f_{v'}(\theta, J, c_B) = f_{v'}(\theta, 1, c_B) - \frac{K_\theta}{2}(J - 1)^2$$
$$= f_{v'}(\theta, 1, c_B) + \frac{1}{2}P(J - 1). \qquad (2.82)$$

The term $P(J - 1)/2$ is the mechanical work done in deforming a unit (reference) volume of crystal from its pressure-free reference state to the actual state of the crystal. This term is equivalent to the elastic strain energy density.

d. Interstitial Solutions

In this section, the thermodynamic treatment for substitutional alloys is extended to interstitial solutions. We consider a crystal with two sublattices denoted as the substitutional and interstitial sublattices. Three mass components are treated; component A, which can be found on either of the sublattices, component B, which is restricted to the substitutional lattice, and component C, which is restricted to the interstitial lattice. Vacant sites are possible on each sublattice and the number of vacant sites, N_V^S and N_V^I, is assumed to be the dependent thermodynamic variables.

a. Free Energy Density. The addition of an A atom to the crystal holding the number of lattice sites constant can occur in two distinct ways. The atom can be added to an interstitial site, removing an interstitial vacancy, or to a substitutional site, removing a substitutional vacancy. In general, each process can lead to a different change in energy and each process must be explicitly treated in the thermodynamic description of the crystal. When the system is in equilibrium, we will show that the energy change associated with each process must be the same. However, situations could arise for which the A atoms occupying the two sublattices might not be in equilibrium, especially if there is a significant activation barrier to the transition of the atom between sublattices.

The change in internal energy is written

$$d\mathcal{E} = \theta d\mathcal{S} - Pd\mathcal{V} + M_{AV}^S d\mathcal{N}_A^S + M_{BV}^S d\mathcal{N}_B^S + \mu_o^c d\mathcal{N}_o^S \\ + M_{AV}^I d\mathcal{N}_A^I + M_{CV}^I d\mathcal{N}_c^I + \mu_o^I d\mathcal{N}_o^I \quad (2.83)$$

where the superscript I and S denote the interstitial and substitutional lattices, respectively. Thus, M_{AV}^S and M_{AV}^I give the change in energy associated with the addition of an A atom to a vacant substitutional site and the addition of an A atom to a vacant interstitial site. The number of interstitial and substitutional sites occurs in a fixed ratio defined by the crystal structure: $\mathcal{N}_o^I = r\mathcal{N}_o^S$ so that $d\mathcal{N}_o^I = rd\mathcal{N}_o^S$. Using this relationship and the definition $\mu_o = \mu_o^c + r\mu_o^I$ gives

$$d\mathcal{E} = \theta d\mathcal{S} - Pd\mathcal{V} + M_{AV}^S d\mathcal{N}_A^S + M_{BV}^S d\mathcal{N}_B^S + \mu_o d\mathcal{N}_o^S \\ + M_{AV}^I d\mathcal{N}_A^I + M_{CV}^I d\mathcal{N}_c^I. \quad (2.84)$$

The term $d\mathcal{N}_o^S$ can be interpreted as the change in the number of unit cells in the crystal, and μ_o is the corresponding energy change when a new unit cell is added to the crystal holding the number of each mass component fixed. Expressions for the Gibbs-Duhem equation and the free energy densities are now obtained using the same arguments as were used for the binary substitutional crystal. First, the internal energy is recognized to be a homogeneous function of the extensive variables $\mathcal{S}$, $\mathcal{V}, \mathcal{N}_A^S, \mathcal{N}_B^S, \mathcal{N}_A^I, \mathcal{N}_C^I$, and $\mathcal{N}_o^S$, and Euler's theorem for homogeneous functions of degree one yields

$$\mathcal{E} = \theta\mathcal{S} - P\mathcal{V} + M_{AV}^S \mathcal{N}_A^S + M_{BV}^S \mathcal{N}_B^S + \mu_o \mathcal{N}_o^S + M_{AV}^I \mathcal{N}_A^I + M_{CV}^I \mathcal{N}_c^I. \quad (2.85)$$

Differentiating Eq. (2.85) and using Eq. (2.84) yields the Gibbs-Duhem equation

$$0 = \mathcal{S}d\theta - \mathcal{V}dP + \mathcal{N}_A^S dM_{AV}^S + \mathcal{N}_B^S dM_{BV}^S + \mathcal{N}_o^S d\mu_o \\ + \mathcal{N}_A^I dM_{AV}^I + \mathcal{N}_c^I dM_{CV}^I. \quad (2.86)$$

An expression for the energy density, $de_{v'}$, is obtained by dividing Eq. (2.85) by the reference volume $\mathcal{V}'$ and then differentiating the resulting expression. Combining with Eq. (2.86) yields

$$de_{v'} = \theta ds_{v'} - PdJ + M^S_{AV}d\rho^{S'}_A + M^S_{BV}d\rho^{S'}_B + M^I_{AV}d\rho^{I'}_A + M^I_{CV}d\rho^{I'}_C. \quad (2.87)$$

The independent variables are $s_{v'}$, J, $\rho^{S'}_A$, $\rho^{S'}_B$, $\rho^{I'}_A$, and $\rho^{I'}_C$, and a fundamental equation has the form $e_{v'} = e_{v'}(s_{v'}, J, \rho^{S'}_A, \rho^{S'}_B, \rho^{I'}_A, \rho^{I'}_C)$.

b. Equilibrium between Crystal and Fluid. The conditions for thermodynamic equilibrium for a hydrostatically stressed crystal possessing interstitial components is obtained in precisely the same way as for the binary substitutional system considered previously. Consider a two-phase system consisting of the crystal and a fluid comprised of the three mass components A, B, and C. Assuming both the crystal and fluid are homogeneous in their thermodynamic states, the energy of the system is simply the sum of the energies of the crystal and the fluid: $\mathcal{E} = \mathcal{E}^c + \mathcal{E}^f$. Perturbing the system under conditions of constant entropy, mass, and volume yields for the change in system energy $\delta\mathcal{E}$:

$$\delta\mathcal{E} = (\theta^c - \theta^f)\delta\mathcal{S}^c - (P^c - P^f)\delta\mathcal{V}^c + \left(M^S_{AV} - \mu^f_A\right)\delta\mathcal{N}^S_A + \left(M^I_{AV} - \mu^f_A\right)\delta\mathcal{N}^I_A$$
$$+ \left(M^S_{BV} - \mu^f_B\right)\delta\mathcal{N}^S_B + \left(M^I_{CV} - \mu^f_C\right)\delta\mathcal{N}^S_C + \mu_o\delta\mathcal{N}^s_o \quad (2.88)$$

where the constraint $\delta\mathcal{N}^S_A + \delta\mathcal{N}^I_A + \delta\mathcal{N}^f_A = 0$ and constraints similar to those appearing in Eq. (2.15) have been used. Because equilibrium obtains when the perturbation in energy vanished, $\delta\mathcal{E} = 0$, each of the terms in parentheses must vanish identically. In addition to the equality of temperature and pressure between the phases, the following conditions are obtained:

$$M^S_{AV} = M^I_{AV} = \mu^f_A \quad (2.89)$$

$$M^S_{BV} = \mu^f_B \quad (2.90)$$

$$M^I_{CV} = \mu^f_C \quad (2.91)$$

and

$$\mu_o = 0. \quad (2.92)$$

The last condition, Eq. (2.92), gives the number of unit cells in the crystal at equilibrium, in a similar manner that the condition $\mu^c_o = 0$ gave the condition on the number of lattice sites at equilibrium for a substitutional alloy. The remaining three conditions determine chemical equilibrium. Equation (2.89) states that at equilibrium the change in energy associated with the addition of an A atom to a vacant interstitial site must be the same as when the A atom is added to a vacant substitutional site, or the diffusion potential of species A on one sublattice must be equal to that on the other sublattice. These equilibrium conditions show that the crystal is best construed as a compendium of several sublattices, each of which might contain one or more component species.

III. Deformation and Stress

This section develops a means for characterizing the elastic deformation of a material when acted on by internal and external forces. The deformation gradient and the strain tensors are established before defining the different stress tensors. These results are used to define various types of eigenstrains and to obtain expressions for the elastic energy density. Initially, no assumptions are made concerning the origin of the forces or the linearity or smallness of the deformation when deriving the equilibrium conditions. These relationships are useful in establishing the thermodynamic equilibrium conditions for single and two-phase crystals under stress. Subsequently, deformation gradients are assumed to be small and linearized relationships for the stress and strain tensors are defined.

While dyadic notation leads to very compact expressions that are valid for both Cartesian and non-Cartesian tensors, it can represent a significant impediment to understanding the development for those not intimately familiar with it. We thus limit ourselves to Cartesian tensors and usually employ indicial notation. Nevertheless, all of the results generalize naturally to general curvilinear coordinate systems. We shall employ the Einstein summation convention: All repeated indices are implicitly summed from 1 to 3 and a comma appearing in a subscript denotes differentiation with respect to a spatial coordinate. Much of the material appearing in this section is from Reference 22.

5. DEFORMATION

a. *Lagrangian Strain Tensor*

When quantifying deformation, it is necessary to define a state, termed the reference state, from which the deformation can be measured. The reference state is defined to be free of deformation, but it is not necessarily free from stress or internal and external forces acting on it. While there is no unique choice for the reference state, a judicious choice of reference state can significantly simplify the solution of a given problem.

Consider the material body shown in Figure 9. The material is assumed initially to be free of all external forces. The vector $\mathbf{x}'$ gives the position of a point or

---

[22] W. C. Johnson, in *Lectures on the Theory of Phase Transformations*, ed. H. I. Aaronson (Warrendale, PA, TMS-AIME, 2000), 35–134.

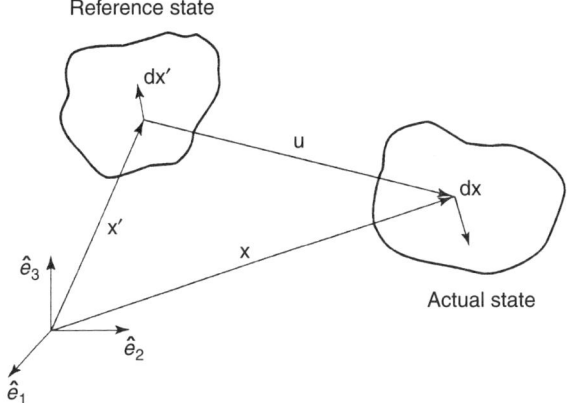

FIG. 9. The vectors **x'** and **x** locate the same material element in the reference and deformed states, respectively, and are connected by the displacement vector **u**. A line element $d\mathbf{x}'$ is rotated and stretched into the line element $d\mathbf{x}$ by the deformation.

volume element in the stress-free material, which is denoted by the reference state in Figure 9. $d\mathbf{x}'$ is an infinitesimal vector located at the position $\mathbf{x}'$, whose direction is arbitrary. When a distribution of forces is applied to the surface of the material, the material can undergo both translation and elastic deformation. During this process, the element of material originally at point $\mathbf{x}'$ is displaced to the point $\mathbf{x}$. The vector $d\mathbf{x}'$ is also displaced, stretched, and rotated and appears in the deformed state as the vector $d\mathbf{x}$. The deformation of the material can be quantified, if a relationship between $d\mathbf{x}'$ and $d\mathbf{x}$ can be found.

For most of the problems to be discussed here, the stress-free state of a particular phase is chosen as the reference state for the measurement of deformation. The position of a material element in the reference state configuration and its position in the deformed state are related by the displacement vector, **u**, as

$$\mathbf{x} = \mathbf{x}' + \mathbf{u}. \tag{3.1}$$

The position of a material element in the deformed state is considered to be a function of its position in the reference state. This relation can be expressed mathematically as

$$\mathbf{x} = \mathbf{x}(\mathbf{x}') \quad \text{or} \quad x_i = x_i(x_1', x_2', x_3') \tag{3.2}$$

where, in the usual Cartesian coordinates, $x_1 = x$, $x_2 = y$ and $x_3 = z$. So long as the deformation does not open up gaps in the material or result in material being folded over upon itself, the inverse relation is also valid.

$$\mathbf{x}' = \mathbf{x}'(\mathbf{x}) \quad \text{or} \quad x_i' = x_i'(x_1, x_2, x_3) \tag{3.3}$$

Equations (3.2) and (3.3) are used to obtain a relationship between the infinitesimal vectors $d\mathbf{x}'$ and $d\mathbf{x}$. Using the chain rule and Eq. (3.2), the components of vectors $d\mathbf{x}$ and $d\mathbf{x}'$ are related by

$$dx_i = \left(\frac{\partial x_i}{\partial x'_j}\right) dx'_j \tag{3.4}$$

or

$$dx_i = F_{ij} dx'_j \tag{3.5}$$

where F_{ij} is the deformation gradient tensor and is defined by[23]

$$F_{ij} = \frac{\partial x_i}{\partial x'_j}. \tag{3.6}$$

The deformation gradient tensor can be considered as an operator that takes the infinitesimal vector $d\mathbf{x}'$ at the point $\mathbf{x}'$ and transforms it into the deformed vector $d\mathbf{x}$ located at the point $\mathbf{x}$. It contains all information on the deformation (stretch and rotation) imparted to $d\mathbf{x}'$ as a result of applied forces. It does not contain information on rigid-body displacements. The deformation gradient can only be used to transform infinitesimal vectors; "longer" vectors could also be "bent" in the deformation process.

Consider a point in the reference state located at $\mathbf{x}'$. When forces are applied, the material in the vicinity of $\mathbf{x}'$ is stretched. The extent of the stretching depends on the displacement direction in the material; two small vectors, $d\mathbf{x}'_1$ and $d\mathbf{x}'_2$, initially of equal length and located at $\mathbf{x}'$, can have different lengths after the deformation, if their initial orientations are different. The strain tensor is constructed from the deformation gradient tensor and is defined so as to provide information on how an infinitesimal vector $d\mathbf{x}'$ located at $\mathbf{x}'$ in the reference state is stretched during the deformation. In defining the strain, only the actual stretching of the material, not rigid-body displacements nor rigid-body rotations, are of concern. In what follows, it is simpler initially to consider the square of the length of the line elements rather than the length itself.

Let dS be the magnitude or length of the material vector $d\mathbf{x}'$ in the undeformed state. Then

$$dS^2 = d\mathbf{x}' \cdot d\mathbf{x}' = dx'_j dx'_j = dx'_j \delta_{jk} dx'_k \tag{3.7}$$

---

[23] L. E. Malvern, *Introduction to the Mechanics of a Continuous Medium*, Prentice-Hall, Inc., Englewood Cliffs, New Jersey (1969).

where δ_{ij} is the Kronecker delta[24]. If ds is the length of line element $d\mathbf{x}$ corresponding to line element $d\mathbf{x}'$ in the deformed state, then

$$ds^2 = d\mathbf{x} \cdot d\mathbf{x} = dx_i dx_i. \tag{3.8}$$

Using Eq. (3.5) to express $d\mathbf{x}$ in terms of $d\mathbf{x}'$ allows Eq. (3.8) to be written as

$$ds^2 = F_{ij} dx'_j F_{ik} dx'_k. \tag{3.9}$$

The "stretch" that $d\mathbf{x}'$ undergoes is related to the difference between Eqs. (3.7) and (3.9):

$$ds^2 - dS^2 = dx'_j F_{ij} F_{ik} dx'_k - dx'_j \delta_{jk} dx'_k \tag{3.10}$$

or

$$ds^2 - dS^2 = dx'_j \left(F_{ij} F_{ik} - \delta_{jk} \right) dx'_k. \tag{3.11}$$

Equation (3.11) can be rewritten as

$$ds^2 - dS^2 = 2 dx'_j E_{jk} dx'_k \tag{3.12}$$

where the strain tensor, E_{jk}, has been defined as

$$E_{jk} = \frac{1}{2} \left(F_{ij} F_{ik} - \delta_{jk} \right). \tag{3.13}$$

When E_{ij} is formulated using the undeformed line elements ($d\mathbf{x}'$), as in Eq. (3.12), it is termed the Lagrangian strain tensor. Other formulations of the strain tensor are also possible. For example, when the term $ds^2 - dS^2$ is written in terms of the deformed configuration ($d\mathbf{x}$), the Eulerian strain tensor is obtained.[23]

Using the definition for the deformation gradient tensor, Eq. (3.6), the strain tensor of Eq. (3.13) can be expressed as

$$E_{jk} = \frac{1}{2}\left[\left(\frac{\partial x_i}{\partial x'_j}\right)\left(\frac{\partial x_i}{\partial x'_k}\right) - \delta_{jk}\right]. \tag{3.14}$$

Replacing the free indices j and k of Eq. (3.14) by the free indices i and j, and the dummy indice i by k, yields the equivalent equation

$$E_{ij} = \frac{1}{2}\left[\left(\frac{\partial x_k}{\partial x'_i}\right)\left(\frac{\partial x_k}{\partial x'_j}\right) - \delta_{ij}\right]. \tag{3.15}$$

---

[24] The Kronecker delta is defined such that $\delta_{ij} = 1$ if $i = j$, and $\delta_{ij} = 0$ if $i \neq j$.

The strain tensor can also be expressed in terms of the displacement vector. Substituting Eq. (3.1) into Eq. (3.6) allows the deformation gradient tensor to be expressed as

$$F_{ij} = \left(\frac{\partial x_i}{\partial x'_j}\right) = \left(\frac{\partial (x'_i + u_i)}{\partial x'_j}\right) = \left(\frac{\partial x'_i}{\partial x'_j}\right) + \left(\frac{\partial u_i}{\partial x'_j}\right) = \delta_{ij} + \left(\frac{\partial u_i}{\partial x'_j}\right). \quad (3.16)$$

Combining Eq. (3.16) with Eq. (3.13) after making the appropriate changes in the free indices gives

$$E_{ij} = \frac{1}{2}\left\{\left[\delta_{ki} + \left(\frac{\partial u_k}{\partial x'_i}\right)\right]\left[\delta_{kj} + \left(\frac{\partial u_k}{\partial x'_j}\right)\right] - \delta_{ij}\right\}. \quad (3.17)$$

Multiplying term by term and noting that $\delta_{ki}\delta_{kj} = \delta_{ij}$, gives for the strain

$$E_{ij} = \frac{1}{2}\left[\left(\frac{\partial u_i}{\partial x'_j}\right) + \left(\frac{\partial u_j}{\partial x'_i}\right) + \left(\frac{\partial u_k}{\partial x'_i}\right)\left(\frac{\partial u_k}{\partial x'_j}\right)\right]. \quad (3.18)$$

It is important to remember that Eq. (3.18), giving the Lagrangian formulation of the strain, is an exact expression and not a second-order approximation to the strain tensor.

b. Small Strain Tensor

For many problems in solid-state phase transformations, the derivatives of the displacement with respect to the material coordinates $\mathbf{x}'$ are small. In such cases, the term containing the product of the displacement gradients is negligible with respect to the linear terms and can be dropped from Eq. (3.18). In this limiting case of the small-strain approximation, the difference between the derivatives with respect to x'_i and x_i can be neglected. The small strain tensor, ϵ_{ij}, is thus defined as

$$\epsilon_{ij} = \frac{1}{2}\left[\left(\frac{\partial u_i}{\partial x'_j}\right) + \left(\frac{\partial u_j}{\partial x'_i}\right)\right] \approx \frac{1}{2}\left[\left(\frac{\partial u_i}{\partial x_j}\right) + \left(\frac{\partial u_j}{\partial x_i}\right)\right]. \quad (3.19)$$

Equation (3.19) indicates that the strain field can be uniquely determined from a known displacement field through differentiation. However, the inverse problem is not so simple: There may be many displacement fields that give the same strain field.

Physical meaning can be imparted to the small strain components by considering the actual change in length of the vector $d\mathbf{x}'$ during deformation. Using Eq. (3.9)

$$ds - dS = [dx'_j F_{ij} F_{ik} dx'_k]^{1/2} - dS. \quad (3.20)$$

If a unit vector, $\mathbf{n}'$, is defined such that it lies in the direction in which $d\mathbf{x}'$ points, then $d\mathbf{x}' = dS\mathbf{n}'$. Since

$$\mathbf{n}' \cdot \mathbf{n}' = n'_j n'_j = n'_j \delta_{jk} n'_k = 1, \qquad (3.21)$$

Equation (3.20) can be written, after adding and subtracting 1 to the term inside the square brackets and using Eq. (3.21), as

$$ds - dS = dS[n'_j F_{ij} F_{ik} n'_k - n'_j \delta_{jk} n'_k + 1]^{1/2} - dS. \qquad (3.22)$$

Introducing the strain tensor from Eq. (3.13) into Eq. (3.22) allows the change in length per-unit-length of line element, or unit extension, to be expressed as

$$\frac{(ds - dS)}{dS} = [1 + 2n'_j E_{jk} n'_k]^{1/2} - 1. \qquad (3.23)$$

Suppose that the unit extension of a line element, initially parallel to the x_1 axis, is to be determined. In this case, $\mathbf{n}' = \hat{e}_1$, where $\hat{e}_1$ is the unit basis vector in the x_1 direction. The components of the unit vector $\mathbf{n}'$ in this case are thus $n'_1 = 1$ and $n'_2 = n'_3 = 0$. Matrix multiplication of $E_{ij} n'_j$ yields the column vector

$$\begin{pmatrix} E_{11} & E_{12} & E_{31} \\ E_{21} & E_{22} & E_{23} \\ E_{31} & E_{32} & E_{33} \end{pmatrix} \begin{pmatrix} 1 \\ 0 \\ 0 \end{pmatrix} = \begin{pmatrix} E_{11} \\ E_{21} \\ E_{31} \end{pmatrix}. \qquad (3.24)$$

Contracting the resulting vector of Eq. (3.24) with the normal n'_i (taking the dot product of the two vectors), $(E_{11}\hat{e}_1 + E_{21}\hat{e}_2 + E_{31}\hat{e}_3) \cdot (\hat{e}_1) = E_{11}$ and Eq. (3.23) becomes

$$\frac{(ds - dS)}{dS} = [1 + 2E_{11}]^{1/2} - 1. \qquad (3.25)$$

If the magnitude of E_{11} is sufficiently small, the unit extension, Eq. (3.25), can be expanded in a Taylor series about $E_{11} = 0$ to give

$$\frac{(ds - dS)}{dS} \approx 1 + \frac{1}{2}(2E_{11}) - 1 = E_{11} = \epsilon_{11} \qquad (3.26)$$

where the small-strain component ϵ_{11} has been substituted for its corresponding term in the Lagrangian formulation E_{11}. Because strains have been assumed to be small, the small-strain component ϵ_{11} can, therefore, be interpreted as the approximate extension, or change in length per unit-length, of a line element initially parallel to the x_1 axis. The same physical interpretation can be given to the other diagonal strain components; ϵ_{22} and ϵ_{33} give the change in length per unit-length of a line element initially parallel to the x_2 and x_3 axes, respectively.

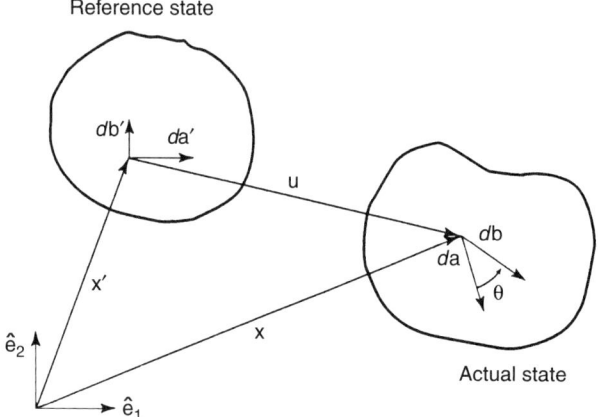

FIG. 10. By examining the relative distortion of two line elements in the $x_1 - x_2$ plane, a physical interpretation of the shear strain can be obtained. Line elements $d\mathbf{a}'$ and $d\mathbf{d}'$ are initially perpendicular. After deformation, the angle between the two line elements is θ.

The physical meaning of the off-diagonal components of the strain tensor can be discerned by considering the relative deformation of two line elements. Figure 10 shows the $x_1 - x_2$ cross-section of a material body in its reference (undeformed) and deformed states. The deformation is assumed to be confined to the $x_1 - x_2$ plane for illustration. Two line elements, $d\mathbf{a}'$ and $d\mathbf{d}'$, are each initially orthogonal and located at $\mathbf{x}'$. $d\mathbf{a}'$ and $d\mathbf{d}'$ are directed along the principal directions $\hat{e}_1$ and $\hat{e}_2$, respectively, such that

$$d\mathbf{a}' = dS_a \hat{e}_1 \quad \text{and} \quad d\mathbf{d}' = dS_b \hat{e}_2 \tag{3.27}$$

where dS_a and dS_b are the magnitudes of the undeformed line elements $d\mathbf{a}'$ and $d\mathbf{d}'$. $d\mathbf{a}$ and $d\mathbf{b}$ are the corresponding line elements in the deformed state. If the magnitudes of $d\mathbf{a}$ and $d\mathbf{b}$ are ds_a and ds_b, respectively, the angle between $d\mathbf{a}$ and $d\mathbf{b}$ in the deformed state, θ, is obtained from the dot product as

$$d\mathbf{a} \cdot d\mathbf{b} = ds_a ds_b \cos\theta = da_k db_k. \tag{3.28}$$

Expressing the line elements in the deformed state with respect to the undeformed state using the deformation gradient tensor, Eq. (3.5), gives

$$ds_a ds_b \cos\theta = F_{ki} da'_i F_{kj} db'_j. \tag{3.29}$$

Using Eq. (3.27) and noting that $F_{ki} da'_i = F_{k1} da'_1 = F_{k1} dS_a$, gives

$$ds_a ds_b \cos\theta = dS_a F_{k1} F_{k2} dS_b. \tag{3.30}$$

The magnitudes ds_a and ds_b can be expressed in terms of dS_a and dS_b, respectively, using Eq. (3.23) (or Eq. (3.25)) as

$$ds_a = dS_a(1 + E_{11}) \quad \text{and} \quad ds_b = dS_b(1 + E_{22}). \tag{3.31}$$

Substituting for ds_a and ds_b and solving Eq. (3.30) for $\cos\theta$ gives

$$\cos\theta = \frac{2E_{12}}{[1 + 2E_{11}]^{1/2}[1 + 2E_{22}]^{1/2}} \tag{3.32}$$

where we have used $F_{k1}F_{k2} = 2E_{12}(= 2E_{21})$ from the definition of strain, Eq. (3.13). Because the line elements $d\mathbf{a}'$ and $d\mathbf{d}'$ were initially orthogonal, the change in angle (ϕ) between $d\mathbf{a}'$ and $d\mathbf{d}'$ caused by the deformation is $\phi = \pi/2 - \theta$. Using the trigonometric identity, $\cos\theta = \cos(\pi/2 - \phi) = \sin\phi$. If the displacement gradient is assumed to be small, the angle ϕ is small and the small-strain approximation can be employed. Expanding $\sin\phi$ in a Taylor series about $\phi = 0$ and the righthand-side of Eq. (3.32) in a Taylor series in the strain components about $E_{ij} = 0$, one obtains, to first-order,

$$\cos\theta = \sin\phi \approx \phi \approx 2E_{12} = 2\epsilon_{12} = 2\epsilon_{21}. \tag{3.33}$$

Thus the off-diagonal terms of the small strain tensor are symmetric and equal to one-half the change in angle owing to the deformation; i.e., $\epsilon_{12} = \epsilon_{21} = \phi/2$.

Either the deformation gradient tensor, F_{ij}, or the strain tensor, E_{ij}, can be used to describe the deformation of the material. The small strain tensor, ϵ_{ij}, is a valid description when the displacement gradients are small. This is a reasonable assumption for many problems in solid-state phase transformations but is certainly not always valid. We will use this physical interpretation of the small strain tensor to motivate expressions for the eigenstrains.

c. Area and Volume Changes

The ratio of the volume of an element in the deformed state to the volume of the corresponding element in the reference state can be approximated by considering the deformation of an initially orthogonal volume element (cuboid). Assume that the cuboid in the reference state has edges defined by the line elements, $d\mathbf{a}'$, $d\mathbf{d}'$, and $d\mathbf{c}'$, which are parallel to the reference axes and which are of length dS_1, dS_2, and dS_3, respectively. The unit extension of the edge originally parallel to the x_1 axis, $dvap$, is given by Eq. (3.25), $ds_1 = (1 + 2E_{11})^{1/2}dS_1$; similar expressions obtain for the other two edges. If a simple deformation is first imagined in which the volume element remains a cuboid, then the volume of the element in the deformed state, dV, is

$$dV = ds_1 ds_2 ds_3 = [(1 + 2E_{11})(1 + 2E_{22})(1 + 2E_{33})]^{1/2} dS_1 dS_2 dS_3. \tag{3.34}$$

Defining J as the ratio of the volumes of the element in the deformed and reference states gives

$$\frac{dV}{dV'} = \frac{ds_1 ds_2 ds_3}{dS_1 dS_2 dS_3} = J = [(1 + 2E_{11})(1 + 2E_{22})(1 + 2E_{33})]^{1/2}. \quad (3.35)$$

If Eq. (3.35) is expanded in a Taylor series in the strain components about $E_{ij} = 0$, and only the leading term in the strain is retained, the small-strain approximation is obtained,

$$\frac{dV}{dV'} = J \approx 1 + E_{11} + E_{22} + E_{33} = 1 + \epsilon_{kk} \quad (3.36)$$

where $\epsilon_{kk} = \epsilon_{11} + \epsilon_{22} + \epsilon_{33}$ is the trace of the strain tensor. In the small-strain approximation, the trace of the strain tensor gives the change in volume per-unit-volume owing to the deformation

$$\frac{dV - dV'}{dV'} \approx \epsilon_{kk}. \quad (3.37)$$

The small strain approximation of Eq. (3.36) can also be obtained directly from Eq. (3.6). We first note that the deformation gradient tensor, F_{ij}, can be expressed as

$$F_{ij} = \frac{\partial x_i}{\partial x'_j} = \frac{1}{2}\left[\frac{\partial x_i}{\partial x'_j} + \frac{\partial x_j}{\partial x'_i}\right] + \frac{1}{2}\left[\frac{\partial x_i}{\partial x'_j} - \frac{\partial x_j}{\partial x'_i}\right]. \quad (3.38)$$

Using the identity, $x_i = x'_i + u_i$, allows Eq. (3.38) to be rewritten as

$$F_{ij} = \delta_{ij} + \frac{1}{2}\left[\frac{\partial u_i}{\partial x'_j} + \frac{\partial u_j}{\partial x'_i}\right] + \frac{1}{2}\left[\frac{\partial u_i}{\partial x'_j} - \frac{\partial u_j}{\partial x'_i}\right]. \quad (3.39)$$

In the limit of small deformations, the difference between the operations $\partial/\partial x'_i$ and $\partial/\partial x_i$ is negligible. Equation (3.38) can thus be approximated in terms of the small strain tensor, ϵ_{ij}, (see Eq. (3.19)), and the antisymmetric rotation tensor, Ω_{ij}, as

$$F_{ij} = \delta_{ij} + \epsilon_{ij} + \Omega_{ij} \quad (3.40)$$

where

$$\Omega_{ij} = \frac{1}{2}\left[\frac{\partial u_i}{\partial x_j} - \frac{\partial u_j}{\partial x_i}\right]. \quad (3.41)$$

Because Ω is antisymmetric, it does not contribute to the determinant $|F_{ij}| = J$, and $J = |\delta_{ij} + \epsilon_{ij}|$ in the small strain limit. Calculating the determinant of the symmetric tensor $\delta_{ij} + \epsilon_{ij}$ using Eq. (3.44), yields Eq. (3.36) to first order in the strain.

In a more general deformation, the angles between the cuboid edges are distorted from right angles. In this case, the volume of the distorted cuboid element becomes

$$dV = (d\mathbf{a} \times d\mathbf{b}) \cdot d\mathbf{c} = \epsilon_{ijk} da_j db_k dc_i = \epsilon_{ijk} F_{jm} F_{kn} F_{il} da'_m db'_n dc'_l \quad (3.42)$$

where ϵ_{ijk} is the permutation tensor. Because the edges of the cuboid in the reference state were chosen parallel to the coordinate axes, Eq. (3.42) simplifies to

$$dV = \epsilon_{ijk} F_{j1} F_{k2} F_{i3} dS_1 dS_2 dS_3 = \det|F_{ij}| dS_1 dS_2 dS_3. \quad (3.43)$$

Thus the change in volume of a small volume element owing to a general deformation F_{ij} is

$$\frac{dV}{dV'} = J = \det|F_{ij}| \quad (3.44)$$

where J is the determinant of the deformation gradient tensor.

As shown by Euler in 1762, if $f(\mathbf{x})$ is a continuous scalar function of position, then the integral of f over the deformed state can be expressed as an integral over the reference state ($\mathbf{x}'$) coordinates as

$$\int_V f(\mathbf{x}) dx_1 dx_2 dx_3 = \int_{V'} f(\mathbf{x}(\mathbf{x}')) J dx'_1 dx'_2 dx'_3 = \int_{V'} f(\mathbf{x}') dx'_1 dx'_2 dx'_3. \quad (3.45)$$

For example, the mass of a region V, $\mathcal{M}$, can be calculated from volume integrals of the density ρ in either the deformed or reference states:

$$\mathcal{M} = \int_V \rho(\mathbf{x}) dx_1 dx_2 dx_3 = \int_{V'} \rho(\mathbf{x}(\mathbf{x}')) J dx'_1 dx'_2 dx'_3$$

$$= \int_{V'} \rho'(\mathbf{x}') dx'_1 dx'_2 dx'_3 \quad (3.46)$$

where ρ' is the density of the material as measured in the reference state.

A relationship between the elements of surface area in the reference and deformed states can also be obtained. Let $d\mathbf{a}'$ and $d\mathbf{d}'$ be line elements that lie in the surface of the reference body and let $d\mathbf{c}' = ds'_c \mathbf{n}'$ be a line element that lies parallel to the surface normal of the reference body $\mathbf{n}'$. $d\mathbf{a}'$ and $d\mathbf{d}'$ delineate an infinitesimal element of surface, which has an area, dS', given by

$$dS' = |d\mathbf{a}' \times d\mathbf{d}'| = (d\mathbf{a}' \times d\mathbf{d}') \cdot \mathbf{n}'. \quad (3.47)$$

Let $d\mathbf{a}$, $d\mathbf{b}$, and $d\mathbf{c}$ be the respective line elements in the deformed state corresponding to $d\mathbf{a}$, $d\mathbf{b}$, and $d\mathbf{c}$. $d\mathbf{a}$ and $d\mathbf{b}$ must still reside in the surface after deformation, but $d\mathbf{c}$ does not necessarily lie parallel to the surface normal $\mathbf{n}$. From Eq. (3.44),

$$J(d\mathbf{a}' \times d\mathbf{d}') \cdot d\mathbf{c}' = (d\mathbf{a} \times d\mathbf{b}) \cdot d\mathbf{c}. \tag{3.48}$$

Using Eq. (3.47) and the corresponding equation for the deformed state, Eq. (3.48) becomes

$$J(dS'n'_j)(ds'_c n'_j) = (dSn_i)(F_{ij} n'_j ds'_c) \tag{3.49}$$

which upon rearrangement yields

$$[Jn'_j dS' - n_i F_{ij} dS]n'_j = 0. \tag{3.50}$$

Equation (3.50) is satisfied only when the term within the brackets vanishes. Thus the relationship between the areas of the surface elements of the deformed and undeformed material is

$$n_j dS = Jn'_i F_{ij}^{-1} dS' \tag{3.51}$$

where F_{ij}^{-1} is the inverse of the deformation gradient tensor. Equation (3.51) is known as Nanson's formula.[23] Equation (3.51) is used to transform a surface integral over the reference state to a surface integral over the actual state. For example,

$$\int_S q_j n_j dS = \int_{S'} q_j Jn'_i F_{ij}^{-1} dS' \tag{3.52}$$

where q_i is some vector function and S and S' are the surface of the crystal as represented in the actual and reference states, respectively.

6. STRESS

If a small cut is made at a given position in the crystal with the unit normal $\mathbf{n}$ to the resultant surface, there is a force exerted by the material into which $\mathbf{n}$ points on the material from which $\mathbf{n}$ points. This force measured per unit area of surface is the traction vector $\mathbf{t}$. Changing the orientation of the cut through the same point results in a different traction vector. The Cauchy stress tensor, σ_{ij}, relates the unit normal to the small cut to the traction vector according to

$$t_i = \sigma_{ji} n_j. \tag{3.53}$$

In the absence of an applied moment, σ_{ij} is a symmetric tensor. Other definitions of the stress are possible. For example, instead of referring all quantities to the actual state of the system as does Eq. (3.53), the traction vector could be expressed as the force per unit area of the reference state. This pseudo-traction vector is designated $\mathbf{t}^o$. The first Piola-Kirchhoff stress tensor, T_{ji}, relates the surface normal for the small cut in the reference state, $\mathbf{n}'$, to the traction vector $\mathbf{t}^o$ according to

$$t_i^o = T_{ji} n_j'. \tag{3.54}$$

Because the first Piola-Kirchhoff stress tensor refers quantities to the reference state, it is especially convenient when formulating the thermodynamics of stressed crystals. The disadvantage of the first Piola-Kirchhoff tensor is that it is nonsymmetric. When the force is rotated along with the material back to the reference state and expressed with respect to a unit area of the reference state, the tensor is symmetric. This is the second Piola-Kirchhoff stress tensor, but the equations of motion are considerably more complicated.[23]

The first Piola-Kirchhoff stress tensor can be related to the Cauchy stress by considering a force $\mathbf{dP}$ acting on a surface element of area dS with normal $\mathbf{n}$. This force can be expressed in terms of either the actual or reference state as

$$T_{ji} n_j' dS' = dP_i = \sigma_{ji} n_j dS \tag{3.55}$$

where dS_o is the area of surface element in the undeformed or reference configuration. Using Nanson's relation, Eq. (3.51), in Eq. (3.55) yields

$$T_{ji} n_j' dS' = J \sigma_{ji} n_k' F_{kj}^{-1} dS'. \tag{3.56}$$

Collecting terms gives

$$\left[T_{ki} - J \sigma_{ji} F_{kj}^{-1} \right] n_k' dS' = 0. \tag{3.57}$$

For Eq. (3.57) to hold requires

$$T_{ki} = J \sigma_{ji} F_{kj}^{-1}. \tag{3.58}$$

Equation (3.58) gives the relationship between the Cauchy and first Piola-Kirchhoff stress tensors in terms of the deformation gradient tensor.

7. Mechanical Equilibrium and Elastic Work

The elastic energy stored in a material body and the conditions for mechanical equilibrium, expressed in terms of derivatives of the components of the stress tensor, are developed in this subsection. These expressions are derived first for the actual state (Eulerian) configuration and then for the reference state (Lagrangian) configuration.

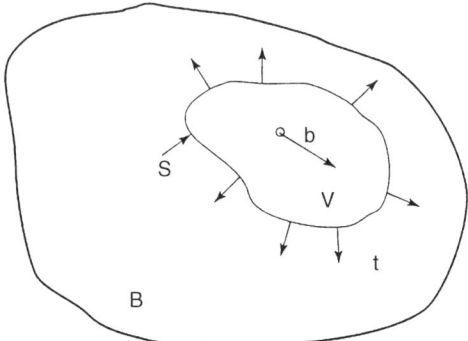

FIG. 11. An arbitrary volume V with surface S is selected within the material B in order to derive the conditions for mechanical equilibrium. **b** is a body force and **t** is the traction acting on surface S.

a. Actual State Representation

Consider the material body B depicted in Figure 11. Choose an arbitrary region V of the material enclosed by the surface S. The region V is acted on by a body force $\mathbf{b}(\mathbf{x})$ (force per unit volume) while a set of tractions acts on the surface S. The region V is in mechanical equilibrium when the sum of all forces acting on the region vanishes. The mathematical expression for the balance of forces when expressed with reference to the deformed state is

$$\int_S t_i dS + \int_V b_i dV = 0. \tag{3.59}$$

Expressing the traction in terms of the Cauchy stress tensor using Eq. (3.53) gives for Eq. (3.59)

$$\int_S \sigma_{ji} n_j dS + \int_V b_i dV = 0. \tag{3.60}$$

Applying the divergence theorem to the surface integral of Eq. (3.60) yields[25]

$$\int_S \sigma_{ji} n_j dS = \int_V \frac{\partial \sigma_{ji}}{\partial x_j} dV. \tag{3.61}$$

Substituting Eq. (3.61) into Eq. (3.60) gives

$$\int_V \left\{ \frac{\partial \sigma_{ji}}{\partial x_j} + b_i \right\} dV = 0. \tag{3.62}$$

---

[25] This equation is an extension of the more familiar form of the divergence theorem applicable to vectors: $\int_V \vec{\nabla} \cdot \mathbf{u} \, dV = \int_S \mathbf{u} \cdot \mathbf{n} dS$ which, in indicial notation, becomes $\int_V \frac{\partial u_i}{\partial x_i} dV = \int_S u_i n_i dS$.

Because the region V can be chosen arbitrarily, Eq. (3.62) is identically satisfied only when the integrand vanishes identically. Mechanical equilibrium is therefore obtained when

$$\frac{\partial \sigma_{ji}}{\partial x_j} + b_i = \sigma_{ji,j} + b_i = 0, \qquad (3.63)$$

where a comma appearing in the subscript implies differentiation with respect to x_j.[26] Equation (3.63) is a partial differential equation which must be satisfied at all points in the material if the material is in mechanical equilibrium.

Assume the material body shown in Figure 11 to be in mechanical equilibrium. If each volume and surface element is imagined to undergo a small, or virtual, displacement,[27] $\delta \mathbf{u}(\mathbf{x})$, then the corresponding virtual work (force multiplied by displacement), δW_{ext}, performed by the surface traction and body forces is

$$\delta W_{ext} = \int_S t_i \delta u_i dS + \int_V b_i \delta u_i dV. \qquad (3.64)$$

The virtual displacement field is arbitrary, so long as the structural integrity of the material is maintained (i.e., no gaps or material overlap occur). Expressing once again the traction vector in terms of the local stress tensor (using Eq. (3.54)) and invoking the divergence theorem, Eq. (3.53), gives

$$\delta W_{ext} = \int_V \left\{ \frac{\partial (\sigma_{ji} \delta u_i)}{\partial x_j} + b_i \delta u_i \right\} dV. \qquad (3.65)$$

Differentiating and rearranging gives

$$\delta W_{ext} = \int_V \{(\sigma_{ji,j} + b_i)\delta u_i + \sigma_{ji}\delta u_{i,j}\} dV. \qquad (3.66)$$

Because the term multiplying δu_i is identically zero when the system is in mechanical equilibrium (see Eq. (3.63)), the work performed by the virtual displacement, δu_i, is

$$\delta W_{ext} = \int_V \sigma_{ji} \delta u_{i,j} dV = \int_V \sigma_{ij} \delta u_{j,i} dV \qquad (3.67)$$

where the second equality follows from the first when the dummy indices are exchanged. Because the stress tensor σ_{ij} is symmetric in i and j, $\sigma_{ij} = \sigma_{ji}$, and

---

[26] The summation convention still applies: $\sigma_{ji,j} = \sigma_{1i,1} + \sigma_{2i,2} + \sigma_{3i,3}$.
[27] The use of $\delta \mathbf{u}$ represents an imaginary displacement performed in such a way that the forces acting on the body are not changed.

Eq. (3.67) yields

$$\delta W_{ext} = \int_V \frac{1}{2}\sigma_{ij}(\delta u_{i,j} + \delta u_{j,i})\, dV \tag{3.68}$$

or, from Eq. (3.19),

$$\delta W_{ext} = \int_V \sigma_{ij}\delta\epsilon_{ij}\, dV \tag{3.69}$$

where $\delta\epsilon_{ij}$ is the virtual strain induced by the virtual displacement.

Equation (3.69) indicates that there is a change in the elastic strain energy of each volume element that is produced by the small displacement δu_i. The change in strain energy density of a volume element, δe_v, is given by

$$\delta e_v = \sigma_{ij}\delta\epsilon_{ij}. \tag{3.70}$$

If the virtual strain is considered infinitesimal, Eq. (3.70) can be written in terms of derivatives as

$$de_v = \sigma_{ij}d\epsilon_{ij}. \tag{3.71}$$

Equation (3.71) connects a small change in the state of strain with a change in the elastic energy density of the volume element. The total elastic energy density of the volume element can thus be obtained by integrating Eq. (3.71) from the unstrained state (the state of zero elastic energy) to the given state of strain. Consequently,

$$\int_0^{e_v} de_v = e_v = \int_0^{\epsilon_{ij}} \sigma_{ij}d\epsilon_{ij}. \tag{3.72}$$

e_v is the elastic energy associated with an infinitesimal volume element and is, therefore, called the elastic strain energy density. (e_v has units of energy per volume.) Integrating the elastic strain energy density over the entire volume of material (i.e., over all volume elements) gives the elastic strain energy of the material.

Integration of Eq. (3.72) in order to obtain the elastic energy stored in a unit volume of crystal requires knowledge of how the stress depends on the state of strain. Such constitutive laws are examined in the following section.

b. Reference State Representation

The mechanical equilibrium conditions and expressions for the elastic strain energy density can also be expressed in terms of quantities of the reference state. In order to calculate the mechanical equilibrium conditions, the sum of the forces acting on the body are set to zero.

$$0 = \int_{S'} t_i^o dS + \int_{V'} b_i^o dV \tag{3.73}$$

where b_i^o is the body force density expressed per unit reference volume ($\mathbf{b} = J^{-1}\mathbf{b}^o$). Expressing the traction vector $\mathbf{t}^o$ in terms of the first Piola-Kirchhoff stress tensor, Eq. (3.54), and using the divergence theorem yields

$$0 = \int_{V'} \left\{ \frac{\partial T_{ji}}{\partial x_j'} + b_i^o \right\} dV. \tag{3.74}$$

Mechanical equilibrium is obtained when the integrand vanishes identically.

$$\frac{\partial T_{ji}}{\partial x_j'} + b_i^o = 0 \tag{3.75}$$

Equation (3.75) gives the condition for mechanical equilibrium in terms of the first Piola-Kirchhoff stress tensor.

The virtual work performed when a system in mechanical equilibrium undergoes a small, virtual displacement can also be expressed in terms of an integral over the reference state as

$$\begin{aligned} \delta W_{ext} &= \int_{S'} t_i^o \delta u_i dS + \int_{V'} b_i^o \delta u_i dV \\ &= \int_{S'} n_j' T_{ji}^o \delta x_i dS + \int_{V'} b_i^o \delta x_i dV \end{aligned} \tag{3.76}$$

Using the divergence theorem on the surface integral of Eq. (3.76) gives

$$\begin{aligned} \delta W_{ext} &= \int_{V'} \left\{ \left[\frac{\partial T_{ji}}{\partial x_j'} + b_i^o \right] \delta x_i + T_{ji} \frac{\partial \delta x_i}{\partial x_j'} \right\} dV \\ &= \int_{V'} T_{ji} \delta F_{ij} dV \end{aligned} \tag{3.77}$$

The first term of the integrand vanishes identically when the system is in mechanical equilibrium. The term δF_{ij} is the virtual change in the deformation gradient tensor owing to a virtual change in the displacement field. The change in the elastic strain energy density, $\delta e_{v'}$, is thus

$$\delta e_{v'} = T_{ji} \delta F_{ij} \quad \text{or} \quad de_{v'} = T_{ji} dF_{ij}. \tag{3.78}$$

When thermodynamic quantities are referred to the reference state, the natural thermodynamic variable to associate with the deformation is the gradient deformation tensor F_{ij}. Its conjugate variable is the first Piola-Kirchhoff stress tensor T_{ij}.

8. CONSTITUTIVE EQUATIONS FOR SMALL STRAIN

The displacement gradient tensor and strain tensor are used to describe the material deformation. The stress tensor is used to describe the forces acting at a point in the material. In order to relate the material deformation to the applied force, however, a constitutive equation for the material must be postulated. In this subsection, several constitutive laws relating deformation to an elastic stress are formulated.

We consider a solid at fixed composition and temperature. If the material is elastic, the deformation is sufficiently small, and the reference state for the measurement of strain is taken as the stress-free state of the material, the relationship between the stress and strain components is linear. Thus,

$$\sigma_{ij} = C_{ijkl}\epsilon_{kl} \tag{3.79}$$

where C_{ijkl} is the elastic stiffness tensor (elastic constants). C_{ijkl} is a fourth-rank tensor that connects the second-rank strain tensor with the second-rank stress tensor. Each component of the stiffness tensor can be viewed as an elastic constant connecting a component of the strain tensor with a component of the stress tensor. The strain tensor can be expressed in terms of the stress components in like manner:

$$\epsilon_{mn} = S_{mnij}\sigma_{ij} \tag{3.80}$$

where S_{mnij} is the elastic compliance tensor and is the inverse of the elastic constants tensor. The compliance and stiffness tensors are inverse tensors and must satisfy

$$S_{mnij}C_{ijkl} = \frac{1}{2}\left(\delta_{km}\delta_{ln} + \delta_{lm}\delta_{kn}\right). \tag{3.81}$$

In a crystal, the 81 individual elastic constants are neither unique nor independent. The symmetry of the elastic strain ($\epsilon_{ij} = \epsilon_{ji}$) and stress ($\sigma_{ij} = \sigma_{ji}$) tensors requires that $C_{ijkl} = C_{ijlk}$ and $C_{ijkl} = C_{jikl}$ so that there are, at most, 36 independent elastic constants. The point group symmetry of the crystal imposes additional constraints on the number of independent elastic constants[28] requiring some of the elastic constants to be identically zero while requiring other components to be equal. The elastic constants are a function of the temperature and composition.

---

[28] J. F. Nye, *Physical Properties of Materials*, Clarendon Press, Oxford (1985).

For an isotropic material, there are only two independent elastic constants: Here we choose one corresponding to the material's resistance to the application of a shear stress (the shear modulus, μ) and the other to the material's resistance to the application of a uniaxial stress (the elastic modulus or Young's modulus, E). Using these elastic constants, the elastic stiffness tensor for an isotropic material can be written as[29]

$$C_{ijkl} = \frac{\mu(E-2\mu)}{(3\mu-E)}\delta_{ij}\delta_{kl} + \mu(\delta_{ik}\delta_{jl} + \delta_{il}\delta_{jk}). \quad (3.82)$$

This relationship is often expressed in terms of Lame's constants as

$$C_{ijkl} = \lambda\delta_{ij}\delta_{kl} + \mu(\delta_{ik}\delta_{jl} + \delta_{il}\delta_{jk}) \quad (3.83)$$

where $\lambda = \mu(E-2\mu)/(3\mu - E) = 2\mu\nu/(1-2\nu)$.

For a cubic material, there are three independent elastic constants, C_{11}, C_{12}, and C_{44}. The stiffness matrix can be expressed as

$$C_{ijkl} = C_{12}\delta_{ij}\delta_{kl} + C_{44}(\delta_{ik}\delta_{jl} + \delta_{il}\delta_{jk}) + (C_{11} - C_{12} - 2C_{44})\delta_{ijkl} \quad (3.84)$$

where the function $\delta_{ijkl} = 1$ if $i = j = k = l$, and is zero otherwise. When $C_{11} = C_{12} + 2C_{44}$, an isotropic system is recovered with the recognition that $C_{12} = \lambda$ and $C_{44} = \mu$.

The components of the compliance tensor can be expressed in terms of the elastic constants by solving Eq. (3.81). For a cubic system,

$$S_{ijkl} = S_{12}\delta_{ij}\delta_{kl} + S_{44}(\delta_{ik}\delta_{jl} + \delta_{il}\delta_{jk}) + (S_{11} - S_{12} - 2S_{44})\delta_{ijkl} \quad (3.85)$$

where

$$S_{11} = \frac{(C_{11} + C_{12})}{(C_{11} + 2C_{12})(C_{11} - C_{12})} \quad (3.86)$$

$$S_{12} = \frac{-C_{12}}{(C_{11} + 2C_{12})(C_{11} - C_{12})} \quad (3.87)$$

and

$$S_{44} = \frac{1}{4C_{44}}. \quad (3.88)$$

For an isotropic material, $S_{11} = S_{12} + 2S_{44}$.

Using a constitutive law connecting the stress and strain fields, the elastic energy stored in each volume element can be calculated by integrating the strain energy

---

[29] Y. C. Fung, *A First Course in Continuum Mechanics*, Prentice Hall, New Jersey (1969).

density of Eq. (3.72). An expression for the strain energy density of an isotropic material under a general state of strain is obtained by combining Eqs. (3.79) and (3.83) and then substituting the result into Eq. (3.72).

$$e_v = \int_0^{\epsilon_{ij}} \sigma_{ij} d\epsilon_{ij} = \int_0^{\epsilon_{ij}} (\lambda \epsilon_{kk} \delta_{ij} + 2\mu \epsilon_{ij}) d\epsilon_{ij}. \qquad (3.89)$$

Breaking the integral into two parts and contracting gives

$$e_v = \int_0^{\epsilon_{kk}} \lambda \epsilon_{kk} d\epsilon_{kk} + \int_0^{\epsilon_{ij}} 2\mu \epsilon_{ij} d\epsilon_{ij}. \qquad (3.90)$$

Integrating each term yields

$$\begin{aligned} e_v &= \frac{1}{2}\lambda \epsilon_{kk}^2 + \mu \epsilon_{ij}\epsilon_{ij} \\ &= \frac{1}{2}[\lambda \epsilon_{kl}\delta_{kl}\epsilon_{ij}\delta_{ij} + 2\mu \epsilon_{ij}\epsilon_{kl}\delta_{ik}\delta_{jl}] \\ &= \frac{1}{2}[\lambda \delta_{ij}\delta_{kl} + \mu(\delta_{ik}\delta_{jl} + \delta_{il}\delta_{jk})]\epsilon_{ij}\epsilon_{kl} = \frac{1}{2}C_{ijkl}\epsilon_{ij}\epsilon_{kl} = \frac{1}{2}\sigma_{ij}\epsilon_{ij} \quad (3.91) \end{aligned}$$

The strain energy density can also be obtained for the general case of an anisotropic crystal. From Eqs. (3.70) and (3.79),

$$de_v = C_{ijkl}\epsilon_{kl} d\epsilon_{ij}. \qquad (3.92)$$

Because the components of the elastic stiffness tensor are constant for a linear system,

$$\int_0^{e_v} de_v = e_v = \int_0^{\epsilon_{ij}} C_{ijkl}\epsilon_{kl} d\epsilon_{ij} = \frac{1}{2}C_{ijkl}\epsilon_{ij}\epsilon_{kl} = \frac{1}{2}\sigma_{ij}\epsilon_{ij} \qquad (3.93)$$

where the strain is measured with respect to the unstressed state of the phase. As shown in the next section, this expression must be modified when eigenstrains are present in the crystal.

9. Eigenstrains

It is often convenient to define the reference state for the measurement of strain in a multi-phase system as the stress-free state of one of the phases. As a consequence, regions within the system with a temperature, composition, or phase structure that differs from those of the reference state experience a non-zero stress, even though the strain is defined to be zero. The strain necessary to transform a volume element of the reference state to its natural, stress-free state at a given composition, structure, and temperature is termed its eigenstrain. The constitutive law for the stress must include those contributions arising from the thermal, compositional, and transformation eigenstrains as well as the contribution arising from the mechanical deformation as measured from the reference state. Because the stress state of a volume element is related to the deformation as measured from the stress-free state of the element, the stress state of a linear elastic system is obtained from the superposition of the stresses arising from the eigenstrains and the deformation of the system. Thus,

$$\sigma_{ij} = C_{ijkl}\left(\epsilon_{kl} - \epsilon_{kl}^p\right) \tag{3.94}$$

where ϵ_{ij}^p is the eigenstrain. The minus sign occurs because the eigenstrain is defined with respect to the reference state of the system. Therefore, a strain of $-\epsilon_{ij}^p$ must be applied to the stress-free element to bring it into coincidence with the reference state. The stress required to produce this eigenstrain is $-C_{ijkl}\epsilon_{kl}^p$.

In the following sections, we express the eigenstrain in terms of the composition, thermal, and phase field variables.

a. Misfit (Transformation) Strain

The misfit or transformation strain, ϵ_{ij}^T, is the eigenstrain used to describe the deformation when two or more phases are present. Two common examples for which a transformation strain is necessary to describe the stress state are the dispersion of a precipitate phase in a matrix or a crystalline film on a substrate. The transformation strain is a measure of the stress incorporated into the reference state when a volume element of the reference state phase is replaced by a volume element of a different phase.

The unit cell of a phase is determined by six lattice parameters, three edge lengths, and three angles specifying the angle between the edges. If a volume element of the reference material (α) is removed and allowed to transform to β phase at constant temperature and composition, the shape and dimensions of the volume element will change. The number of variables required to describe this change depends on the point group symmetry of each phase: A cubic crystal requires only one lattice parameter while a triclinic crystal would require all six.

In Section III.2, the diagonal components of the small strain tensor were shown to give the relative length change of a volume element and the off-diagonal components were a measure of the change in angle between cell edges. As such, the diagonal components of the transformation strain tensor can be used to describe the change in edge length of a transformed volume element and the off-diagonal components to describe changes in the angles. For example, consider two phases, α and β, which possess a cubic crystal structure with lattice parameters a^α and a^β, respectively (see Fig. 12). Let stress-free α be used as the reference state for the measurement of strain. Remove a unit cell of α from the reference state and allow

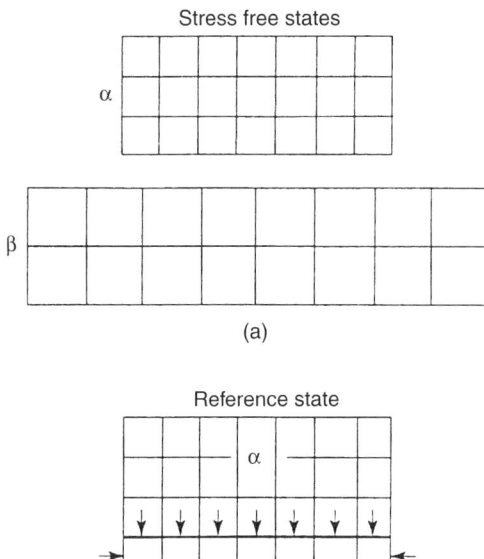

FIG. 12. In their unstressed states, the α and β phases, here assumed to be rectangular plates, possess different lattice parameters. If unstressed α is chosen as the reference state for deformation, forces must be applied to the β phase to bring the two lattices into coincidence, as shown in the coherent structure of (b). (Arrows represent constraining forces acting on β.)

it to transform to β. The transformation required to bring the β lattice back into coincidence with the α lattice requires no shear deformation, so the off-diagonal components of the eigenstrain tensor vanish. The diagonal terms of the eigenstrain tensor represent an extension of the cell edges, and we have from Eq. (3.26)

$$\epsilon_{11}^T = \epsilon_{22}^T = \epsilon_{33}^T = (a^\beta - a^\alpha)/a^\alpha. \tag{3.95}$$

In tensor form the transformation tensor becomes

$$\epsilon_{ij}^T = \frac{(a^\beta - a^\alpha)}{a^\alpha}\delta_{ij} = \epsilon^T \delta_{ij} \tag{3.96}$$

where $\epsilon^T = (a^\beta - a^\alpha)/a^\alpha$ is the dilatational misfit.

If the cubic phase transforms to a tetragonal phase, two distinct diagonal components of the transformation strain tensor are required to describe the change in lengths of the unit cell. However, there are three different ways to describe the transformation strain. If the tetragonal axis is directed along the x_3 direction,

$\epsilon_{11}^T = \epsilon_{22}^T \neq \epsilon_{33}^T$. If the tetragonal axis is directed along the x_1 direction, $\epsilon_{11}^T \neq \epsilon_{22}^T = \epsilon_{33}^T$. If the tetragonal axis is directed along the x_2 direction, $\epsilon_{22}^T \neq \epsilon_{11}^T = \epsilon_{33}^T$. Thus there are three different representations of the transformation matrix, each corresponding to a specific crystallographic variant of the new phase. In describing the microstructure of a system, it is often necessary to consider all three variants. Extension to the transformation strain of crystals with less symmetry proceeds in the same manner, with the off-diagonal components representing changes in the angle between cell edges (measured in radians).

The eigenstrain is not restricted to two phases possessing a common crystal structure, for it is possible to relate, in a continuum sense, a volume element of the reference crystal to the shape and size the volume element would have when transformed to the crystal structure of the new phase. The transformation strain is still defined as above.

b. Thermal Strain

A non-uniform temperature field will also induce an eigenstrain, ϵ_{ij}^θ, as all six lattice parameters can depend in different ways on the temperature. The reference state for measurement of deformation is chosen to be the lattice of the homogeneous crystal at a given reference temperature, θ_r. If an element of material is removed from the reference state and its temperature is changed to θ at constant composition and without change of phase, the changes in the dimensions and shape of the element can be related directly to the components of the thermal strain tensor in precisely the same way as the changes in the unit cell dimensions are related to the transformation strain.

Once again assume a cubic material where an increase in temperature induces a change in the lattice parameter, a. Expanding the lattice parameter in a Taylor series in the temperature about the lattice parameter found in the reference state, the lattice parameter at the absolute temperature θ is approximated as

$$a(\theta) = a(\theta_r) + \left(\frac{\partial a}{\partial \theta}\right)_{\theta=\theta_r} (\theta - \theta_r) + \frac{1}{2}\left(\frac{\partial^2 a}{\partial \theta^2}\right)_{\theta=\theta_r} (\theta - \theta_r)^2 + \cdots. \quad (3.97)$$

The relative change in lattice parameter induced by the temperature difference gives the diagonal components of the thermal strain tensor ϵ_{ij}^θ (using Eq. (3.97)) as

$$\epsilon^\theta = \epsilon_{11}^\theta = \epsilon_{22}^\theta = \epsilon_{33}^\theta = \frac{[a(\theta) - a(\theta_r)]}{a(\theta_r)}$$

$$= \alpha_1^\theta (\theta - \theta_r) + \frac{1}{2}\alpha_2^\theta (\theta - \theta_r)^2 + \cdots \quad (3.98)$$

where

$$\alpha_1^\theta = \frac{1}{a(\theta_r)}\left(\frac{\partial a}{\partial \theta}\right)_{\theta=\theta_r} \quad \text{and} \quad \alpha_2^\theta = \frac{1}{a(\theta_r)}\left(\frac{\partial^2 a}{\partial \theta^2}\right)_{\theta=\theta_r}. \quad (3.99)$$

The second-order terms are retained as the lattice parameter often depends strongly on temperature. For a cubic material, no shear strains are introduced by the temperature change so the thermal strain or eigenstrain, ϵ_{ij}^θ, can be written in tensor form as

$$\epsilon_{ij}^\theta = \frac{[a(\theta) - a(\theta_r)]}{a(\theta_r)}\delta_{ij} = \epsilon^\theta \delta_{ij} = \alpha^\theta(\theta - \theta_r)\delta_{ij} \quad (3.100)$$

where α^θ is the temperature-dependent coefficient of thermal expansion. For a system where a second-order expansion of the lattice parameters with temperature is accurate, $\alpha = \alpha_1^\theta + \alpha_2^\theta(\theta - \theta_r)$. Because the stress at a point results from the superposition of the thermal and mechanical strains,

$$\sigma_{ij} = C_{ijkl}\left[\epsilon_{kl} - \epsilon_{kl}^\theta \delta_{kl}\right] = C_{ijkl}[\epsilon_{kl} - \alpha^\theta(\theta - \theta_r)\delta_{kl}]. \quad (3.101)$$

When the temperature field is non-uniform, the stress at a point depends on both the mechanical deformation, measured through ϵ_{ij}, and the local temperature, θ.

For non-cubic systems, additional components of the thermal strain tensor are present. In such cases, the thermal expansion coefficient becomes a second-rank tensor, α_{ij}^θ. It is related to the thermal strain tensor by $\epsilon_{ij}^\theta = \alpha_{ij}^\theta(\theta - \theta_r)$.

c. Compositional Strain

a. Binary Substitutional Alloy with No Vacancies. When the lattice parameters of a crystal are a function of composition, a change in the local composition engenders a non-zero compositional eigenstrain ϵ_{ij}^c in precisely the same fashion as a temperature change engenders a thermal eigenstrain. Consider first a binary homogeneous crystal with cubic symmetry comprised of *substitutional* components A and B with no vacancies (see Fig. 13). We take the mole fraction of component B to be the independent composition variable. Remove a volume element of the crystal and change its composition from c_r to c, where c is the mole fraction of component B, at constant temperature and without change of phase. This change in composition is accomplished by an exchange of A and B atoms. If the composition of the reference state is c_r, the lattice parameter at composition c, $a(c)$, can be estimated by expanding the lattice parameter in a Taylor series about the reference state composition.

$$a(c) = a(c_r) + \left(\frac{\partial a}{\partial c}\right)_{c=c_r}(c - c_r) + \frac{1}{2}\left(\frac{\partial^2 a}{\partial c^2}\right)_{c=c_r}(c - c_r)^2 + \cdots \quad (3.102)$$

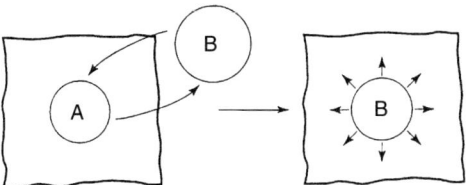

FIG. 13. An exchange of atoms possessing different partial molar volumes will induce a deformation in the surrounding crystal. This deformation is depicted schematically when a B atom replaces a smaller A atom, distorting the surrounding lattice planes. The arrows depict the displacement of the surrounding lattice.

The relative change in lattice parameter of the cubic material induced by the composition change gives the diagonal components of the compositional strain tensor ϵ_{ij}^c (using Eq. (3.97)) as

$$\epsilon^c = \epsilon_{11}^c = \epsilon_{22}^c = \epsilon_{33}^c = \frac{[a(c) - a(c_r)]}{a(c_r)}$$

$$= \epsilon_1^c(c - c_r) + \frac{1}{2}\epsilon_2^c(c - c_r)^2 + \cdots \quad (3.103)$$

where

$$\epsilon_1^c = \frac{1}{a(c_r)}\left(\frac{\partial a}{\partial c}\right)_{c=c_r} \quad \text{and} \quad \epsilon_2^c = \frac{1}{a(c_r)}\left(\frac{\partial^2 a}{\partial c^2}\right)_{c=c_r}. \quad (3.104)$$

If ϵ_2^c is neglected, the crystal is said to obey Vegard's law, i.e., the lattice parameter depends linearly on the composition. ϵ_1^c is also termed a solute expansion coefficient in analogy to the thermal expansion coefficient for thermally induced strains. The second-order term gives the deviation from Vegard's law. Its retention can be important, especially when dealing with small particles where surface stress effects are important.[30] Thus, for a cubic system, the compositional strain tensor is expressed as

$$\epsilon_{ij}^c = \epsilon^c(c)\delta_{ij}. \quad (3.105)$$

Analogous to the case of thermally induced strain, Eq. (3.101), the stress induced at a point by a composition change at the same point can be written as

$$\sigma_{ij} = C_{ijkl}[\epsilon_{kl} - \epsilon^c\delta_{kl}]. \quad (3.106)$$

If a composition change does not distort the lattice uniformly in all directions, the compositional strain must be expressed as a tensor, rather than a scalar. In such

---

[30] W. C. Johnson, *Acta Mater.* **49**, 3463 (2001).

cases,

$$\sigma_{ij} = C_{ijkl}[\epsilon_{kl} - \epsilon^c_{kl}]. \tag{3.107}$$

The dependence of ϵ^c_{kl} on composition can also be nonlinear.

b. Binary Substitutional Alloy with Vacancies. The compositional strain tensor for a binary substitutional alloy with vacancies can be constructed by extension from the binary system. For simplicity, consider a binary alloy with cubic symmetry. Let the lattice parameter be a function of the two independent mole fractions c_A and c_B, $a(c_A, c_B)$. This implies that the solute expansion coefficient measures the lattice parameter change with an exchange of an A atom with a vacancy. This is the analog in the compositional strain case to the diffusion potential for the change in energy associated with an exchange of an atom with a vacancy. If there are no vacancies then the remaining dependent composition variable could be a third mass component. Choosing c^r_A and c^r_B to be the reference composition, the lattice parameter for a crystal of composition (c_A, c_B) can be approximated to first order in the composition as

$$a(c_A, c_B) = a(c^r_A, c^r_B) + \left(\frac{\partial a}{\partial c_A}\right)(c_A - c^r_A) + \left(\frac{\partial a}{\partial c_B}\right)(c_B - c^r_B) + \cdots \tag{3.108}$$

where the partial derivatives are evaluated at the reference composition (c^r_A, c^r_B). The relative change in lattice parameter owing to the change in composition is

$$\frac{a(c_A, c_B) - a(c^r_A, c^r_B)}{a(c^r_A, c^r_B)} = \epsilon^c_A(c_A - c^r_A) + \epsilon^c_B(c_B - c^r_B) \tag{3.109}$$

where

$$\epsilon^c_i = \frac{1}{a(c^r_A, c^r_B)}\left(\frac{\partial a}{\partial c_i}\right)_{c_A=c^r_A, c_B=c^r_B} \quad i = A, B. \tag{3.110}$$

The composition strain is written

$$\epsilon^c_{ij}(c_A, c_B) = \epsilon^c(c_A, c_B)\delta_{ij} = [\epsilon^c_A(c_A - c^r_A) + \epsilon^c_B(c_B - c^r_B)]\delta_{ij}. \tag{3.111}$$

c. Interstitial Alloy. Comparable expressions for the compositional strain can also be obtained for interstitial solutions. However, some care must be exercised in writing such expressions, even for cubic crystals. In a simple cubic crystal, there is one type of interstitial site. Addition of an atom to this site at the center of the unit cell induces a uniform dilatation so that the compositional strain tensor has the same form as for a substitutional solution. In the face-centered cubic crystal, there exist two different interstitial sites. Addition of an interstitial atom to either a tetragonal or octahedral site induces a dilatational strain, but the magnitude of the compositional strain will depend on which site is occupied. In most cases, only one type of site is occupied and there is no difficulty defining the compositional strain as in Eq. (3.103). If both sites are occupied, the compositional strain depends on the concentration of interstitials in each interstitial site and cannot be expressed just in terms of the total interstitial concentration.

The addition of an interstitial atom to a BCC crystal, however, induces a tetragonal distortion. The octahedral sites are located at the center of the cube edge, and if a specific edge site is occupied, the composition-induced crystal distortion along the direction of the cube edge differs from the other two directions. For example, if the interstitial site lying along a cell edge parallel to the x_3 axis is occupied, a compositional strain tensor with components $\epsilon^c_{11} = \epsilon^c_{22} \neq \epsilon^c_{33}$ is produced. If the interstitial atoms are distributed uniformly on the three sublattices comprising all interstitial sites, the average compositional strain reduces to a dilatational field with $\epsilon^c = \epsilon^c_{kk}/3$. Representing the compositional strain as a dilatational tensor is permissible in hydrostatically stressed crystals. However, for a non-hydrostatically stressed crystal, the distinction is important in that the distribution of interstitials will no longer be uniform among the various interstitial sublattices, thereby giving rise to a position-dependent, tetragonal compositional strain tensor.

As an example, consider a binary BCC crystal in which one atom species occupies the substitutional lattice and the other the interstitial lattice. Allow the edges of the cubic unit cell to lie parallel to the Cartesian coordinate system. Identify the interstitial sublattices as (1), (2), and (3) according to whether the two nearest neighbor substitutional atoms lie in the x_1, x_2, and x_3 directions, respectively. If $\rho^{(i)}$ is the number density of interstitials occupying sublattice (i) as measured in the reference state and ρ_o is the number density of lattice sites on each sublattice, the interstitial mole fraction occupying sublattice (i), $y^{(i)}$, is defined as

$$y^{(i)} = \rho^{(i)}/\rho_o \quad (i) = (1), (2), (3) \qquad (3.112)$$

Note that the number density of interstitial sites is $3\rho_o$. Because there are six interstitial sites and two substitutional sites per unit cell, ρ_o also gives the density of substitutional sites. Let y be the fraction of occupied interstitial sites so that $y = (y^{(1)} + y^{(2)} + y^{(3)})/3$. For simplicity, we refer to the composition vector $\mathbf{y} = (y^{(1)}, y^{(2)}, y^{(3)})$.

The compositional strain associated with an interstitial atom occupying sublattice (i) is $e^{c(i)}_{kl}$. In general, $e^{c(i)}_{kl}$ is a function of composition. If we assume that the lattice parameter depends linearly[31] on the interstitial composition $y^{(i)}$, the

---

[31] A nonlinear dependence of the lattice parameter on composition is treated in the same way. However, the resulting equations for the diffusion potential will not decouple.

compositional strain is also a linear function of $y^{(i)}$,

$$e_{kl}^{c(i)} = \eta_{kl}^{(i)}\left(y^{(i)} - y_o\right) \tag{3.113}$$

where

$$\eta_{kl}^{(1)} = \begin{pmatrix} \eta_c & 0 & 0 \\ 0 & \eta_a & 0 \\ 0 & 0 & \eta_a \end{pmatrix} \quad \eta_{kl}^{(2)} = \begin{pmatrix} \eta_a & 0 & 0 \\ 0 & \eta_c & 0 \\ 0 & 0 & \eta_a \end{pmatrix} \quad \eta_{kl}^{(3)} = \begin{pmatrix} \eta_a & 0 & 0 \\ 0 & \eta_a & 0 \\ 0 & 0 & \eta_c \end{pmatrix} \tag{3.114}$$

η_c and η_a are a measure of relative extensions of the lattice owing to the interstitial along the tetragonal and perpendicular axes, respectively. The compositional strain associated with a particular volume element, ϵ_{kl}^c, is the sum of the compositional strain contributions arising from the interstitial atoms on each sublattice.

$$\epsilon_{kl}^c = e_{kl}^{c(1)} + e_{kl}^{c(2)} + e_{kl}^{c(3)} = \eta_{kl}^{(1)}\left(y^{(1)} - y_o\right) + \eta_{kl}^{(2)}\left(y^{(2)} - y_o\right)$$
$$+ \eta_{kl}^{(3)}\left(y^{(3)} - y_o\right). \tag{3.115}$$

If the interstitial sublattice sites are equally occupied, as would be expected for a stress-free crystal or a hydrostatically stressed crystal, the compositional strain is isotropic or dilatational in nature. If the crystal is subjected to nonhydrostatic stresses, however, the distribution of interstitials among the sublattices would no longer be expected to be uniform and the compositional strain of a volume element could exhibit orthorhombic symmetry.

d. Elastic Energy Density with Eigenstrains

The presence of an eigenstrain induces stresses into the reference state of the crystal. As such, the elastic energy of a crystal volume element does not vanish in the reference state and the elastic energy density given by Eq. (3.93) is not valid.

In order to calculate the elastic energy density for the system with eigenstrains, the work done in deforming the volume element from its stress-free state to the reference state must be included. The strain state of the stress-free volume element is ϵ_{ij}^p, and we let ϵ_{ij} be the final strain state. Then from Eqs. (3.71) and (3.94),

$$de_v = \sigma_{ij} d\epsilon_{ij} = C_{ijkl}\left(\epsilon_{kl} - \epsilon_{ij}^p\right) d\epsilon_{ij}. \tag{3.116}$$

Integration from the stress-free state to the final elastic state yields

$$\int_0^{e_v} de_v = \int_{\epsilon_{ij}^p}^{\epsilon_{ij}} C_{ijkl}\left(e_{kl} - \epsilon_{kl}^p\right) de_{ij}. \tag{3.117}$$

Integration of Eq. (3.117) is achieved by making the change of variable $f_{ij} = e_{ij} - \epsilon_{ij}^p$. This gives

$$e_v = \int_0^{\epsilon_{ij}-\epsilon_{ij}^p} C_{ijkl} f_{kl} df_{ij} = \frac{1}{2} C_{ijkl} f_{kl} f_{ij} \Big|_0^{\epsilon_{ij}-\epsilon_{ij}^p}, \quad (3.118)$$

which yields for the elastic strain energy density with eigenstrains

$$e_v = \frac{1}{2} C_{ijkl} \left(\epsilon_{ij} - \epsilon_{ij}^p\right)\left(\epsilon_{kl} - \epsilon_{kl}^p\right) = \frac{1}{2} \sigma_{ij} \left(\epsilon_{ij} - \epsilon_{ij}^p\right). \quad (3.119)$$

IV. Thermodynamics of a Single-Phase System

As discussed in the section on hydrostatically stressed crystals, a central issue in formulating a thermodynamics of crystals is the necessity of differentiating shape changes owing to mechanical deformation from those resulting from mass accretion. This difference in a system with shear stresses is illustrated by the simple example of a solid immersed in a liquid phase in which the solid is soluble. In Figure 14(a), the solid is in equilibrium with the liquid. The crystal, after some small variation in the condition of the two-phase system, is shown in Figure 14(b). In determining the equilibrium conditions, it is necessary to determine if the configuration in Figure 14(b) is a result of deformation of the solid or of mass exchange with the fluid. Separating these two variations is not possible when only the macroscopic shape is given. However, when the solid is crystalline, consideration of the crystalline lattice does permit these two types of variations to be differentiated. Shown in Figure 15 is a crystal with the same macroscopic shape of the crystal shown in Figure 14(b), but with the lattice inscribed. In Figure 15(A) the new shape of the crystal is a result of deformation, as the number and contiguity of lattice sites in the crystal does not change. In Figure 15(B), the number of lattice sites increases to give the new shape without subsequent deformation, and thus mass exchange with the fluid is responsible for the change in shape of the crystal. This example illustrates the difference between shape changes owing to mass accretion and changes resulting from mechanical deformation. The crystalline lattice can be used to differentiate between the two processes.[21]

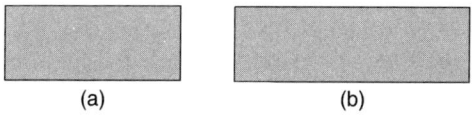

FIG. 14. A region of crystal of rectangular shape in equilibrium with a fluid is depicted in (a). The new shape of the crystal after a variation in the system is shown in (b). It is not possible to determine if the change in shape from (a) to (b) is a result of accretion of atoms or of elastic deformation.

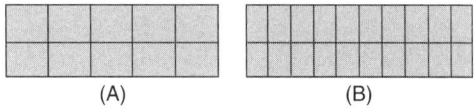

FIG. 15. The same region of crystal as in Fig. 14(b) is shown with representative lattice planes of a tetragonal system sketched in. In (A), the change in the initial shape of the crystal in Fig. 14(a) is due to elastic deformation, as the unit cell is stretched to be nearly square. In (B), the shape change is a result of the addition of several unit cells (accretion) to the original shape. The lattice allows the origin of the shape change to be identified.

In general, the presence of a lattice is not required in order to differentiate between mechanical deformation and mass exchange. It is simply necessary for a grid of some sort to exist that permits these two processes to be differentiated. The silicate network in a glass, or a polymer immersed in a solvent in which the solvent can be absorbed, are both physical examples of a network that could be used to measure mass accretion. The lattice, however, imposes constraints on mass and defect exchange *within* the crystal and must be incorporated into the thermodynamic description of the non-hydrostatically stressed crystal.

The change in shape of a crystal for which the number and contiguity of the lattice sites remains constant is due to deformation, or the application of force to the crystal. Such mechanical forces are well known and accepted. Changes in the shape of the crystal as a result of the addition of lattice planes can also be viewed as a response of the system to the application of *configurational* forces.[13,32,33] Configurational forces have also been termed chemical stresses.[34] Such configurational forces are *not* forces in the usual mechanical sense. The driving force for interface migration and the "force-balance" of interfacial energies at a trijunction are examples of configurational forces from materials science. In mechanics, configuration forces are used to examine the effects of stress on defects in crystal such as cracks, but the connection to more general changes in the structure of the lattice, such as an interface between two phases, is usually not made. Realizing that deformational and configurational forces give rise to distinct changes in a crystal will avoid confusion in the development of the equilibrium conditions for a crystal. Because a fluid lacks a network, these two forces are the same in a fluid.

The thermodynamic equilibrium conditions are obtained using Gibbs' variational approach.[20] This is a powerful and broadly applicable tool that allows

---

[32] J. D. Eshelby, in *Inelastic Behavior of Solids*, eds. M. F. Kanninen, W. F. Alder, A. R. Rosenfeld, and R. I. Jaffee (McGraw-Hill, New York, 1970), 77.

[33] M. E. Gurtin, in *Configurational Forces as Basic Concepts of Continuum Physics*, eds. J. E. Marsden and L. Sirovich (Springer Verlag, 2000), Applied Mathematical Sciences.

[34] P. Noziéres, in *Solids Far From Equilibrium*, ed. C. Godréche, Cambridge University Press (1991), 1–154.

very general relationships among the thermodynamic variables to be formulated. Its starting point is the combined form of the first two laws of thermodynamics, which Gibbs stated as[20]

> For the equilibrium of any isolated system, it is necessary and sufficient that in all possible variations in the state of the system which do not alter its entropy, the variation of its (internal) energy shall either vanish or be positive.

Gibbs' variational principle thus states that an isolated system is in equilibrium if the internal energy of the system is a minimum. It also provides a method for finding the equilibrium conditions of a system regardless of the imposed experimental conditions. An arbitrary region of the system is imagined to be isolated from its surroundings. If this region is in thermodynamic equilibrium, then arbitrary perturbations in the thermodynamic state of the system (such as changes in the local composition, entropy, or strain) that maintain a constant entropy in the region result in a change in internal energy for the region that is either positive or zero. It is the internal energy that is defined by the first two laws of thermodynamics. Other free energies, such as the Helmholtz and Gibbs free energies, are derived quantities.

We consider crystals in which there is only one type of substitutional site and one type of interstitial site. More precisely, we consider a crystal in which atoms occupy two different sublattices S and I. As for the hydrostatic crystal, the superscript denotes the sublattice. In addition, atoms of a given species are found on only one type of site.[35] The only defect considered is vacancies, which can exist on both types of sites. These restrictions can be relaxed in order to treat more complex crystals with multiple sublattices.[36,37]

As mentioned previously, the crystalline lattice introduces constraints on the variations of the chemical component densities. If the number density of substitutional sites in the reference state is ρ'_o, then

$$\rho_o^{S'} = \rho_A^{S'} + \rho_B^{S'} + \rho_V^{S'} \tag{4.1}$$

where $\rho_A^{S'}$ and $\rho_B^{S'}$ are the number density of substitutional components A and B, respectively, in the reference state, and $\rho_V^{S'}$ is the number density of substitutional vacancies. Similarly, for the interstitial component,

$$\rho_o^{I'} = \rho_C^{I'} + \rho_V^{I'} \tag{4.2}$$

where $\rho_C^{I'}$ is the number density of interstitial component C and $\rho_V^{I'}$ is the density of interstitial vacancies. In many interstitial alloys, the number of interstitial vacancies

---
[35] The treatment of crystals for which an atom can occupy two sublattices was elucidated for the hydrostatic case.
[36] W. C. Johnson, *Acta metall. mater.* **77**, 1581 (1994).
[37] W. W. Mullins and R. F. Sekerka, *J. Chem. Phys.* **82**, 5192 (1986).

far exceeds the density of interstitial atoms, and thus this constraint has often been neglected in past treatments.[18,21] Because this condition is not true in general, we treat this constraint explicitly. Due to these two constraints, the internal energy per volume in the reference state, $e_{v'}$, is a function of only two of the three densities, $\rho_A^{S'}$, $\rho_B^{S'}$, $\rho_V^{S'}$ and one of the two densities $\rho_C^{I'}$ and $\rho_V^{I'}$. The number of lattice sites per unit volume is a constant. This does not contradict the assumptions used previously to derive the equilibrium conditions in a hydrostatically stressed crystal. In that case we dealt with the system energy, not the energy per unit volume. For the present, we choose $e_{v'}(\rho_A^{S'}, \rho_B^{S'}, \rho_C^{I'})$. Nevertheless, one must remember that vacancies are present and, for example, the derivative

$$\left.\frac{\partial e_{v'}}{\partial \rho_A^{S'}}\right|_{\rho_B^{S'}, \rho_C^{I'}} \tag{4.3}$$

implies an implicit variation in the density of vacancies on the substitutional sublattice. If there are no substitutional vacancies in the system, $\rho_A^{S'}$ and $\rho_B^{S'}$ are not independent variables and it is not possible write Eq. (4.3); i.e. it is not possible to vary the concentration of component A keeping the concentration of component B constant. Alternate choices for the independent composition variables for the energy such as $e_{v'}(\rho_A^{S'}, \rho_V^{S'}, \rho_C^{I'})$ are advantageous when an expression for the chemical potential of a vacancy is required or when substitutional vacancies must be treated explicitly, as during certain diffusional processes. Similarly, the internal energy density cannot be a function of both $\rho_C^{I'}$ and $\rho_o^{I'}$. We use $\rho_C^{I'}$ as the independent concentration variable.

Equations (4.1) and (4.2) are local conditions satisfied at every point (or within every volume element) of the crystal. In addition, the quantities $\rho_o^{S'}$ and $\rho_o^{I'}$ are constants that depend only on the choice of the reference state for a given crystal. Larché and Cahn referred to these conditions as the "network constraint".[21]

A network constraint is not present in fluids. In a simple binary fluid, the energy density is of the form $e_v^f(s_v, \rho_A, \rho_B)$ where s_v is the entropy per unit volume. Although this form appears similar to that written previously, there is no defect structure of the fluid and,

$$\left.\frac{\partial e_v^f}{\partial \rho_A^f}\right|_{s_v, \rho_B^f} \tag{4.4}$$

has a clear, physical interpretation.

In general, the thermodynamic state of a crystal is heterogeneous; the composition, strain, and entropy are functions of position. As a consequence, the system entropy, internal energy, and other extensive properties must be expressed as an integral of a density over the volume of the system. For example, if e_v is the internal energy density, then the internal energy of the system, $\mathcal{E}$, is

$$\mathcal{E} = \int_V e_v dV. \tag{4.5}$$

Because the volume of an element of solid is affected by its state of strain, expressing a density as per-unit-volume of the actual or deformed state is usually not very convenient for elastically stressed crystals; when the state of strain is varied, the density of each of the mass components changes as well. Instead, densities expressed per-unit-volume of the reference state are often used. Thus,

$$\mathcal{E} = \int_{V'} e_{v'} dV \tag{4.6}$$

where, in accord with Eq. (3.45), $e_{v'} = Je_v$ and $J = \det F$. The internal energy contained in the volume V of the actual state is identical to the internal energy contained in the corresponding volume V' of the reference state as they describe the same material.

It is reasonable to assume that the internal energy of any volume element of the crystal depends on its entropy, concentration of each component, and state of deformation.[38] The deformation can be represented by one of the strain tensors but, if the internal energy is referred to the reference state, it is more convenient to use the deformation gradient tensor, at least initially. Therefore, we assume the internal energy density is a function of the following variables:

$$e_{v'} = e_{v'}\left(s_{v'}, F_{ij}, \rho_A^{S'}, \rho_B^{S'}, \rho_C^{I'}\right) \tag{4.7}$$

where $s_{v'}$ is the entropy density and F_{ij} is the deformation gradient tensor. For the hydrostatically stressed crystal, only the determinant of F was needed to describe the mechanical state.

From Eq. (4.7), a small variation (or perturbation) of the local entropy, deformation state or composition field results in a change in internal energy of a volume element given by

$$\delta e_{v'} = \frac{\partial e_{v'}}{\partial s_{v'}}\delta s_{v'} + \frac{\partial e_{v'}}{\partial F_{ij}}\delta F_{ij} + \frac{\partial e_{v'}}{\partial \rho_A^{S'}}\delta \rho_A^{S'} + \frac{\partial e_{v'}}{\partial \rho_B^{S'}}\delta \rho_B^{S'} + \frac{\partial e_{v'}}{\partial \rho_C^{I'}}\delta \rho_C^{I'} \tag{4.8}$$

---

[38] Of course, the internal energy density can depend on other variables as well. For example, in an ionic crystal, the internal energy is also a function of the electric displacement and the various atomic species can have different charge states.

where the summation convention is again being used.

$$\frac{\partial e_{v'}}{\partial F_{ij}} \delta F_{ij} = \sum_{i=1}^{3} \sum_{j=1}^{3} \frac{\partial e_{v'}}{\partial F_{ij}} \delta F_{ij} \qquad (4.9)$$

From the second law of thermodynamics, we associate

$$\frac{\partial e_{v'}}{\partial s_{v'}} = \theta \qquad (4.10)$$

where θ is the absolute temperature. Because the energy density is reckoned per unit volume of the reference state, Eq. (3.78) gives

$$\frac{\partial e_{v'}}{\partial F_{ij}} = T_{ji} \qquad (4.11)$$

where T_{ji} is the first Piola-Kirchhoff stress tensor.[23]

The compositional derivatives of the internal energy density are the diffusion potentials[21]

$$\frac{\partial e_{v'}}{\partial \rho_A^{S'}} = M_{AV}^S$$
$$\frac{\partial e_{v'}}{\partial \rho_B^{S'}} = M_{BV}^S \qquad (4.12)$$

and

$$\frac{\partial e_{v'}}{\partial \rho_C^{I'}} = M_{CV}^I. \qquad (4.13)$$

As for the treatment of hydrostatically stressed crystals, the subscript on the diffusion potential, AV for example, is an explicit statement that, for a crystal, the concentration derivative involves an exchange of species. In this case, M_{AV}^S gives the change in energy density owing to an increase in the number density of A and a corresponding decrease in the number density of vacancies V.

The variation in the internal energy density owing to small perturbations in each of the independent variables is therefore given by

$$\delta e_{v'} = \theta \delta s_{v'} + T_{ji} \delta F_{ij} + M_{AV}^S \delta \rho_A^{S'} + M_{BV}^S \delta \rho_B^{S'} + M_{CV}^I \delta \rho_C^{I'}. \qquad (4.14)$$

The variation of the internal energy of a crystal region V, $\delta \mathcal{E}$, is obtained by integrating the change in the internal energy density, $\delta e_{v'}$, over the reference volume V'.

Thus,

$$\delta \mathcal{E} = \int_{V'} \delta e_{v'} dV = \int_{V'} \{\theta \delta s_{v'} + T_{ji} \delta F_{ij} + M_{AV}^S \delta \rho_A^{S'} + M_{BV}^S \delta \rho_B^{S'} + M_{CV}^I \delta \rho_C^{I'}\} dV. \quad (4.15)$$

10. EQUILIBRIUM

In this subsection, the conditions for thermodynamic equilibrium in a multicomponent, single-phase crystal are derived using Gibbs' statement of the first two laws of thermodynamics. Equilibrium is obtained when the internal energy of the crystal is a minimum subject to the constraints that the crystal entropy, volume, and number of atoms of each component be constant. We assume that the crystal contains no defects other than vacancies, thus there are no sources of vacancies within the crystal. We proceed by choosing an arbitrary region of the crystal, V. This region is imagined to be isolated from its surroundings. Small perturbations in the various thermodynamic fields are then made holding the total entropy and number of atoms of each component fixed. Because the system is isolated, the variation of the displacement vector along the surface of V must vanish. The equilibrium conditions are those that lead to no change in the energy of the system owing to the small perturbations.

Three kinds of perturbations in the thermodynamic state of the system are allowed. First, changes in the entropy density of the volume elements composing V can be made. Such entropy changes can result, for example, from the flow of heat between neighboring volume elements. Second, changes in the component concentrations of the volume elements accrue owing to the exchange of mass between the volume elements. Third, the state of deformation can be varied locally.

In keeping with the equilibrium principle, the perturbation in the local entropy density must be accomplished holding the total entropy of the crystal, $\mathcal{S}$, fixed. Because

$$\mathcal{S} = \int_{V'} s_{v'} dV, \quad (4.16)$$

constant entropy requires

$$\delta \mathcal{S} = \int_{V'} \delta s_{v'} dV = 0. \quad (4.17)$$

Because the region is required to be isolated, the perturbations in the compositions and displacement field of the volume elements are not completely arbitrary. For example, the displacement along the surface of the isolated region must vanish, otherwise the perturbation would do work against an external force and the system would not be isolated. Because mass cannot be exchanged with the surroundings, the total number of atoms of each component in the region must remain constant. This mass constraint is expressed by

$$\mathcal{N}_A = \int_{V'} \rho_A^{S'} dV$$

$$\mathcal{N}_B = \int_{V'} \rho_B^{S'} dV \qquad (4.18)$$

$$\mathcal{N}_C = \int_{V'} \rho_C^{I'} dV$$

where $\mathcal{N}_i$ is the total number of atoms of species i contained in V. This leads to the following constraints on the composition perturbations

$$\delta\mathcal{N}_A = \int_{V'} \delta\rho_A^{S'} dV = 0$$

$$\delta\mathcal{N}_B = \int_{V'} \delta\rho_B^{S'} dV = 0 \qquad (4.19)$$

$$\delta\mathcal{N}_C = \int_{V'} \delta\rho_C^{I'} dV = 0$$

The constraints imposed on the perturbation of the internal energy can be included in the variational procedure by means of Lagrange multipliers.[22,23] Introducing the multipliers θ_L for the constraint on the entropy, λ_A and λ_B for each of the substitutional components, and λ_C for the interstitial component, a free energy $\mathcal{E}^*$ can be defined as

$$\mathcal{E}^* = \mathcal{E} - \theta_L \mathcal{S} - \lambda_A \mathcal{N}_A - \lambda_B \mathcal{N}_B - \lambda_C \mathcal{N}_C. \qquad (4.20)$$

The minimization of $\mathcal{E}^*$ is equivalent to the minimization of $\mathcal{E}$ subject to the constraints of constant entropy and mass. The first variation, or perturbation, of $\mathcal{E}^*$ is

$$\delta\mathcal{E}^* = \delta\mathcal{E} - \theta_L \delta\mathcal{S} - \lambda_A \delta\mathcal{N}_A - \lambda_B \delta\mathcal{N}_B - \lambda_C \delta\mathcal{N}_C. \qquad (4.21)$$

Substituting Eqs. (4.15), (4.17), and (4.19) into Eq. (4.21) yields

$$\delta\mathcal{E}^* = \int_{V'} \{(\theta - \theta_L)\delta s_{v'} + T_{ji}\delta F_{ij} + (M_{AV}^S - \lambda_A)\delta\rho_A^{S'} + (M_{BV}^S - \lambda_B)\delta\rho_B^{S'} + (M_{CV}^I - \lambda_A)\delta\rho_C^{I'}\}dV. \qquad (4.22)$$

The variations or perturbations appearing in Eq. (4.22) are still not independent, because the components of δF_{ij} cannot be varied independently. The perturbation of the displacement gradient tensor gives, from Eq. (3.1),

$$\delta F_{ij} = \frac{\partial}{\partial x'_j}(\delta x_i) = \frac{\partial}{\partial x'_j}(\delta x'_i + \delta u_i) = \frac{\partial \delta u_i}{\partial x'_j}. \tag{4.23}$$

Making use of the identity

$$\frac{\partial}{\partial x'_j}(T_{ji} \delta u_i) = T_{ji} \frac{\partial \delta u_i}{\partial x'_j} + \frac{\partial T_{ji}}{\partial x'_j} \delta u_i, \tag{4.24}$$

the volume integral containing the stress term appearing in Eq. (4.22) becomes, after using Eq. (4.23),

$$\int_{V'} T_{ji} \delta F_{ij} dV = \int_{V'} \frac{\partial}{\partial x'_j}[T_{ji} \delta u_i] dV - \int_{V'} \frac{\partial T_{ji}}{\partial x'_j} \delta u_i dV. \tag{4.25}$$

Applying the divergence theorem, Eq. (3.61), to Eq. (4.25) gives

$$\int_{V'} T_{ji} \delta F_{ij} dV = \int_{S'} T_{ji} \delta u_i n'_j dS - \int_{V'} \frac{\partial T_{ji}}{\partial x'_j} \delta u_i dV. \tag{4.26}$$

Equation (4.26) shows that the nine variations corresponding to δF_{ij} are not independent, but rather can be expressed in terms of the three independent variations δu_i.

Substituting Eq. (4.26) into Eq. (4.22) gives for the variation in the free energy $\mathcal{E}^*$

$$\delta \mathcal{E}^* = \int_{V'} \left\{ (\theta - \theta_L) \delta s_{v'} - \frac{\partial T_{ji}}{\partial x'_j} \delta u_i - (M_{AV}^S - \lambda_A) \delta \rho_A^{S'} - (M_{BV}^S - \lambda_B) \delta \rho_B^{S'} \right.$$

$$\left. - (M_{CV}^I - \lambda_C) \delta \rho_C^{I'} \right\} dV + \int_{S'} T_{ji} n'_j \delta u_i dS. \tag{4.27}$$

In order to determine the equilibrium conditions, the variation in the system free energy, $\delta \mathcal{E}^*$, is set equal to zero in Eq. (4.27). In keeping with the equilibrium criterion that the region of the system under consideration must be isolated from its surroundings, the surface integral in Eq. (4.27) must vanish as the perturbation in the displacement on the surface of the region must be zero at every point, otherwise work is done against external forces. Because all variations appearing in the volume integral of Eq. (4.27) are now independent, $\delta \mathcal{E}^* = 0$ for an arbitrary region of the system only when the coefficient of each term appearing in the volume integrand of Eq. (4.27) vanishes. This yields the equilibrium conditions

$$\theta = \theta_L, \tag{4.28}$$

$$\frac{\partial T_{ji}}{\partial x'_j} = 0, \tag{4.29}$$

$$M^S_{AV} = \lambda_A \tag{4.30}$$

$$M^S_{BV} = \lambda_B \tag{4.31}$$

and

$$M^I_{CV} = \lambda_C. \tag{4.32}$$

The first condition, Eq. (4.28), is the condition for thermal equilibrium requiring that the temperature field, $\theta(\mathbf{x})$, be constant and uniform everywhere in the system. If the temperature is not everywhere uniform, then the system is not in equilibrium and heat flow will occur. The second condition, Eq. (4.29), is the condition for mechanical equilibrium, which states that the divergence of the first Piola-Kirchhoff stress tensor vanishes in the absence of body forces. The remaining terms refer to chemical equilibrium and state that the diffusion potential of the substitutional and interstitial species must be uniform at equilibrium. If the crystal were composed of only pure A and vacancies, the preceding equations would still hold.

11. Stress-Dependence of Diffusion Potential

General relationships for the stress (and composition) dependence of the chemical potential can be determined formally using Maxwell relations and the first two laws of thermodynamics. We consider two cases that illustrate the approach used to obtain the expression for the diffusion potential. The first system is a binary substitutional alloy in which vacancies can be ignored. The second is a binary interstitial alloy with a body-centered cubic crystal structure. This second example illustrates the importance of considering multiple sublattices.

a. Substitutional Binary Crystal

As before, we take the crystal to be a binary $(A - B)$ substitutional alloy with vacancies. No assumption is made initially on the crystal symmetry or the functional dependence of the lattice parameter and elastic constants on composition and temperature. In the section on hydrostatically stressed crystals the energy density was taken to be a function of $s_{v'}$, J, ρ'_A, ρ'_B and the total differential was given in Eq. (2.22). If the crystal is under nonhydrostatic stress, we take the energy density to be a function of $s_{v'}$, F_{ij}, ρ'_A, and ρ'_B. Thus the work term, $-PdJ$ in Eq. (2.22), is replaced by $T_{ji}dF_{ij}$ for the nonhydrostatically stressed crystal with the result,

$$de_{v'} = \theta ds_{v'} + T_{ji}dF_{ij} + M_{AV}d\rho'_A + M_{BV}d\rho'_B. \tag{4.33}$$

As written, the variables θ, T_{ji}, M_{AV}, and M_{BV} are functions of the independent variables $s_{v'}$, F_{ij}, ρ'_A, and ρ'_B. A change in variable is obtained by defining a new function, $g_{v'}$ (using a Legendre transform), as[21]

$$g_{v'} = e_{v'} - \theta s_{v'} - T_{ji} F_{ij}. \tag{4.34}$$

$g_{v'}$ is a state function whose value depends only on the current thermodynamic state and is independent of the path taken to reach this state. Differentiating Eq. (4.34),

$$dg_{v'} = de_{v'} - \theta ds_{v'} - s_{v'} d\theta - T_{ji} dF_{ij} - F_{ij} dT_{ji}, \tag{4.35}$$

and using Eq. (4.33) gives

$$dg_{v'} = -s_{v'} d\theta - F_{ij} dT_{ji} + \rho'_o M_{AV} dc_A + \rho'_o M_{BV} dc_B \tag{4.36}$$

where $c_i = \rho'_i / \rho'_o$ for $i = A, B$. As written, $g_{v'}$ is a function of θ, T_{ji}, c_A, and c_B. If the small-strain approximation is used for the elastic deformation, then

$$dg_{v'} = -s_{v'} d\theta - \epsilon_{ij} d\sigma_{ij} + \rho'_o M_{AV} dc_A + \rho'_o M_{BV} dc_B \tag{4.37}$$

with $g_{v'} = g_{v'}(\theta, \sigma_{ij}, c_A, c_B)$.

Because $g_{v'}$ is a state function of the variables θ, σ_{ij}, c_A, and c_B, $dg_{v'}$ is an exact differential and one can write

$$dg_{v'}(\theta, \sigma_{ij}, c_A, c_B) = \left(\frac{\partial g_{v'}}{\partial \theta}\right) d\theta + \left(\frac{\partial g_{v'}}{\partial \sigma_{ij}}\right) d\sigma_{ij}$$
$$+ \left(\frac{\partial g_{v'}}{\partial c_A}\right) dc_A + \left(\frac{\partial g_{v'}}{\partial c_B}\right) dc_B. \tag{4.38}$$

In performing the partial differentiation, it is implicit that the independent thermodynamic variables not involved in the differentiation are held constant. Comparing Eq. (4.38) with Eq. (4.37) allows the following associations to be made:

$$\left(\frac{\partial g_{v'}}{\partial \theta}\right) = -s_{v'}, \quad \left(\frac{\partial g_{v'}}{\partial \sigma_{ij}}\right) = -\epsilon_{ij}, \quad \left(\frac{\partial g_{v'}}{\partial c_A}\right) = \rho'_o M_{AV}, \quad \left(\frac{\partial g_{v'}}{\partial c_B}\right) = \rho'_o M_{BV}. \tag{4.39}$$

Because the order of differentiation can be exchanged for an exact differential, the following equality must hold:

$$\frac{\partial}{\partial c_k}\left(\frac{\partial g_{v'}}{\partial \sigma_{ij}}\right) = \frac{\partial}{\partial \sigma_{ij}}\left(\frac{\partial g_{v'}}{\partial c_k}\right) \tag{4.40}$$

for $k = A$ or B. Substituting the relationships of Eq. (4.39) into Eq. (4.40) gives the following Maxwell relations:

$$-\left(\frac{\partial \epsilon_{ij}}{\partial c_A}\right)_{\theta, c_B \sigma_{ij}} = \rho'_o \left(\frac{\partial M_{AV}}{\partial \sigma_{ij}}\right)_{\theta, c_A, c_B \sigma_{kl \neq ij}}$$

$$-\left(\frac{\partial \epsilon_{ij}}{\partial c_B}\right)_{\theta, c_A, \sigma_{ij}} = \rho'_o \left(\frac{\partial M_{BA}}{\partial \sigma_{ij}}\right)_{\theta, c_A, c_B \sigma_{kl \neq ij}}.$$
(4.41)

The subscripts appearing in Eq. (4.41) indicate explicitly those quantities held constant during the partial differentiation. The term $\sigma_{kl \neq ij}$ means that all components of the stress tensor are held constant except the component σ_{ij}. Equation (4.41) is quite general with no particular assumptions on material behavior other than the validity of employing the small-strain approximation. To proceed further, a constitutive law connecting the stress and strain must be invoked. If the temperature is assumed to remain constant, stresses are induced when the material is deformed or a composition change occurs, and Eq. (3.107) can be used as a stress-strain constitutive law. Therefore,

$$\sigma_{ij} = C_{ijkl}[\epsilon_{kl} - \epsilon^c_{kl}]$$
(4.42)

where the compositional strain, ϵ^c_{kl} and

$$\frac{\sigma_{ij}}{\partial x_j} = \frac{\partial}{\partial x_j}\{C_{ijkl}[\epsilon_{kl} - \epsilon^c_{kl}]\} = 0$$
(4.43)

is the equilibrium condition.

In order to combine Eq. (4.42) with the Maxwell relation, Eq. (4.41), it is first necessary to solve for the strain in terms of the stress. This is accomplished by contracting each side of Eq. (4.42) with the elastic compliance tensor, S_{ijkl}.

$$S_{mnij}\sigma_{ij} = S_{mnij}C_{ijkl}[\epsilon_{kl} - \epsilon^c_{kl}].$$
(4.44)

Using Eq. (3.81), Eq. (4.44) can be simplified to

$$S_{mnij}\sigma_{ij} = \frac{1}{2}[\delta_{mk}\delta_{nl} + \delta_{ml}\delta_{nk}][\epsilon_{kl} - \epsilon^c_{kl}] = \epsilon_{mn} - \epsilon^c_{mn}.$$
(4.45)

Solving Eq. (4.45) for the strain gives

$$\epsilon_{mn} = \epsilon^c_{mn} + S_{mnij}\sigma_{ij}.$$
(4.46)

The compositional derivatives of the strain at constant temperature and stress become

$$\left(\frac{\partial \epsilon_{mn}}{\partial c_A}\right)_{\theta,\sigma_{ij}} = \left(\frac{\partial \epsilon_{mn}^c}{\partial c_A}\right)_{\theta,\sigma_{ij}} + \left(\frac{\partial S_{mnij}}{\partial c_A}\right)_{\theta,\sigma_{ij}} \sigma_{ij}$$

$$\left(\frac{\partial \epsilon_{mn}}{\partial c_B}\right)_{\theta,\sigma_{ij}} = \left(\frac{\partial \epsilon_{mn}^c}{\partial c_B}\right)_{\theta,\sigma_{ij}} + \left(\frac{\partial S_{mnij}}{\partial c_B}\right)_{\theta,\sigma_{ij}} \sigma_{ij}.$$

(4.47)

Substituting Eq. (4.47) into Eq. (4.41) gives

$$\rho'_o \left(\frac{\partial M_{AV}}{\partial \sigma_{ij}}\right)_{\theta,c_A,c_B,\sigma_{mn}\neq ij} = -\left(\frac{\partial \epsilon_{ij}^c}{\partial c_A}\right)_{\theta,\sigma_{mn}} - \left(\frac{\partial S_{ijkl}}{\partial c_A}\right)_{\theta,\sigma_{mn}} \sigma_{kl}$$

$$\rho'_o \left(\frac{\partial M_{BV}}{\partial \sigma_{ij}}\right)_{\theta,c_A,c_B,\sigma_{mn}\neq ij} = -\left(\frac{\partial \epsilon_{ij}^c}{\partial c_B}\right)_{\theta,\sigma_{mn}} - \left(\frac{\partial S_{ijkl}}{\partial c_B}\right)_{\theta,\sigma_{mn}} \sigma_{kl}.$$

(4.48)

$M_{AV}(\theta, P, c_a, c_B)$ and $M_{BV}(\theta, P, c_a, c_B)$ are the diffusion potentials of an unstressed volume element at temperature θ, pressure P, and composition c_A, c_B. The diffusion potentials of the volume element at the same composition and temperature but under stress σ_{ij} are obtained by integrating each of Eqs. (4.48).

$$\int_{M_{AV}(\theta,P,c_A,c_B)}^{M_{AV}(\theta,\sigma_{ij},c_A,c_B)} \rho'_o dM_{BA} = -\int_{-P\delta_{ij}}^{\sigma_{ij}} \left(\frac{\partial \epsilon_{ij}^c}{\partial c_A}\right) d\tau_{ij} - \int_{-P\delta_{ij}}^{\sigma_{ij}} \left(\frac{\partial S_{ijkl}}{\partial c_A}\right) \tau_{kl} d\tau_{ij}$$

$$\int_{M_{BV}(\theta,P,c_A,c_B)}^{M_{BV}(\theta,\sigma_{ij},c_A,c_B)} \rho'_o dM_{BA} = -\int_{-P\delta_{ij}}^{\sigma_{ij}} \left(\frac{\partial \epsilon_{ij}^c}{\partial c_B}\right) d\tau_{ij} - \int_{-P\delta_{ij}}^{\sigma_{ij}} \left(\frac{\partial S_{ijkl}}{\partial c_B}\right) \tau_{kl} d\tau_{ij}.$$

(4.49)

Because the two composition derivatives are independent of the stress state, integration yields

$$M_{AV}(\theta, \sigma_{ij}, c_A, c_B) = M_{AV}(\theta, P, c_A, c_B) - \frac{1}{\rho'_o}\left(\frac{\partial \epsilon_{ij}^c}{\partial c_A}\right)(\sigma_{ij} + P\delta_{ij})$$

$$-\frac{1}{2\rho'_o}\left(\frac{\partial S_{ijkl}}{\partial c_A}\right)(\sigma_{ij}\sigma_{kl} - P^2\delta_{ij}\delta_{kl})$$

(4.50)

$$M_{BV}(\theta, \sigma_{ij}, c_A, c_B) = M_{BV}(\theta, P, c_A, c_B) - \frac{1}{\rho'_o}\left(\frac{\partial \epsilon_{ij}^c}{\partial c_B}\right)(\sigma_{ij} + P\delta_{ij})$$

$$-\frac{1}{2\rho'_o}\left(\frac{\partial S_{ijkl}}{\partial c_B}\right)(\sigma_{ij}\sigma_{kl} - P^2\delta_{ij}\delta_{kl})$$

Equation (4.50) gives the stress dependence of the diffusion potential when the small strain approximation is valid. If the elastic constants are independent of composition, and we make the reasonable approximation for most materials systems $P = 0$, Eq. (4.50) simplifies to

$$M_{AV}(\theta, \sigma_{ij}, c_A, c_B) = M_{AV}(\theta, 0, c_A, c_B) - \frac{1}{\rho'_o}\left(\frac{\partial \epsilon^c_{ij}}{\partial c_A}\right)\sigma_{ij}$$
$$M_{BV}(\theta, \sigma_{ij}, c_A, c_B) = M_{BV}(\theta, 0, c_A, c_B) - \frac{1}{\rho'_o}\left(\frac{\partial \epsilon^c_{ij}}{\partial c_B}\right)\sigma_{ij}$$
(4.51)

If the crystal is isotropic or has cubic symmetry, the compositional strain is given by Eq. (3.108) and the diffusion potential becomes

$$M_{AV}(\theta, \sigma_{ij}, c_A, c_B) = M_{AV}(\theta, 0, c_A, c_B) - \frac{1}{\rho'_o}\epsilon^c_A \sigma_{kk}$$
$$M_{BV}(\theta, \sigma_{ij}, c_A, c_B) = M_{BV}(\theta, 0, c_A, c_B) - \frac{1}{\rho'_o}\epsilon^c_B \sigma_{kk}.$$
(4.52)

These two equations can be used to determine the equilibrium distribution of atoms and vacancies in systems under nonhydrostatic stress. Because equilibrium demands $M_{AV}(\theta, \sigma_{ij}, c_A, c_B)$ and $M_{BV}(\theta, \sigma_{ij}, c_A, c_B)$ must be constant, if the trace of the stress is nonuniform, $M_{AV}(\theta, 0, c_A, c_B)$ and $M_{BV}(\theta, 0, c_A, c_B)$ must be nonuniform. This implies that the concentration of A, B atoms and vacancies must be nonuniform.

In many applications the concentrations of the substitutional atoms is of more interest than vacancies. Thus, to focus on just the concentration of substitutional atoms, we subtract the two equations Eq. (4.52) to yield

$$M_{BA}(\theta, \sigma_{ij}, c_A, c_B) = M_{BA}(\theta, 0, c_A, c_B) - \frac{1}{\rho'_o}\left(\epsilon^c_B - \epsilon^c_A\right)\sigma_{kk} \quad (4.53)$$

where $M_{BA} = M_{BV} - M_{AV}$. Because each solute expansion coefficient involves the exchange with a vacancy, the difference in solute expansion coefficients measures the change on an exchange of an A and B atom, ϵ^c, thus

$$M_{BA}(\theta, \sigma_{ij}, c_A, c_B) = M_{BA}(\theta, 0, c_A, c_B) - \frac{1}{\rho'_o}\epsilon^c \sigma_{kk}. \quad (4.54)$$

In the section on hydrostatically stressed solids we defined $M_{BA}(\theta, 0, c_A, c_B) = \mu^V_B(\theta, 0, c_A, c_B) - \mu^V_A(\theta, 0, c_A, c_B)$. If the concentration of vacancies is very low, the chemical potential of component A or B does not depend strongly on the vacancy concentration. Thus only one mole fraction can be an independent variable and to a good approximation, $\mu_A^V(\theta, 0, c_A, c_B) = \mu_A^V(\theta, 0, c_A), \mu_B^V(\theta, 0, c_A, c_B) = \mu_B^V(\theta, 0, c_A)$ and

$$M_{BA}(\theta, \sigma_{ij}, c_B) = M_{BA}(\theta, 0, c_B) - \frac{1}{\rho'_o}\epsilon^c \sigma_{kk}. \tag{4.55}$$

This is a very convenient expression to calculate the composition field in a binary substitutional alloy.

12. Stress-induced Solute Redistribution

In this section, we present three examples illustrating how compositional strains and external sources of stress can interact to induce solute redistribution in a binary alloy. We begin with a substitutional alloy in contact with a large, fluid reservoir. A uniform external stress is imposed on the crystal and the change in alloy composition is calculated. The second example considers a large substitutional crystal in which a nonuniform stress field induces solute redistribution. The third example considers an interstitial BCC alloy for which three different interstitial sublattices are present. In addition to a change in the net composition of a volume element, the stresses induce a redistribution of the interstitial between the different interstitial sites. The concept of open-system elastic constants[21] is introduced to treat the last two problems.

a. Example 1: Stress-induced Composition Change

As a simple example of how application of an elastic stress can induce a change in the equilibrium composition of an alloy, consider an unstressed $(A - B)$ binary crystal in equilibrium with a fluid containing the two components A and B (Figure 16). The crystal is assumed to be cubic and the compositional strain to depend linearly on composition. The elastic constants are assumed independent of composition. The chemical potentials of the two components in the fluid are μ_A^f and μ_B^f and the composition of component B of the unstressed crystal in equilibrium with the fluid is $c = c_o$. To focus on the composition of the chemical components we shall assume that the vacancy concentration is small. Chemical equilibrium between the hydrostatically stressed crystal under zero pressure and fluid requires, upon subtracting the two equations of Eq. (2.22),

$$M_{BA}(\theta, c_o) = \mu_B^v - \mu_A^v = \mu_B^f - \mu_A^f. \tag{4.56}$$

where c is the concentration of component B. Equation (4.56) states that the simultaneous addition of a B atom to and removal of an A atom from the crystal produces the same change in free energy as the addition of a B to and the removal

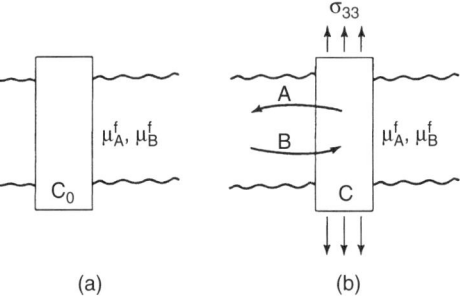

FIG. 16. (a) The unstressed crystal, initially of composition c_o, is in equilibrium with a large fluid reservoir of chemical potential μ_A^f and μ_B^f. (b) Application of a uniaxial stress ($\sigma_{app} = \sigma_{33}$) to the crystal changes the diffusion potential of the crystal and induces mass flow between the crystal and fluid reservoir. The equilibrium composition of the crystal under stress is c.

of an A from the fluid. If this were not the case, mass exchange between the crystal and fluid would occur.

If a uniaxial stress, $\sigma_{app} = \sigma_{33}$, is applied to the crystal in the x_3 direction *while the chemical potentials of A and B in the fluid remain fixed*, a change in composition of the crystal will result in order to bring the crystal back into chemical equilibrium with the fluid. The applied stress changes the diffusion potential M_{BA} in the crystal such that the condition for chemical equilibrium, Eq. (4.56), is no longer satisfied and the resulting gradient in chemical (diffusion) potential will induce mass flow between the crystal and fluid reservoir. In the presence of the applied stress, chemical equilibrium in this cubic system requires, from Eq. (4.52),

$$M_{BA}(\theta, c, \sigma_{ij}) = M_{BA}(\theta, c) - \frac{\epsilon^c}{\rho'_o}\sigma_{kk} = \mu_B^f - \mu_A^f \qquad (4.57)$$

where the composition c that satisfies Eq. (4.57) is different from the initial composition c_o. Because the chemical potentials of the fluid are unchanged, Eq. (4.56) can be substituted into Eq. (4.57) to give

$$M_{BA}(\theta, c) - \frac{\epsilon^c}{\rho'_o}\sigma_{kk} = M_{BA}(\theta, c_o) \qquad (4.58)$$

Equation (4.58) is, in general, a transcendental equation for the composition c. However, in the limit $|(c - c_o)/c_o| \ll 1$, an expression for the composition shift can be obtained from Eq. (4.58) by expanding the diffusion potential in a Taylor's series to first order in the composition about the initial composition c_o,

$$M_{BA}(\theta, c) = M_{BA}(\theta, c_o) + \left(\frac{\partial M_{BA}}{\partial c}\right)_{c=c_o}(c - c_o) + \cdots. \qquad (4.59)$$

Using Eq. (2.74) and recognizing that $J = 1$ for a stress-free system gives

$$M_{BA}(\theta, c) = M_{BA}(\theta, c_o) + \frac{1}{\rho'_o}\left(\frac{\partial^2 f_v}{\partial c^2}\right)_{c=c_o}(c - c_o) = M_{BA}(\theta, c_o) + \frac{f''_v}{\rho'_o}(c - c_o) \quad (4.60)$$

where $f''_v = \partial^2 f_v/\partial c^2$ evaluated at $c = c_o$. Substituting Eq. (4.60) into Eq. (4.58) and solving yields

$$\frac{f''_v}{\rho'_o}(c - c_o) = \frac{\epsilon^c}{\rho'_o}\sigma_{kk} = \frac{\epsilon^c}{\rho'_o}\sigma_{app} \quad (4.61)$$

where the only source of stress is the applied uniaxial stress, σ_{app}, so that $\sigma_{kk} = \sigma_{app}$.

An alternate derivation can be obtained by substituting the chemical potentials into Eq. (4.58) to give for the equilibrium condition

$$\mu_B^V(\theta, c) - \mu_A^V(\theta, c) - \frac{\epsilon^c}{\rho'_o}\sigma_{kk} = \mu_B^V(\theta, c_o) - \mu_A^V(\theta, c_o). \quad (4.62)$$

The chemical potentials are expanded in a Taylor's series to first order about the composition c_o, and

$$\mu_i^V(c) = \mu_i^V(c_o) + \left.\frac{\partial \mu_i^V}{\partial c}\right|_{c=c_o}(c - c_o) \quad (4.63)$$

for $i = A, B$. The derivatives of the chemical potentials are not independent and are conveniently expressed in terms of derivatives of the Helmholtz or Gibbs free energies (see Eq. (2.53) and Eq. (2.54)). The Gibbs free energy per atomic site, g_o, can also be used as the Helmholtz and Gibbs free energies are equivalent in the stress-free case. Using Eq. (2.53) and Eq. (2.54), but expressed as the Helmholtz free energy per atom site f_o, Eq. (4.63) in Eq. (4.62) yields

$$\frac{d^2 f_o}{dc^2}(c - c_o) = \frac{\epsilon^c}{\rho'_o}\sigma_{kk}. \quad (4.64)$$

Equation (4.64) is identical to Eq. (4.61) as $\rho'_o f_o = f_v$.

The magnitude of the composition change induced by application of an external stress, Eq. (4.61), depends on three terms; the compositional strain (ϵ^c), the applied stress (σ_{app}), and the curvature of the Helmholtz free energy density with respect to composition in the absence of stress (f''_v). The product $\epsilon^c \sigma_{app}$ is a measure of the elastic work term. When ϵ^c is large, more elastic energy can be relieved when one type of atom is replaced by another at a given applied stress. Changing the composition, however, engenders a change in the chemical free energy and the elasticity-induced composition change can only proceed until the decrease in elastic energy is exactly balanced by the increase in chemical energy resulting from the composition change. f_v'' is an inverse measure of the allowed composition change. If f_v'' is large, corresponding to a strong curvature of the free energy, small changes in the composition result in large changes in the chemical free energy. If f_v'' is small and the free energy curve is a relatively weak function of the composition, larger changes in composition give rise to progressively smaller increases in the free energy. Thus if the alloy composition is near a spinodal where f_v'', the elastic stress-induced changes are very large. This tradeoff between elastic and chemical energy is the source of the kinetic instability of alloy thin films grown under conditions where f_v'' is small.[39,40] The magnitude of the elasticity-induced composition changes always depends on the trade-off between the elastic and chemical energies and can range from essentially zero up to about 10 $at\%$.

b. Example 2: Open-System Elastic Constants

An example illustrating the interaction between elastic stress and composition concerns solute redistribution around such stress centers as dislocations and second-phase precipitates.[41-43] Consider an infinite, and initially homogeneous, binary substitutional crystal of composition c_o with cubic (or isotropic) elastic constants. The compositional strain is dilatational in nature and we take it to depend linearly on the composition according to Eq. (3.105). The reference state for the measurement of strain is the stress-free crystal of composition c_o. A dislocation or second-phase particle is introduced into the crystal and its stress field induces solute redistribution. It is expected that larger solute atoms would segregate to regions of tension in the crystal while smaller solute atoms would segregate to regions of compression, both in order to reduce the elastic energy of the system. Solute redistribution results in a concomitant increase in the chemical energy of the system, however, and will continue until the decrease in elastic energy is exactly offset by the increase in chemical energy. Equilibrium is achieved once the diffusion potential, M_{BA}, is uniform throughout the crystal. Because the elastic field is position dependent, the equilibrium composition will also be position dependent. The magnitude of solute redistribution cannot be determined by direct application of Eq. (4.61) with the stress field of the stress center substituted for the stress term on the righthand side of the equation, because a change in composition (from c_o) induces a compositional strain that partially offsets the stress field of the stress center. Stresses thus arise from two sources in this case, the stress center and

---

[39] J. E. Guyer and P. W. Voorhees, *Phys. Rev. Lett.* **74**, 4031 (1995).
[40] V. G. Malyshkin and V. A. Shchukin, *Semiconductors* **27**, 1062 (1994).
[41] W. C. Johnson and P. W. Voorhees, *Metall. Trans. A* **16A**, 337 (1985).
[42] D. Barnett, G. Wang, and W. Nix, *Acta metall.* **30**, 2035 (1982).
[43] A. H. Cottrell and M. A. Jaswon, *Proc. Roy. Soc.* **A199**, 104 (1949).

the heterogeneous composition field. This is in contrast to the previous example, where it was not necessary to consider the stress arising from the compositional strain as the stress state was imposed on the homogeneous crystal. This is rarely the case in physical situations.

The stress and composition fields are coupled by two equations: the expressions for the stress dependence of the diffusion potential, Eq. (4.50), where we must also use the stress-strain constitutive equation of Eq. (3.106), and the mechanical equilibrium condition

$$\frac{\partial \sigma_{ij}}{\partial x_j} = \frac{\partial}{\partial x_j}\{C_{ijkl}[\epsilon_{kl} - \epsilon^c(c - c_o)\delta_{kl}]\}, \qquad (4.65)$$

where c is the local composition and we have taken the ϵ^c as the compositional strain measuring the change in the lattice parameter with an exchange of A and B atoms, defined previously.

We shall assume that far from the source of stress, the crystal composition remains unperturbed at c_o and solute redistribution can be considered to occur at constant diffusion potential or under open-system conditions. Here, the crystal is assumed to be sufficiently large so as to act as a chemical potential reservoir. Mass is considered to flow toward or away from the source of stress such that the diffusion potential remains constant in the stressed region. Equilibrium is obtained when the diffusion potential at all points near the stress center assumes a value equal to the diffusion potential far from the stress center. For an isotropic system in which the elastic constants do not depend on composition, this requires

$$M_{BA}(\theta, c_o) = M_{BA}(\theta, c) - \frac{\epsilon^c}{\rho'_o}\sigma_{kk}. \qquad (4.66)$$

Equations (4.65) and (4.66) must be solved simultaneously for the composition and mechanical fields. In general, this must be done numerically. However, an analytical solution accurate to first order in the change in composition can be obtained by using the approximation to Eq. (4.66) given by Eq. (4.61). Using $e^c_{kl} = \epsilon^c(c - c_o)\delta_{kl}$, in Eq. (3.106) and Eq. (4.61) gives

$$\sigma_{ij} = C_{ijkl}\left[\epsilon_{kl} - \frac{(\epsilon^c_1)^2}{f''_v}\sigma_{nn}\delta_{kl}\right]. \qquad (4.67)$$

Contracting each side of Eq. (4.67) with the elastic compliance tensor, S_{mnij}, yields

$$S_{mnij}\sigma_{ij} = S_{mnij}C_{ijkl}\left[\epsilon_{kl} - \frac{(\epsilon^c)^2}{f''_v}\sigma_{pp}\delta_{kl}\right]. \qquad (4.68)$$

Because S_{ijkl} is the inverse of C_{ijkl} Eq. (4.68) can be simplified to

$$S_{mnij}\sigma_{ij} = \frac{1}{2}[\delta_{km}\delta_{ln} + \delta_{kn}\delta_{lm}]\left[\epsilon_{kl} - \frac{(\epsilon^c)^2}{f_v''}\sigma_{pp}\delta_{kl}\right]$$

$$= \epsilon_{mn} - \frac{(\epsilon^c)^2}{f_v''}\sigma_{pp}\delta_{mn}. \tag{4.69}$$

Solving for the strain field and using $\sigma_{pp} = \sigma_{jj} = \sigma_{ij}\delta_{ij}$ gives

$$\epsilon_{mn} = \left[S_{mnij} + \frac{(\epsilon^c)^2}{f_v''}\delta_{ij}\delta_{mn}\right]\sigma_{ij} = S^o_{mnij}\sigma_{ij} \tag{4.70}$$

where a modified, or open-system compliance tensor, S^o_{mnij}, has been defined as

$$S^o_{mnij} = S_{mnij} + \frac{(\epsilon^c)^2}{f_v''}\delta_{ij}\delta_{mn}. \tag{4.71}$$

Equation (4.70) is a modified constitutive law connecting the stress and strain. The modified compliance tensor, S^o_{mnij}, contains both elastic and thermodynamic information about the system. It permits the actual stress field in the system, arising from both the stress center (e.g., dislocation) and solute redistribution, to be determined to first-order in composition change. The elastic solution pertaining to the system without solute redistribution is used, but the actual elastic constants are replaced by the modified elastic compliances of Eq. (4.71). A set of modified elastic constants, C^o_{ijkl}, connecting the strain to the stress tensor can also be defined by the relation

$$\sigma_{ij} = C^o_{ijkl}\epsilon_{kl}. \tag{4.72}$$

The components of C^o_{ijkl} are found by solving $C^o_{ijkl}S^o_{klmn} = [\delta_{im}\delta_{jn} + \delta_{in}\delta_{jm}]/2$ as in Eq. (3.81). The modified elastic constants for an isotropic system, μ^o (shear modulus), ν^o (Poisson's ratio), and E^o (Young's modulus), are defined by

$$\mu^o = \mu, \tag{4.73}$$

$$\nu^o = \frac{\nu - 2\mu(1+\nu)(\epsilon^c)^2/f_v''}{1 + 2\mu(1+\nu)(\epsilon^c)^2/f_v''}, \tag{4.74}$$

and

$$E^o = \frac{E}{1 + (\epsilon^c)^2 E/f_v''}. \tag{4.75}$$

Because solute redistribution occurs under open-system conditions (constant diffusion potential) the modified elastic constants are referred to as open-system elastic constants.[21]

In order to illustrate the power of the open-system elastic constants, we consider solute redistribution in the vicinity of an edge dislocation directed along the positive x_3 axis with Burger's vector b directed along the x_2 axis. The uniform composition of the isotropic crystal in the absence of stress is c_o, and the reference state for the measurement of strain is the stress-free crystal at composition c_o. The trace of the stress field owing to the isolated edge dislocation in a uniform composition field without solute redistribution is given by[44]

$$\sigma_{kk} = \frac{-\mu b(1+\nu)\sin\phi}{\pi(1-\nu)r} \qquad (4.76)$$

where ϕ is the angle measured from the x_2 axis in the direction of the x_1 axis and $r^2 = x_1^2 + x_2^2$. The actual stress field in the vicinity of the dislocation accounting for the compositional strains that arise owing to solute redistribution is, according to Eq. (4.72), obtained by replacing the actual elastic constants of the homogeneous system by the open-system elastic constants of Eqs. (4.73) and (4.74). The trace of the stress is then given by

$$\sigma_{kk} = \frac{-\mu^o b(1+\nu^o)\sin\phi}{\pi(1-\nu^o)r}. \qquad (4.77)$$

Substituting Eq. (4.77) into Eq. (4.61) gives the composition field in the vicinity of the dislocation,

$$c(r,\phi) = c_o - \frac{\mu^o b(1+\nu^o)\epsilon^c \sin\phi}{f_v'' \pi(1-\nu^o)r} \qquad (4.78)$$

where c_o is the initial (uniform) composition of component B in the crystal. From Eq. (4.77), the trace of the stress is negative (the region is under compression) when $\sin\phi > 0$, corresponding to a point for which $x_1 > 0$. If $\epsilon^c > 0$; i.e., the partial molar volume of component B is greater than that of component A, then Eq. (4.78) predicts that regions of the crystal under compression will be depleted of component B while regions under tension will be enhanced in component B. This result is expected intuitively as the strain energy can be partially relieved by replacing a larger atom with a smaller atom in a region of compression. The magnitude of the solute redistribution depends on the curvature of the free-energy density versus composition curve f_v''. If f_v'' is large, small changes in the composition from c_o give a significant change in the chemical energy. Thus, for a given

---

[44] J. P. Hirth and J. Lothe, *Theory of Dislocations*, John Wiley and Sons, New York (1982).

stress field, the magnitude of the composition shift from the initial composition c_o will be small.

The net solute enhancement in the vicinity of the dislocation is obtained by integrating the change in composition over the volume surrounding the dislocation. Solute enhancement per unit length of dislocation is thus

$$\int_0^\infty \int_0^{2\pi} (c - c_o) r \, d\phi \, dr = \int_0^\infty \frac{-\mu^o b(1 + \nu^o)\epsilon^c}{f_v'' \pi (1 - \nu^o) r} r \, dr \int_0^{2\pi} \sin \phi \, d\phi = 0. \quad (4.79)$$

Thus, to first order in the change in composition from c_o, for the case of a straight-edge dislocation in an isotropic matrix, there is no *net* solute enhancement around the dislocation. Regions enriched in solute are compensated by regions depleted in solute.

Precipitates, or second-phase particles, are sources of stress that can also lead to solute redistribution in the surrounding matrix phase. For the case of an isolated misfitting spherical precipitate in an isotropic matrix, Eshelby[45] showed that the trace of the stress field in the matrix vanishes, $\sigma_{kk} = 0$. This means that, for a spherical precipitate in an isotropic matrix, there is no solute enhancement in the vicinity of the precipitate if the solute expansion tensor is purely dilatational, as it is in the case considered previously. For an anisotropic matrix, the trace of the stress in the matrix does not vanish, and solute redistribution in the vicinity of the spherical precipitate would be expected.[46]

The interaction of two spherical precipitates, however, can lead to solute redistribution. As an example, Figure 17. shows the solute redistribution arising from the interaction between two elastically spherical inhomogeneities subjected to a uniaxial stress field, σ_{app} applied in the x_3 direction.[41] The far-field composition of the infinite matrix is c_o and the reference state for the measurement of strain is assumed to be the stress-free crystal of composition c_o. The transformation strain of the two spherical regions is assumed to vanish, but the elastic constants of the two regions are different from those of the surrounding matrix phase. Solute redistribution in the vicinity of the inhomogeneities is calculated using the open-system elastic constants[47] developed previously and, therefore, accounts for the stress fields engendered by the inhomogeneities as well as by the compositional inhomogeneity. Contour lines are lines of constant concentration determined by solving the elasticity problem using the open-system elastic constants. The isoconcentrates

---

[45] J. D. Eshelby, *Proc. Roy. Soc.* **A252**, 561 (1959).
[46] J. Piller, M. K. Miller, and S. S. Brenner, in *Proc. 29th Int. Field Emmisions Symp.*, eds. H. O. Andren and H. Norden (Goteborg, Sweden, 1982).
[47] Note that the application of the uniaxial stress in the absence of the two inhomogeneities changes the diffusion potential, but does not lead to a change in solute composition.

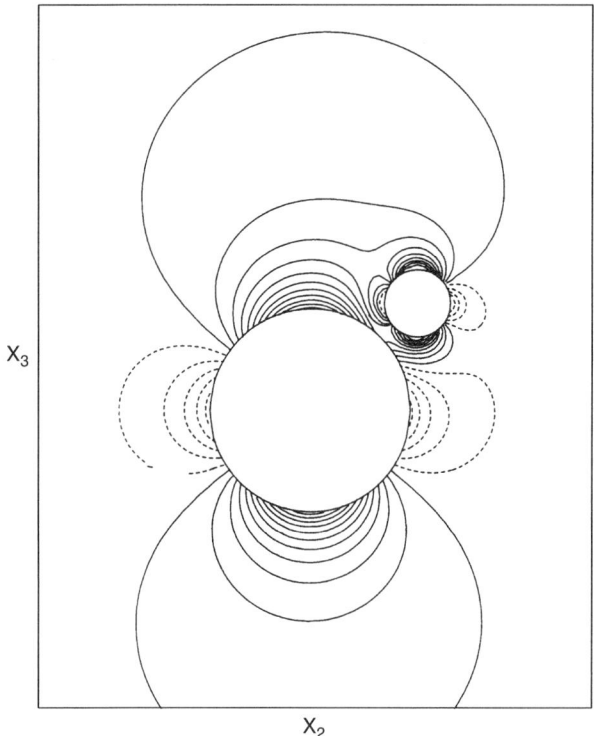

FIG. 17. Solute redistribution in the vicinity of two spherical inhomogeneities subjected to a uniaxial stress field σ_{app} in the x_3 direction is depicted for an isotropic matrix [41]. The solid lines represent regions of solute enhancement when $Z_a > 0$. The contour lines are in intervals of 0.062.

shown have been normalized by Z_a where

$$Z_a = \frac{\epsilon^c \sigma_{app}(1 + \nu^o)(\mu^\beta - \mu^\alpha)}{\mu^\alpha(1 - \nu^o)f_v''}. \tag{4.80}$$

Solute enhancement occurs when $Z_a > 0$ in the regions of the solid lines and depletion in the regions of dashed lines. Changing the sign of the compositional strain ϵ^c, the applied stress, or the relative hardness of the phases changes the sign of Z_a. The contour lines are in intervals of 0.062.

c. Example 3: Interstitial BCC Solution

This example is used to illustrate how a nonhydrostatic stress can introduce a redistribution of interstitial atoms between individual sublattices as well as a net change in the local concentration field[47a]. We consider a binary alloy with cubic symmetry and a compositional strain that is linear in the interstitial composition. We take component B to be the interstitial and assume no solute redistribution occurs on the substitutional lattice. The compositional strain is given by Eq. (3.113) where $y^{(i)} = \rho_B^{(i)}/\rho_o$ is the fraction of occupied interstitial sites on sublattice (i). (See description in section on compositional strains.) Let $y = (y^{(1)} + y^{(2)} + y^{(3)})/3$ be the total fraction of occupied interstitial sites and y_o be the fraction of occupied interstitial sites in the stress-free crystal far from the source of stress. $\mathbf{y} = (y^{(1)}, y^{(2)}, y^{(3)})$ is the vector of interstitial compositions. Chemical equilibrium in the stressed region requires the diffusion potential of B on each sublattice (i), $M_{BV}^{(i)}(\mathbf{y}, \sigma_{ij})$, to equal the stress-free far-field diffusion potential $M_{BV}^{(i)}(\mathbf{y}_o, \sigma_{ij})$.

a. Stress Dependence of Diffusion Potentials. The stress dependence of the interstitial diffusion potentials on each sublattice is obtained as in Section IV. We begin with the combined form of the first two laws of thermodynamics for the interstitial system assuming small strain elasticity;

$$de_{v'} = \theta ds_{v'} + \sigma_{ij}d\epsilon_{ij} + M_{BV}^{(1)}d\rho_B^{(1)} + M_{BV}^{(2)}d\rho_B^{(2)} + M_{BV}^{(3)}d\rho_B^{(3)}, \quad (4.81)$$

where the densities are referred to the reference state. Defining the state function,

$$g_{v'} = e_{v'} - \theta s_{v'} - \epsilon_{ij}\sigma_{ij} \quad (4.82)$$

gives rise to a Maxwell relation that connects the stress and diffusion potential of the interstitial on site (i).

$$-\left(\frac{\partial \epsilon_{ij}}{\partial y^{(1)}}\right)_{\theta, y^{(2)}, y^{(3)}, \sigma_{kl}} = \rho_o' \left(\frac{\partial M_{BV}^{(1)}}{\partial \sigma_{ij}}\right)_{\theta, \mathbf{y}, \sigma_{kl} \neq ij} \quad (4.83)$$

Assuming the elastic constants are independent of composition $\mathbf{y}$ gives

$$\left(\frac{\partial \epsilon_{ij}}{\partial y^{(1)}}\right)_{\theta, y^{(2)}, y^{(3)}, \sigma_{kl}} = \left(\frac{\partial \epsilon_{ij}^c}{\partial y^{(1)}}\right)_{\theta, y^{(2)}, y^{(3)}, \sigma_{kl}} = \eta_{ij}^{(1)} \quad (4.84)$$

where the compositional strain as given by Eq. (3.113) has been used. Substituting Eq. (4.84) into Eq. (4.83) and integrating yields

$$M_{BV}^{(1)}(\mathbf{y}, \sigma_{kl}) = M_{BV}^{(1)}(\mathbf{y}, 0) - \frac{1}{\rho_o'}\eta_{kl}^{(1)}\sigma_{kl}. \quad (4.85)$$

[47a] W. C. Johnson and J. Y. Hut Metall. Trans., **34a**, 2819 (2003).

Similar expressions hold for $M^{(2)}$ and $M^{(3)}$ so that, in general,

$$M_{BV}^{(i)}(\mathbf{y}, \sigma_{kl}) = M_{BV}^{(i)}(\mathbf{y}, 0) - \frac{1}{\rho_o'}\eta_{kl}^{(i)}\sigma_{kl}. \tag{4.86}$$

The precise form of $M^{(i)}$ depends on the stress-free solution thermodynamics of the crystal. Assuming an ideal solution, the stress-free Helmholtz free energy density can be written as[48]

$$f_{v'}(\theta, \mathbf{y}) = \rho_o'\left(y^{(1)} + y^{(2)} + y^{(3)}\right)h_I + \rho_o'k\theta \sum_{i=1}^{3}\left[\left(1 - y^{(i)}\right)\ln\left(1 - y^{(i)}\right)\right.$$
$$\left. + y^{(i)}\ln y^{(i)}\right] \tag{4.87}$$

where h_I is an enthalpy per interstitial atom and is independent of the interstitial site on which the atom resides. The stress-free diffusion potential is

$$M_{BV}^{(i)}(\mathbf{y}, 0) = \frac{1}{\rho_o'}\frac{\partial f_v}{\partial y^{(i)}} = h_I + k\theta \ln\left(\frac{y^{(i)}}{1 - y^{(i)}}\right) \tag{4.88}$$

The stress-dependent diffusion potential is, from Eqs. (4.86) and (4.88),

$$M_{BV}^{(i)}(\mathbf{y}, \sigma_{kl}) = h_I + k\theta \ln\left(\frac{y^{(i)}}{1 - y^{(i)}}\right) - \frac{1}{\rho_o'}\eta_{kl}^{(i)}\sigma_{kl}. \tag{4.89}$$

b. *Open-system Elastic Constants.* Chemical equilibrium requires for each sublattice (i)

$$M_{BV}^{(i)}(\mathbf{y}^o, 0) = M_{BV}^{(i)}(\mathbf{y}, 0) - \frac{1}{\rho_o'}\eta_{kl}^{(i)}\sigma_{kl} \quad \text{for } (i) = (1), (2), (3). \tag{4.90}$$

Substituting for the diffusion potentials and solving for $y^{(i)}$ gives

$$\frac{y^{(i)}}{1 - y^{(i)}} = \frac{y_o}{1 - y_o}\exp\left[\frac{\eta_{kl}^{(i)}\sigma_{kl}}{\rho_o'k\theta}\right]. \tag{4.91}$$

Equation (4.91) is the exact expression relating the interstitial composition $\mathbf{y}$ to the stress state. This term can be linearized when

$$\left|\frac{\eta_{kl}^{(i)}\sigma_{kl}}{\rho_o'k\theta}\right| \ll 1, \quad (i) = (1), (2), (3) \tag{4.92}$$

---

[48] C. H. P. Lupis, *Chemical Thermodynamics of Materials*, Elsevier, Amsterdam (1983).

or equivalently,

$$\left| \frac{(y^{(i)} - y_o)}{y_o} \right| \ll 1 \quad (i) = (1), (2), (3). \tag{4.93}$$

Expanding the exponential term in Eq. (4.91) to first order yields

$$\Delta y^{(i)} = y^{(i)} - y_o = \frac{y_o(1 - y_o)}{\rho'_o k\theta} \eta^{(i)}_{kl} \sigma_{kl}. \tag{4.94}$$

The open-system elastic constants can be obtained for the interstitial alloy by substituting the linearized form of the composition change, Eq. (4.94), into the constitutive relations, Eq. (3.106) and Eq. (3.115). The linearized compositional strain can be expressed as

$$\epsilon^c_{kl}(\mathbf{y}) = \frac{y_o(1 - y_o)}{\rho'_o k\theta} N_{klmn} \sigma_{mn} \tag{4.95}$$

where we define the fourth-rank compositional strain tensor,

$$N_{klmn} = \eta^{(1)}_{kl} \eta^{(1)}_{mn} + \eta^{(2)}_{kl} \eta^{(2)}_{mn} + \eta^{(3)}_{kl} \eta^{(3)}_{mn}. \tag{4.96}$$

Substituting Eq. (4.95) into the stress-strain constitutive relation, Eq. (3.106), yields

$$\sigma_{ij} = C_{ijkl} \left[E_{kl} - \frac{y_o(1 - y_o)}{\rho'_o k\theta} N_{klmn} \sigma_{mn} \right]. \tag{4.97}$$

Contracting each side of Eq. (4.97) with the compliance tensor, S_{pqij}, and rearranging gives

$$S^o_{pqij} \sigma_{ij} = E_{pq} \tag{4.98}$$

where the open-system elastic compliance tensor for this interstitial system, S^o_{pqij}, is defined as

$$S^o_{pqij} = S_{pqij} + \frac{y_o(1 - y_o)}{\rho'_o k\theta} N_{pqij}. \tag{4.99}$$

For the BCC crystal with interstitials, using Eq. (3.114) in Eq. (4.99) gives the nonzero components of the compliance tensor $S^o_{44} = S_{44}$,

$$S^o_{12} = S_{12} + \frac{y_o(1 - y_o)\eta_a(\eta_a + 2\eta_c)}{\rho'_o k\theta} \tag{4.100}$$

and

$$S_{11}^o = S_{11} + \frac{y_o(1-y_o)(2\eta_a^2 + \eta_c^2)}{\rho_o' k\theta}. \quad (4.101)$$

The components of the open-system elastic constants are $C_{44}^o = C_{44}$,

$$C_{12}^o = \frac{-S_{12}^o}{(S_{11}^o + 2S_{12}^o)(S_{11}^o - S_{12}^o)} \quad (4.102)$$

and

$$C_{11}^o = \frac{S_{11}^o + S_{12}^o}{(S_{11}^o + 2S_{12}^o)(S_{11}^o - S_{12}^o)}. \quad (4.103)$$

c. Estimates of Compositional Change. In order to assess the validity of the open-system elastic constants approximation for estimating stress-induced solute redistribution, we compare three different approximations for the elastically induced composition change for a homogeneously stressed crystal with the exact solution of the mechanical and thermodynamic equilibrium conditions. In order to make the problem tractable, we consider a nonhydrostatically stressed, but uniformly deformed, crystal in contact with a chemical potential reservoir. The reservoir establishes a composition of y^o in the crystal in the absence of stress. The crystallographic axes are aligned with the coordinate axes. Traction boundary conditions are imposed along the x_1 and x_2 directions such that $\sigma_{11} = \sigma_{22} = 0$. A strain $\epsilon_{33} = 0.01$ is imposed in the x_3 direction. We use the elastic constants of BCC iron:[44] $C_{11} = 24.2 \times 10^{11}$, $C_{12} = 14.65 \times 10^{11}$, and $C_{44} = 11.2 \times 10^{11}$ ergs/cm³; and $\sigma_{11} = 7.6 \times 10^{-13}$, $S_{12} = -2.87 \times 10^{-13}$, and $S_{44} = 2.23 \times 10^{-13}$ cm³/erg. For calculation purposes, we choose the compositional strain of hydrogen in BCC iron[49] with $\eta_a = -0.11$ and $\eta_c = 0.37$, $\rho_o = 8.5 \times 10^{22}$ sites/cm³, and $\theta = 400$K.

Figures 18 and 19 show the dependence of the scaled composition on each sublattice site, $y^{(1)}/y_o$ and $y^{(3)}/y_o$, on the far-field composition y_o for the imposed strain $\epsilon_{33} = 0.01$. Figure 20. gives the net (scaled) concentration change of the interstitial, y/y_o. The exact solution, obtained by numerical solution of the mechanical and thermodynamic equilibrium conditions, is given by the solid curve. Three common approximations to the equilibrium compositions are also shown. The approximation for which the compositional self-stress is ignored is given by the dashed line with two dots. The approximation based on the assumption that the compositional strain can be treated as a dilatation (i.e., the tetragonal nature of the compositional strain is ignored) is depicted by the dot-dash curve. The open system elastic constants approach, for which the solutions are accurate to first order in the composition change, is given by the dashed curve.

---

[49] Q. Bai, W. Chu, and C. Hsiao, *Scripta metall.* **21**, 613 (1987).

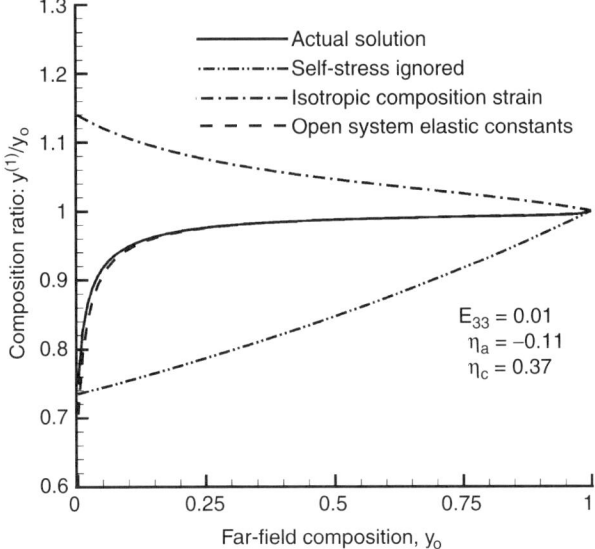

FIG. 18. The scaled composition of interstitial on sublattice site (1) is plotted as a function of the far-field composition, $y^{(1)47a}$.

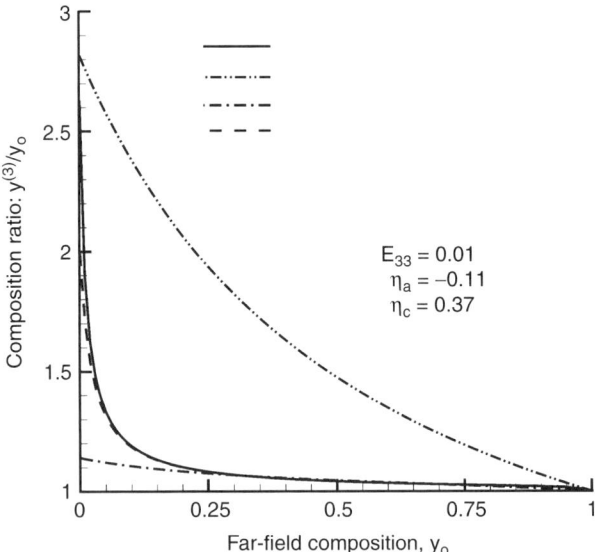

FIG. 19. The scaled composition of interstitial on sublattice site (3) is plotted as a function of the far-field composition, $y^{(1)47a}$.

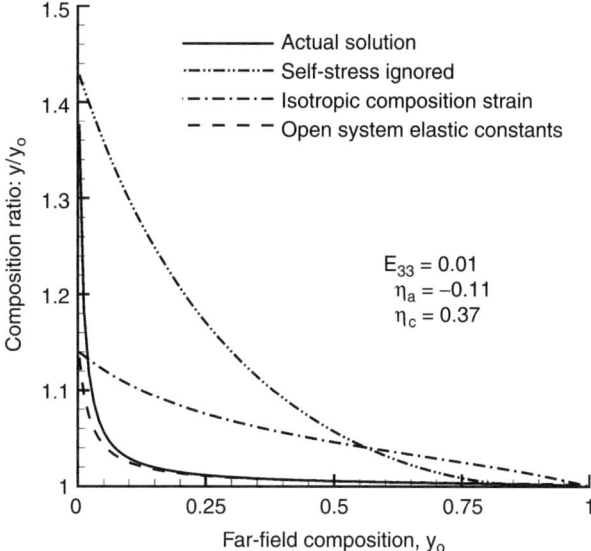

FIG. 20. The net scaled composition of interstitial is plotted as a function of the far-field composition, y_0[47a].

Application of the positive strain $\epsilon_{33} = 0.01$ induces a tensile stress in the x_3 direction. The elastic energy associated with this tensile stress and with the compositional strain of the interstitials can be reduced by having the interstitial occupy sublattice (3), thereby aligning the positive component of the compositional strain with the tensile field. Similarly, the tensile stress tends to reduce the occupation of sublattices (1) and (2). This behavior is apparent from the actual solutions shown in Figures 18 and 19.

When the compositional self-stress is ignored, significant error can be introduced into the calculation of the elastically induced composition change. This approximation systematically underestimates the composition change induced by the imposed strain on sublattice (1) and overestimates the composition change on sublattice (3). As the distribution of the interstitials on the sublattices is changed, there is a change in the stress-free dimensions of the crystal. Enhancement of interstitials on sublattice (3) induces a net (positive) displacement in the x_3 direction. This stress-free compositional strain reduces the force required to impart the given strain E_{33} and, thereby, reduces the net stress, σ_{33}. As a consequence, the change in composition on a given sublattice will not be as extreme as predicted when the compositional strain is ignored.

The approximation based on the assumption of a dilatational misfit strain can also misrepresent the direction and magnitude of the composition shift. For the calculation shown here, the dilatational strain is given by $(\eta_c + 2\eta_a)/3 = 0.05$.

Consequently, imposing a tensile field on the crystal yields a net increase in the interstitial composition. Because the compositional strain is taken to be isotropic, the enhancement occurs uniformly on all three sublattices. However, the tetragonal nature of the compositional strain breaks the degeneracy of the interstitial sublattices, and a decrease in the composition on sublattice (1) occurs rather than the increase predicted by this approximation. Likewise, the increase in composition on sublattice (3) predicted for a dilatational compositional strain underestimates the actual increase, especially for smaller far-field compositions. The isotropic strain significantly overestimates the total composition increase for far-field compositions $y_o > 0.05$ and underestimates it for smaller far-field compositions $y_o < 0.05$.

The open-system elastic constants approach is seen to give a very good approximation to the exact solution except for very small far-field compositions. This approximation gives the same trends as the exact solution for the composition changes on all sublattices for the far-field compositions shown.

d. Free Energy Densities

In this section, we present some useful expressions for the stress and strain dependence of the Helmholtz free energy density, $f_{v'}$, and the thermodynamic potential, $\omega_{v'}$, which are valid for a linear elastic crystal in the small strain approximation. These quantities are used frequently to establish boundary conditions for problems in diffusion and microstructural evolution. For concreteness, we take the crystal to be a binary ($A - B$) substitutional alloy with vacancies. No assumption is made on the crystal symmetry or the functional dependence of the lattice parameter and elastic constants on composition or temperature. The approach presented here can be extended directly to more complicated crystal structures with atoms occupying multiple sublattices.

a. Helmholtz Free Energy Density. The Helmholtz free energy density can be defined as a Legendre transform of the internal energy density,

$$f_{v'} = e_{v'} - \theta s_{v'}. \tag{4.104}$$

Using Eq. (4.33) and making the small strain substitution for the elastic work term, $T_{ji} dF_{ij} = \sigma_{ij} d\epsilon_{ij}$, the differential of $f_{v'}$ can be written as

$$df_{v'} = -s_{v'} d\theta + \sigma_{ij} d\epsilon_{ij} + M_{AV} d\rho'_A + M_{BV} d\rho'_B. \tag{4.105}$$

In the small strain approximation, the natural variables to associate with $f_{v'}$ are $s_{v'}$, ϵ_{ij}, ρ'_A, and ρ'_B. The composition can also be used as an independent variable using the mole fraction $c_i = \rho'_i/\rho'_o$ for $i = A, B$. This gives, for the differential $df_{v'}$,

$$df_{v'} = -s_{v'} d\theta + \sigma_{ij} d\epsilon_{ij} + \rho'_o M_{AV} dc_A + \rho'_o M_{BV} dc_B. \tag{4.106}$$

The dependence of $f_{v'}$ on strain is obtained through integration of Eq. (4.106) under conditions of constant temperature and composition. In order to complete this integration, a reference state for the measurement of strain must be defined. Once again, we choose the reference state as a homogeneous crystal at temperature θ^o and composition $\rho_A^{o'}$ and $\rho_B^{o'}$ (or c_A^o and c_B^o). Integration of Eq. (4.106) from the state of zero strain to the state of strain ϵ_{ij} yields

$$\int_{f_{v'}(\theta,\rho'_A,\rho'_B,0)}^{f_{v'}(\theta,\rho'_A,\rho'_B,\epsilon_{ij})} df_{v'} = \int_0^{\epsilon_{ij}} \sigma_{ij} de_{ij} \qquad (4.107)$$

or

$$f_{v'}(\theta, \rho'_A, \rho'_B, \epsilon_{ij}) = f_{v'}(\theta, \rho'_A, \rho'_B, 0) + \int_0^{\epsilon_{ij}} \sigma_{ij} de_{ij} \qquad (4.108)$$

where $f_{v'}(\theta, \rho'_A, \rho'_B, 0)$ is the free energy density in the (zero strain) reference state. (This is not the zero-stress state, unless $\theta = \theta^o$, $c_A = c_A^o$, and $c_B = c_B^o$.) In order to evaluate the remaining integral, a constitutive equation for the stress must be introduced for the linear elastic crystal. Assuming the composition and temperature dependence of the lattice parameter can be expressed in terms of a general eigenstrain $\epsilon_{ij}^P(\theta, c_A, c_B)$, the constitutive law for the stress can be expressed as (see Eq. (3.94))

$$\sigma_{ij} = C_{ijkl}(\theta, c_A, c_B)\left(\epsilon_{kl} - \epsilon_{kl}^P(\theta, c_A, c_B)\right). \qquad (4.109)$$

Substituting Eq. (4.109) into Eq. (4.108) and integrating gives the change in free energy when a homogeneous element of material at temperature θ and composition $\rho_A^{o'}, \rho_B^{o'}$ is deformed from a state of zero strain to a strain ϵ_{ij}.

$$f_{v'}(\theta, \rho'_A, \rho'_B, \epsilon_{ij}) = f_{v'}(\theta, \rho'_A, \rho'_B, 0) + \frac{1}{2} C_{ijkl} \epsilon_{ij} \epsilon_{kl} - C_{ijkl} \epsilon_{ij} \epsilon_{kl}^P \qquad (4.110)$$

This expression for the strain dependence of $f_{v'}$ is not always useful. An expression for the reference term $f_{v'}(\theta, \rho'_A, \rho'_B, 0)$ must still be obtained, and this term does not, in general, correspond to a stress-free state. (If all eigenstrains vanish, then the reference state is stress free.) Therefore, it is often more convenient to express $f_{v'}$ in terms of the stress tensor, so that when the stress vanishes, the elastic energy also vanishes.

There are at least two approaches that can be used to calculate the dependence of $f_{v'}$ on stress. The first approach entails integrating the increment of elastic work, $\sigma_{ij} d\epsilon_{ij}$, from the strain state corresponding to the stress-free state to the desired strain state and then expressing the strain tensor in terms of the stress tensor using the constitutive law for the stress. This was the approach taken in the last section to calculate the strain energy density for a system with eigenstrains.

Because the stress vanishes, $\sigma_{ij} = 0$, when $\epsilon_{ij} = \epsilon_{ij}^p$ the change in the free energy density on deforming from the stress-free state to the stress state of interest becomes

$$\int_{f_{v'}(\theta,\rho_A',\rho_B',\sigma_{ij}=0)}^{f_{v'}(\theta,\rho_A',\rho_B',\sigma_{ij})} df_{v'} = \int_{f_{v'}(\theta,\rho_A',\rho_B',\epsilon_{ij}^p)}^{f_{v'}(\theta,\rho_A',\rho_B',\epsilon_{ij})} df_{v'} = \int_{\epsilon_{ij}^p}^{\epsilon_{ij}} \sigma_{ij} d\epsilon_{ij}. \quad (4.111)$$

Using Eqs. (3.116)–(3.119), we obtain

$$f_{v'}(\theta, \rho_A', \rho_B', \sigma_{ij}) = f_{v'}(\theta, \rho_A', \rho_B', \sigma_{ij} = 0)$$
$$+ \frac{1}{2} C_{ijkl} (\epsilon_{ij} - \epsilon_{ij}^p)(\epsilon_{kl} - \epsilon_{kl}^p) \quad (4.112)$$

Inverting the constitutive law for the stress yields $\epsilon_{kl} - \epsilon_{kl}^p = S_{klij}\sigma_{ij}$. Substituting this expression into Eq. (4.112) yields the following expression for the stress dependence of the Helmholtz free energy density:

$$f_{v'}(\theta, \rho_A', \rho_B', \sigma_{ij}) = f_{v'}(\theta, \rho_A', \rho_B', \sigma_{ij} = 0) + \frac{1}{2} S_{ijkl}\sigma_{ij}\sigma_{kl}. \quad (4.113)$$

The standard state term $f_{v'}(\theta, \rho_A', \rho_B', \sigma_{ij} = 0)$ is for a stress-free system and can be expressed as a function of just the three variables θ, ρ_A', and ρ_B' (for example, see Eq. (2.73)). A formal expression for the standard state term $f_{v'}(\theta, \rho_A', \rho_B', \sigma_{ij} = 0)$ can also be obtained directly from Eq. (2.63). Setting $P = 0$ and denoting the diffusion potential M_{AV}^o as the diffusion potential under zero-stress conditions evaluated at the temperature (θ) and composition (ρ_A', ρ_B') of interest, gives

$$f_{v'}(\theta, \rho_A', \rho_B', \sigma_{ij} = 0) = M_{AV}^o \rho_A' + M_{BV}^o \rho_B' + \mu_V^v \rho_o' \quad (4.114)$$

where μ_V^v is the vacancy chemical potential. Substituting Eq. (4.114) into Eq. (4.113) yields

$$f_{v'}(\theta, \rho_A', \rho_B', \sigma_{ij}) = M_{AV}^o \rho_A' + M_{BV}^o \rho_B' + \mu_V^v \rho_o' + \frac{1}{2} S_{ijkl}\sigma_{ij}\sigma_{kl}. \quad (4.115)$$

Sometimes it is convenient to use a homogeneous system at constant pressure P^o as the standard state. In this case, the free energy density of the system at $\sigma_{ij} = -P^o \delta_{ij}$ is

$$f_{v'}(\theta, \rho_A', \rho_B', -P^o\delta_{ij}) = f_{v'}(\theta, \rho_A', \rho_B', \sigma_{ij} = 0) + \frac{1}{2} S_{kkii}(P^o)^2. \quad (4.116)$$

Subtracting Eq. (4.116) from Eq. (4.113) gives

$$f_{v'}(\theta, \rho'_A, \rho'_B, \sigma_{ij}) = f_{v'}(\theta, \rho'_A, \rho'_B, -P^o\delta_{ij}) + \frac{1}{2}S_{ijkl}\sigma_{ij}\sigma_{kl}$$
$$- \frac{1}{2}S_{kkii}(P^o)^2. \qquad (4.117)$$

The Helmoholtz free energy density at pressure P^o can be obtained from Eq. (2.63). For small strains, $J = (1 + \epsilon_{kk})$. If M_{AV}^P denotes the diffusion potential for a system under hydrostatic pressure P^o (or $\sigma_{ij} = -P^o\delta_{ij}$), then

$$f_{v'}(\theta, \rho'_A, \rho'_B, -P^o\delta_{ij}) = -P^o(1 + \epsilon_{kk}) + M_{AV}^P\rho'_A + M_{BV}^P\rho'_B + \mu_V^v\rho'_o. \qquad (4.118)$$

Substituting Eq. (4.118) into Eq. (4.117) gives

$$f_{v'}(\theta, \rho'_A, \rho'_B, \sigma_{ij}) = -P^o(1 + \epsilon_{kk}) + \frac{1}{2}S_{ijkl}\sigma_{ij}\sigma_{kl} - \frac{1}{2}S_{iikk}(P^o)^2$$
$$+ M_{AV}^P\rho'_A + M_{BV}^P\rho'_B + \mu_V^v\rho'_o \qquad (4.119)$$

Equation (4.117) is an expression for the stress dependence of the Helmholtz free energy density when the standard state is at pressure P^o. For a linear elastic crystal, the stress is independent of the referential state for the measurement of strain, and Eq. (4.117) is valid for a reference state as described previously, or for a reference state chosen such that the strain vanishes for the homogeneous crystal at pressure P^o.

The second approach to determining the stress dependence of the Helmholtz free energy density entails defining a new energy density for which the natural variable is the stress. This approach was used to construct Maxwell relations for calculating the stress dependence of the diffusion potential. The appropriate free energy $g_{v'}$, was given by Eq. (4.34) which, in the small strain limit, becomes $g_{v'} = e_{v'} - \theta s_{v'} - \sigma_{ij}\epsilon_{ij} = f_{v'} - \sigma_{ij}\epsilon_{ij}$. Because the natural independent variables of $g_{v'}$ are θ, ρ'_A, ρ'_B, and σ_{ij}, the stress dependence of $g_{v'}$ is obtained by the following integration:

$$\int_{g_{v'}(\theta,\rho'_A,\rho'_B,0)}^{g_{v'}(\theta,\rho'_A,\rho'_B,\sigma_{ij})} dg_{v'} = -\int_0^{\sigma_{ij}} \epsilon_{ij}ds_{ij}. \qquad (4.120)$$

Expressing the strain in terms of the stress in Eq. (4.120) and integrating yields

$$g_{v'}(\theta, \rho'_A, \rho'_B, \sigma_{ij}) = g_{v'}(\theta, \rho'_A, \rho'_B, 0) - \int_0^{\sigma_{ij}} [\epsilon_{ij}^P + S_{ijkl}s_{kl}]ds_{ij}$$
$$= g_{v'}(\theta, \rho'_A, \rho'_B, 0) - \frac{1}{2}S_{ijkl}\sigma_{ij}\sigma_{kl} - \epsilon_{ij}^P\sigma_{ij} \qquad (4.121)$$

Recognizing that $f_{v'} = g_{v'} + \sigma_{ij}\epsilon_{ij}$, Eq. (4.113) is obtained immediately when $\sigma_{ij}\epsilon_{ij}$ is added to Eq. (4.121).

b. Thermodynamic Potential. In this section, we present expressions for the grand canonical energy density, $\omega_{v'}$, of a nonhydrostatically stressed crystal using the results obtained for $f_{v'}$. Like $f_{v'}$, it is simplest to express $\omega_{v'}$ in terms of the stress rather than its natural variable, the strain tensor. By definition,

$$\omega_{v'}(\theta, M_{AV}, M_{BV}, \sigma_{ij}) = f_{v'}(\theta, \rho'_A, \rho'_B, \sigma_{ij}) - M_{AV}\rho'_A - M_{BV}\rho'_B \quad (4.122)$$

where M_{AV} and M_{BV} are the diffusion potentials for a system at temperature θ, composition ρ'_A and ρ'_B, and stress state σ_{ij}. To evaluate $\omega_{v'}$ at zero pressure we use the Helmholtz free energy density at zero pressure (Eq. (4.115)),

$$\omega_{v'}(\theta, M_{AV}, M_{BV}, \sigma_{ij}) = \mu_V^v \rho'_o + \frac{1}{2}S_{ijkl}\sigma_{ij}\sigma_{kl} + (M_{AV}^o - M_{AV})\rho'_A$$
$$+ (M_{BV}^o - M_{BV})\rho'_B. \quad (4.123)$$

The density $\omega_{v'}$ can also be referred to a standard state at a pressure P^o instead of the stress-free state. In this case we use Eq. (4.122) and the expression for the Helmholtz free energy at a reference P^o, Eq. (4.119), to yield

$$\omega_{v'}(\theta, M_{AV}, M_{BV}, \sigma_{ij}) = -P^o(1 + \epsilon_{kk}) + \mu_V^v \rho'_o + \frac{1}{2}S_{ijkl}\sigma_{ij}\sigma_{kl} - \frac{1}{2}S_{iikk}(P^o)^2$$
$$+ (M_{AV}^p - M_{AV})\rho'_A + (M_{BV}^p - M_{BV})\rho'_B \quad (4.124)$$

V. Capillary and Interfacial Properties

13. Introduction

Surfaces and interfaces play a major role in the evolution of microstructure and the properties of materials.[50] Thermodynamic properties usually change rapidly in the vicinity of an interface, and interfaces display properties much different from those of the contiguous phases. It is at interfaces where phase transformations actually occur, and a viable description of the thermodynamic properties of interfaces is of central importance in the description of phase transformations. Many texts treat quite nicely the thermodynamic properties of fluid–fluid interfaces.[48] However, research dating from the time of Gibbs has shown that interfaces involving at least one phase that is a crystalline display properties that are quite distinct from

---

[50] D. A. Porter and K. E. Easterling, *Phase Transformations in Metals and Alloys*, Chapman and Hall, Londan (1992).

fluid–fluid interfaces in many respects. In order to illustrate these differences and to set the background for the ensuing material on crystal–fluid and crystal–crystal equilibrium, we develop a thermodynamic description of a spherical, crystalline particle in contact with a fluid. This treatment also has important implications for the thermodynamic properties of nanoparticles, as capillary effects are dominant at small particle sizes.

Specifically, we treat isotropic, hydrostatically stressed spherical crystals, for which there is both a surface energy and a surface stress. Large deformations are possible, although we invoke the small-strain assumption when the resulting equations are applied to specific examples. We begin with a brief discussion of thermodynamic excess or superficial properties of the interface and identify the surface stress with the excess tangential force that is present as a result of the interface. We then obtain the conditions for thermodynamic equilibrium between a small spherical crystal in contact with a fluid. Finally, the equilibrium conditions are used to show the relative contributions of surface stress and surface energy to several thermodynamic properties, including the diffusion and chemical potentials, as a function of particle size. The constraints of a spherical geometry and a hydrostatic stress within the crystal are relaxed later in the monograph.

14. SURFACE EXCESS QUANTITIES

The physical (observed) volume of a crystal can be changed in two distinct ways. First, lattice sites can be accreted to the crystal for a fixed state of mechanical deformation. Second, mechanical forces can be applied to the crystal, resulting in a change in its deformation state for a fixed configuration of lattice sites. The change in energy associated with the accretion term is different from the change in mechanical energy associated with the crystal deformation and, as was done for the treatment of a hydrostatically stressed crystal, each energy term must be considered explicitly in the thermodynamic formulation of crystals. Observation of the physical volume or shape of the crystal alone does not allow these two processes to be readily differentiated, as a small increase in volume could be a result of either a mechanical expansion of the crystal (a change in its strain state) or the addition (or rearrangement) of lattice sites. Therefore, it is convenient to express the thermodynamic extensive quantities with respect to a stress-free reference state where these two processes can be readily differentiated.

Similarly, the physical area of the crystal can be changed by either the accretion (or local rearrangement) of lattice sites at the surface or by the mechanical deformation of the surface for a fixed arrangement of lattice sites. Each of these thermodynamic processes can be expressed with respect to either the reference or actual states of the system. The conjugate thermodynamic variable associated with the mechanical deformation of the surface at constant configuration of the lattice sites is the Piola-Kirchhoff surface stress tensor, $\hat{T}$, when expressed in terms of the reference state. Creating a new interface by the accretion or rearrangement of lattice sites while the thermodynamic state of the crystal remains constant yields the scalar conjugate variable, γ', an energy density (energy/area) that, like an energy density for the bulk crystal, can be expressed with respect to either the reference state of the crystal (Lagrangian state indicated by the prime superscript) or the actual state of the crystal (Eulerian state).

On an atomistic scale, it is usually not possible to identify a precise position for an interface. This ambiguity arises from the observation that most thermodynamic densities change continuously, although over very short distances, from their value in one phase to their value in the other phase, rather than discontinuously at some geometrically sharp interface. In his treatment of interfaces, Gibbs defined a surface of discontinuity as the "non-homogeneous film which separates homogeneous or nearly homogeneous masses."[20] Thermodynamic densities, such as the composition, entropy, and energy densities, change continuously through this region, from their values in one homogeneous phase to their corresponding values in the adjacent homogeneous phase. In contrast, the intensive variables conjugates to these thermodynamic densities, which would include the temperature and chemical potentials, were shown to be constant throughout both homogeneous phases and the surface of discontinuity.

For a variety of reasons, Gibbs found it useful to represent the physical system with its surface of discontinuity in terms of a hypothetical system consisting of two homogeneous phases separated by a geometrical surface called the dividing surface. Associated with the imaginary dividing surface are thermodynamic excess or superficial quantities that are introduced to ensure that the imaginary system accurately reflects the mass, entropy, and energy of the actual system. The superficial quantities are non-physical, as they depend on the precise position of the imaginary dividing surface. There is usually no reason other than convenience to establish the position of the dividing surface. However, Gibbs also showed that various combinations of the superficial quantities can yield physically meaningful and experimentally measurable variables that are independent of the dividing surface position.

A system consisting of a spherical crystal surrounded by a low density fluid or gas has the advantage that a dividing surface can often be placed unambiguously at the boundary of the crystal in the actual system. This definition allows locations the precise position of the interface in the reference state configuration and establishment it the number of unit cells in the crystal.

There are situations, however, where the boundary between the crystal and surrounding fluid is not so distinct. For example, a surface of discontinuity of a nanometer or more can exist when a crystal is in contact with its melt. Figure 21(a) depicts schematically the radial dependence of a thermodynamic density $\phi_v(r)$. (For example, $\phi_v(r)$ could represent the entropy or mass density.) $\phi_v(r)$ achieves uniform values within each phase at distances removed from the surface of

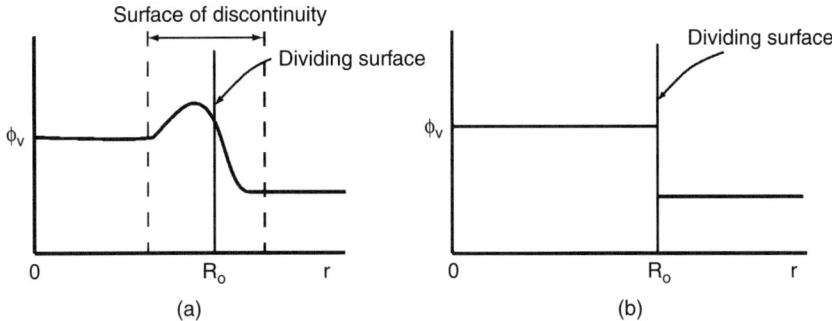

FIG. 21. (a) depicts the physical system with a continuous profile of a thermodynamic density ϕ through the surface of discontinuity bounded by the dashed, vertical lines. (b) is the corresponding imaginary system with the dividing surface located at $r = R$ separating two homogeneous phases with densities ϕ_v^s and ϕ_v^f.

discontinuity. Within the surface of discontinuity, which contains portions of both the crystal and fluid, the density changes rapidly with radial position. Figure 21(b) depicts the actual system as it is represented in the corresponding imaginary system. The density fields are treated as being uniform up to the surface from both phases and a superficial quantity, ϕ_a, is associated with the surface. As seen from the figure, displacing the dividing surface would change the value of the superficial quantity. The imaginary system has representations in both the actual and reference states.

The superficial density of any component in the radially symmetric system is defined as

$$\phi_a^\Sigma = \frac{1}{R^2} \int_0^{R_{ex}} \phi_v r^2 dr - \frac{1}{R^2} \int_0^R \phi_v^s r^2 dr - \frac{1}{R^2} \int_R^{R_{ex}} \phi_v^f r^2 dr \qquad (5.1)$$

where ϕ_v^s and ϕ_v^f are the ϕ densities in the crystal and fluid, respectively, and R_{ex} is the external radius of the system. The superficial density expressed with respect to the reference state, $\phi_{a'}^\Sigma$, is

$$\phi_{a'}^\Sigma = \phi_a^\Sigma \frac{\mathcal{A}^s}{\mathcal{A}^{s'}} = \phi_a^\Sigma \frac{R^2}{R_o^2} = \phi_a^\Sigma \hat{J}. \qquad (5.2)$$

$\mathcal{A}^{s'}$ is the surface area of the crystal as measured in the reference state and $\hat{J} = \mathcal{A}^s/\mathcal{A}^{s'}$. The superficial entropy and mass densities of components A and B are designated as s_a^Σ, Γ_A, and Γ_B, respectively.

The surface stress is defined to be the superficial tangential stress that arises owing to the presence of the surface. It has units of force per unit length. The surface stress can be determined by an integration either over the reference state

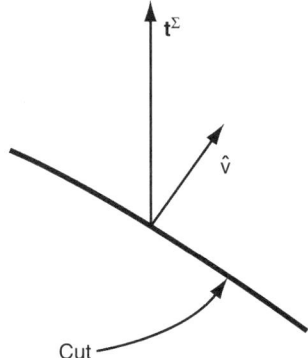

FIG. 22. The superficial traction vector, t^Σ, is defined as the force per unit length exerted by the material into which $\hat{v}$ points on the material from which $\hat{v}$ points. $\hat{v}$ is a unit vector which resides in the tangent plane to the surface at the point of interest.

configuration or over the actual state of the system. When expressed with respect to the Lagrangian formulation, the excess tangential stress is denoted as $\hat{\mathbf{T}}$, when expressed with respect to the actual state, $\hat{\sigma}$.

The superficial traction vector, t^Σ, has a direct analog with the (bulk) traction vector, t. Imagine a small cut being made in the surface of the crystal. Perpendicular to this cut and in the tangent plane of the surface, draw a unit vector $\hat{v}$, as in Figure 22. The superficial traction vector, t^Σ, is defined as the force vector per unit length exerted by the surface material into which $\hat{v}$ points on the surface material from which $\hat{v}$ points. In general, the superficial traction vector need not point in the direction of $\hat{v}$, although it does so for the radially symmetric system considered here. The surface stress tensor relates the unit vector $\hat{v}$ to the superficial traction vector in the same way as the stress tensor relates the bulk unit normal $\mathbf{n}$ to the traction vector t in Eq. (3.54). In the Eulerian description,

$$\hat{t}_i = \hat{\sigma}_{ji} v_j \tag{5.3}$$

where $\hat{\sigma}_{ij}$ is the Cauchy surface stress tensor. The traction vector in the Lagrangian description, $\hat{t}_i^o$, is the actual force acting on the element of length in the surface per unit reference length. Thus,

$$\hat{t}_i^o = \hat{T}_{ji} v'_j. \tag{5.4}$$

For the isotropic, radially symmetric system being treated,

$$\hat{\sigma}_{ji} = \hat{f} \hat{\delta}_{ij} \tag{5.5}$$

and

$$\hat{T}_{ji} = \hat{T}\hat{\delta}_{ij} \qquad (5.6)$$

where $\hat{\delta}_{ij}$ is the Kronecker delta function, which acts only on the surface coordinates. For the sphere, it is a 2×2 tensor that acts on the θ and ϕ components. Thus $\hat{f}_{\theta\theta} = \hat{f}_{\phi\phi} = \hat{f}$.

The relationship between $\hat{T}$ and $\hat{f}$ for the radially symmetric system is obtained by considering the superficial traction vector. The actual force acting on the element of length in the surface is the same regardless of its representation in the Eulerian or Lagrangian description. Thus,

$$\hat{t}_i d\ell = \hat{t}_i^o d\ell' \qquad (5.7)$$

where $d\ell$ and $d\ell'$ are the elements of length in the actual state and its corresponding length in the reference state, respectively. Substituting for the tractions in terms of the surface stress tensors yields

$$\hat{\sigma}_{ji} v_j d\ell = \hat{T}_{ji} v'_j d\ell' \qquad (5.8)$$

or

$$\hat{f} v_i d\ell = \hat{T} v'_i d\ell'. \qquad (5.9)$$

For the radially symmetric deformation of the sphere, the tangential normals $\hat{v}$ and $\hat{v}'$, are equal. In addition, $d\ell = R d\theta_s$ and $d\ell' = R_o d\theta_s$ where θ_s is a parameter that increases along the arclength of the sphere. Substituting these expressions into Eq. (5.9) yields the following relationship between $\hat{f}$ and $\hat{T}$:

$$\left[\hat{f}\left(\frac{R}{R_o}\right) - \hat{T} \right] R_o d\theta_s = 0 \quad \text{or} \quad \hat{T} = \hat{f}\left(\frac{R}{R_o}\right) \qquad (5.10)$$

With analogy to the bulk crystal, the superficial energy for the surface can be considered to be a function of the local superficial entropy, $s_{a'}^\Sigma$, the superficial concentrations of each species, Γ'_i, and the surface deformation tensor, $\hat{T}_{ij}$. Gibbs also showed that the superficial energy can depend on the curvature of the dividing surface.[51] For the sphere, we need only consider the mean curvature κ'. Thus, $e_{a'}^\Sigma = e_{a'}^\Sigma(s_{a'}^\Sigma, \hat{F}_{ij}, \Gamma'_A, \Gamma'_B, \Gamma'_C, \kappa')$. Differentiating $e_{a'}^\Sigma$ and associating each derivative with the appropriate conjugate variable yields the Gibbs equation for the surface,

$$de_{a'}^\Sigma = \theta^\Sigma ds_{a'}^\Sigma + \hat{T}_{ji} d\hat{F}_{ij} + \mu_A^\Sigma d\Gamma'_A + \mu_B^\Sigma d\Gamma'_B + \mu_C^\Sigma d\Gamma'_C + K d\kappa', \qquad (5.11)$$

---

[51] See Gibbs,[20] pp. 225–229.

where the coefficient K is defined by

$$K = \frac{\partial e_{a'}^\Sigma}{\partial \kappa'}. \tag{5.12}$$

The surface deformation gradient tensor $\hat{F}_{ij}$ is that portion of the deformation gradient tensor that lies in the surface of the crystal; i.e., it is a projection of the bulk deformation gradient tensor F_{ij} onto the tangent plane of the crystal surface.[52] Here, $\hat{F}_{ij}$ is an isotropic tensor given by

$$\hat{F}_{ij} = \frac{R}{R_o}\hat{\delta}_{ij}. \tag{5.13}$$

The mechanical work term simplifies to

$$\hat{T}_{ji} d\hat{F}_{ij} = \hat{T}\hat{\delta}_{ij} d\left(\frac{R}{R_o}\right)\hat{\delta}_{ij} = 2\hat{T} d\left(\frac{R}{R_o}\right). \tag{5.14}$$

Noting that

$$d\hat{J} = d\left(\frac{A^s}{A^{s'}}\right) = d\left(\frac{R}{R_o}\right)^2 = \frac{2R}{R_o} d\left(\frac{R}{R_o}\right) \tag{5.15}$$

allows the mechanical work term to also be expressed in terms of the magnitude of the isotropic Cauchy surface stress $\hat{f}$ as

$$\hat{T}_{ji} d\hat{F}_{ij} = 2\hat{T} d\left(\frac{R}{R_o}\right) = 2\hat{T}\left(\frac{R_o}{2R}\right) d\hat{J} = \hat{f} d\hat{J}. \tag{5.16}$$

The Gibbs equation for the surface, Eq. (5.11), can thus be written using this alternate formulation of the mechanical work term as

$$de_{a'}^\Sigma = \theta^\Sigma ds_{a'}^\Sigma + \hat{f} d\hat{J} + \mu_A^\Sigma d\Gamma_A' + \mu_B^\Sigma d\Gamma_B' + \mu_C^\Sigma d\Gamma_C' + K' d\kappa'. \tag{5.17}$$

The question remains as to the position of the dividing surface. Gibbs showed that there is a position of the dividing surface for which the coefficient K' must vanish. Furthermore, he showed that this position "will in general be sensibly coincident with the physical surface of discontinuity." Therefore, we follow Gibbs and henceforth assume that the position of the dividing surface is chosen so that $K' = 0$. As such, the term $K' d\kappa'$ appearing in Eq. (5.17) vanishes. This convention defines precisely how to calculate the excess properties. It is possible to show that if the interface thickness is much smaller that the radius of the particle, this extra term is indeed small. Thus, for example, in solid-vapor systems the effects of

---

[52] M. E. Gurtin and I. Murdoch, *Arch. Rational Mech. Anal.* **57**, 291 (1975).

neglecting this term are negligible for all but the smallest particles. As such, the differential of the excess energy associated with the surface becomes

$$de_{a'}^{\Sigma} = \theta^{\Sigma} ds_{a'}^{\Sigma} + \hat{f} d\hat{j} + \mu_A^{\Sigma} d\Gamma_A' + \mu_B^{\Sigma} d\Gamma_B' + \mu_C^{\Sigma} d\Gamma_C'. \quad (5.18)$$

15. Thermodynamic Equilibrium Conditions

We consider a spherical crystal with dividing surface located at radius R_o as measured in the reference state, surrounded by a fluid large in extent. Both the crystal and fluid are ternary solutions, and the crystal is isotropic in all aspects. Although the various thermodynamic fields do not necessarily remain homogeneous in the vicinity of the surface of discontinuity, all fields are assumed to retain their radial symmetry. Two sublattices are associated with the crystal, which we refer to as the interstitial and substitutional sublattices. Species A, B, and vacancies are found exclusively on the substitutional sublattice, while only component C and vacancies are found on the interstitial sublattice. Small strain deformation is not assumed. The reference state for the measurement of deformation is taken to be a stress-free, infinite crystal of composition $\mathbf{c}_o = (c_A^{so}, c_B^{so}, c_C^{lo})$, where c^s and c^l are the indicated mole fraction on the substitutional and interstitial sublattices, respectively. Thus, $c_A^{so} + c_B^{so} + c_V^{so} = 1$ and $c_C^{lo} + c_V^{lo} = 1$.

From Eq. (2.84), the crystal energy obeys

$$d\mathcal{E}^s = \theta^s d\mathcal{S}^s - P^s d\mathcal{V}^s + M_{AV}^s d\mathcal{N}_A^s + M_{BV}^s d\mathcal{N}_B^s$$
$$+ M_{CV}^s d\mathcal{N}_C^I + \mu_o d\mathcal{N}_o^s \quad (5.19)$$

where $\mathcal{N}_o^s$ is the number of unit cells in the crystal, defined as one interstitial and one substitutional site per cell, and μ_o is the corresponding energy change when a unit cell is added to the crystal holding the number of mass components fixed. The addition of a unit cell is equivalent to adding a vacancy to each sublattice (see subsection 4 of Section II). P^s expresses the force acting on the crystal per unit actual area. Expressed as an energy density with respect to the reference state, Eq. (5.19) becomes

$$de_{v'}^s = \theta^s ds_{v'}^s - P^s dJ + M_{AV}^s d\rho_A^{s'} + M_{BV}^s d\rho_B^{s'} + M_{CV}^s d\rho_C^{l'}. \quad (5.20)$$

The differential $d\mathcal{N}_o^s$ does not appear in Eq. (5.20) for the energy density. The density of lattice sites on both sublattices in the reference configuration remains constant so that $d\rho_o^s = d\rho_o^I = d\mathcal{N}_o^s = 0$.

For a large crystal, $\mathcal{E}^s$ is a homogeneous function to first order in its independent variables. Using Euler's theorem and Eq. (5.19) yields

$$\mathcal{E}^s = \theta^s \mathcal{S}^s - P^s \mathcal{V}^s + M_{AV} \mathcal{N}_A^s + M_{BV} \mathcal{N}_B^s + M_{CV} \mathcal{N}_C^s + \mu_o \mathcal{N}_o^s. \quad (5.21)$$

The thermodynamic potential, or grand canonical potential, ω, is defined for the crystal as

$$\omega^s = \mathcal{E}^s - \theta^s \mathcal{S}^s - M_{AV}\mathcal{N}_A^s - M_{BV}\mathcal{N}_B^s - M_{CV}\mathcal{N}_C^I$$
$$= -P^s \mathcal{V}^s + \mu_o \mathcal{N}_o^s. \qquad (5.22)$$

The respective densities for the thermodynamic potential with respect to the actual and reference states, ω_v and $\omega_{v'}$, are

$$\omega_v^s = -P^s + \mu_o \rho_o^s \qquad (5.23)$$

and

$$\omega_{v'}^s = J\omega_v^s = -P^s J + \mu_o \rho_o^{s'} \qquad (5.24)$$

For bulk systems, we showed that $\mu_o = 0$ at equilibrium. This is not necessarily true for a small crystal and this term must be retained until its value is determined.

From Eq. (2.1), we have the Gibbs equation for the fluid

$$d\mathcal{E}^f = \theta^f d\mathcal{S}^f - P^f d\mathcal{V}^f + \mu_A^f d\mathcal{N}_A^f + \mu_B^f d\mathcal{N}_B^f + \mu_C^f d\mathcal{N}_C^f \qquad (5.25)$$

with the corresponding expression for the density of the grand canonical potential, ω_v^f,

$$\omega_v^f = e_v^f - \theta^f s_v^f - \mu_A^f \rho_A^f - \mu_B^f \rho_B^f - \mu_C^f \rho_C^f = -P^f \qquad (5.26)$$

The energy of the system, $\mathcal{E}$, is the sum of the crystal, fluid, and surface energies

$$\mathcal{E} = \mathcal{E}^s + \mathcal{E}^\Sigma + \mathcal{E}^f = e_{v'}^s \mathcal{V}^{s'} + e_{a'}^\Sigma \mathcal{A}^{s'} + e_v^f \mathcal{V}^f \qquad (5.27)$$

where the energy of the crystal and surface are expressed with respect to the reference state and energy of the fluid with respect to the actual state. The second expression for the system energy is its representation in the imaginary system in which the thermodynamic densities of each phase are taken to be uniform up to the dividing surface. For a superficial density such as $e_{a'}^\Sigma$, the superscript Σ denotes the surface and the subscript a' identifies the energy density as an energy per unit reference area. Likewise, the system entropy and number of mass components can be expressed as

$$\mathcal{S} = \mathcal{S}^s + \mathcal{S}^\Sigma + \mathcal{S}^f = s_{v'}^s \mathcal{V}^{s'} + s_{a'}^\Sigma \mathcal{A}^{s'} + s_v^f \mathcal{V}^f \qquad (5.28)$$

and

$$\mathcal{N}_A = \mathcal{N}_A^s + \mathcal{N}_A^\Sigma + \mathcal{N}_A^f = \rho_A^{s'} \mathcal{V}^{s'} + \Gamma_A' \mathcal{A}^{s'} + \rho_A^f \mathcal{V}^f \qquad (5.29)$$

$$\mathcal{N}_B = \mathcal{N}_B^s + \mathcal{N}_B^\Sigma + \mathcal{N}_B^f = \rho_B^{s'} \mathcal{V}^{s'} + \Gamma_B' \mathcal{A}^{s'} + \rho_B^f \mathcal{V}^f \qquad (5.30)$$

$$\mathcal{N}_C = \mathcal{N}_C^s + \mathcal{N}_C^\Sigma + \mathcal{N}_C^f = \rho_C^{I'} \mathcal{V}^{s'} + \Gamma_C' \mathcal{A}^{s'} + \rho_C^f \mathcal{V}^f \qquad (5.31)$$

where Γ'_i is the superficial number density of component i. We again use Gibbs' variational approach to establish the conditions for thermodynamic equilibrium. This requires minimizing the energy of the system subject to the constraints of constant entropy, mass, and volume. Introducing Lagrange multipliers θ_L for the entropy constraint and λ_A, λ_B, and λ_C for the mass constraints, the thermodynamic potential for the entire system, $\mathcal{E}^*$, is defined as

$$\mathcal{E}^* = \mathcal{E} - \theta_L \mathcal{S} - \lambda_A \mathcal{N}_A - \lambda_B \mathcal{N}_B - \lambda_C \mathcal{N}_C. \tag{5.32}$$

Extremizing $\mathcal{E}^*$ is mathematically equivalent to minimizing $\mathcal{E}$ with the imposed constraints; the constraint on the volume is treated implicitly. Using Eqs. (5.27)–(5.31), the variation in $\mathcal{E}^*$, $\delta\mathcal{E}^*$, owing to small perturbations in the thermodynamic state of the crystal, fluid, and interface becomes

$$\begin{aligned}\delta\mathcal{E}^* &= \{\delta e^s_{v'} - \theta_L \delta s^s_{v'} - \lambda_A \delta\rho^{s'}_A - \lambda_B \delta\rho^{s'}_B - \lambda_C \delta\rho^{l'}_C\}\mathcal{V}^{s'} + \omega^s_{v'}\delta\mathcal{V}^{s'} \\ &+ \{\delta e^\Sigma_{a'} - \theta_L \delta s^\Sigma_{a'} - \lambda_A \delta\Gamma'_A - \lambda_B \delta\Gamma'_B - \lambda_C \delta\Gamma'_C\}\mathcal{A}^{s'} + \gamma'\delta\mathcal{A}^{s'} \\ &+ \{\delta e^f_v - \theta_L \delta s^f_v - \lambda_A \delta\rho^f_A - \lambda_B \delta\rho^f_B - \lambda_C \delta\rho^f_C\}\mathcal{V}^f + \omega^f_v \delta\mathcal{V}^f \end{aligned} \tag{5.33}$$

where γ' is the superficial grand canonical density defined as

$$\gamma' = e^\Sigma_{a'} - \theta_L s^\Sigma_{a'} - \lambda_A \Gamma'_A - \lambda_B \Gamma'_B - \lambda_C \Gamma'_C. \tag{5.34}$$

The outer boundary of the system must remain rigid in order to ensure the system remains isolated and does no work on the surroundings. This gives for the constant volume constraint, $\delta\mathcal{V} = \delta(\mathcal{V}^f + \mathcal{V}^s) = 0$, so that

$$\delta\mathcal{V}^f = -\delta\mathcal{V}^s = -\delta(J\mathcal{V}^{s'}) = -\mathcal{V}^{s'}\delta J - J\delta\mathcal{V}^{s'} = -\mathcal{V}^{s'}\delta J - J\mathcal{A}^{s'}\delta R_o. \tag{5.35}$$

The last expression follows from $\mathcal{V}^{s'} = 4\pi R_o^3/3$ as

$$\delta\mathcal{V}^{s'} = 4\pi R_o^2 \delta R_o = \mathcal{A}^{s'}\delta R_o. \tag{5.36}$$

Equation (5.35) is the mathematical representation of the concept discussed earlier that there are two distinct ways in which the volume of a crystal can be changed: first, by deforming the sphere at a constant number of unit cells, as represented by the variation δJ and, second, by the addition or removal of lattice sites at constant deformation, as given by the variation of the size of the particle in the reference state δR_o.

Similarly, a variation in the reference volume of the crystal leads to changes in the reference area of the crystal. Because $\mathcal{A}^{s'} = 4\pi R_o^2$,

$$\delta\mathcal{A}^{s'} = 8\pi R_o \delta R_o = 4\pi R_o^2 \left(\frac{2\delta R_o}{R_o}\right) = \mathcal{A}^{s'}\left(\frac{2\delta R_o}{R_o}\right). \tag{5.37}$$

Substituting for the variation in the energy densities and using Eqs. (5.18), (5.20), and (5.35)–(5.37) yields for the variation in $\mathcal{E}^*$

$$\delta\mathcal{E}^* = \left\{(\theta^s - \theta_L)\delta s_{v'}^s + (M_{AV} - \lambda_A)\delta\rho_A^{s'} + (M_{BV} - \lambda_B)\delta\rho_B^{s'}\right.$$
$$\left. + (M_{CV} - \lambda_C)\delta\rho_C^{I'}\right\}\mathcal{V}^{s'}$$
$$+ \left\{(\theta^f - \theta_L)\delta s_v^f + (\mu_A^f - \lambda_A)\delta\rho_A^f + (\mu_B^f - \lambda_B)\delta\rho_B^f\right.$$
$$\left. + (\mu_C^f - \lambda_C)\delta\rho_C^f\right\}\mathcal{V}^f$$
$$+ \left\{(\theta^\Sigma - \theta_L)\delta s_{a'}^\Sigma + (\mu_A^\Sigma - \lambda_A)\delta\Gamma_A' + (\mu_B^\Sigma - \lambda_B)\delta\Gamma_B'\right.$$
$$\left. + (\mu_C^\Sigma - \lambda_C)\delta\Gamma_C'\right\}\mathcal{A}^{s'}$$
$$+ \left\{-P^s - \omega_v^f + \frac{2\hat{f}}{R}\right\}\mathcal{V}^{s'}\delta J + \left\{\omega_{v'}^s - J\omega_v^f + \frac{2\gamma'}{R_o}\right\}\mathcal{A}^{s'}\delta R_o \quad (5.38)$$

where the following relationship has been used:

$$\hat{f}\mathcal{A}^{s'}\delta J = \hat{f}\left(4\pi R_o^2\right)\left(\frac{2R\delta R}{R_o^2}\right) = \frac{2\hat{f}}{R}\left(\frac{4\pi R_o^3}{3}\right)\left(\frac{R^2\delta R}{R_o^3}\right)$$
$$= \frac{2\hat{f}}{R}\mathcal{V}^{s'}\delta J. \quad (5.39)$$

Equilibrium is obtained when the variation in $\mathcal{E}^*$ vanishes. This occurs only when the coefficients multiplying each of the independent variations is identically zero. The first three lines of Eq. (5.38) pertain to perturbations in the thermodynamic state of the crystal, fluid, and surface, respectively. They yield the condition on thermal equilibrium;

$$\theta^s = \theta^\Sigma = \theta^f = \theta_L, \quad (5.40)$$

which requires a uniform temperature field. The conditions for chemical equilibrium are

$$M_{AV}^S = \mu_A^\Sigma = \mu_A^f = \lambda_A$$
$$M_{BV}^S = \mu_B^\Sigma = \mu_B^f = \lambda_B \quad (5.41)$$
$$M_{CV}^I = \mu_C^\Sigma = \mu_C^f = \lambda_C.$$

Chemical equilibrium requires an equality of the chemical potentials of the fluid and surface with the corresponding diffusion potential of the crystal. In keeping with previous usage, the diffusion potentials are defined as the change in energy associated with placing the component species on a vacant lattice site and are accomplished at a constant number of lattice sites. Physically, it means that the

THE THERMODYNAMICS OF ELASTICALLY STRESSED CRYSTALS 105

energy change accompanying the addition of an A atom to the fluid, the energy change accompanying the addition of an A atom to the interface region (surface of discontinuity), and the energy change accompanying the addition of an A atom to a vacant substitutional site must all be equivalent at equilibrium. It is also possible to consider the exchange of two species at constant number of lattice sites. For example, if the exchange of a B atom with an A atom is considered, equilibrium necessitates

$$M^s_{BV} - M^s_{AV} = M^s_{BA} = \mu^\Sigma_B - \mu^\Sigma_A = \mu^f_B - \mu^f_A. \quad (5.42)$$

The two expressions appearing in the last line of Eq. (5.38) concern the energy changes associated with the two physically different mechanisms by which the volume and surface area of the crystal can be changed. The term multiplying δJ, which corresponds to a perturbation in the displacement field, refers to the mechanical work done when the deformation state of the crystal is changed. Mechanical equilibrium requires a balance of mechanical forces exerted by the crystal, fluid, and surface,

$$P^s = -\omega^f_v + \frac{2\hat{f}}{R} \quad \text{or} \quad P^s = P^f + \frac{2\hat{f}}{R}, \quad (5.43)$$

where the second condition follows from Eq. (5.26). P^s and P^f are the normal forces per unit actual area exerted by the crystal and fluid, respectively. $2\hat{f}/R$ is the normal force per unit of actual surface area exerted by the excess tangential forces on the surface. This mechanical equilibrium condition can also be expressed with respect to the reference state. Multiplying Eq. (5.43) by $\hat{J} = R^2/R_o^2$ and using Eq. (5.10),

$$P^s \hat{J} = P^f \hat{J} + \frac{2\hat{f}}{R}\hat{J} = P^f \hat{J} + \frac{2\hat{T}}{R_o}. \quad (5.44)$$

$P^s \hat{J}$ can be expressed in terms of the first Piola-Kirchhoff stress tensor using Eq. (3.58), $\sigma_{ij} = -P^s \delta_{ij}$, and recognizing that, at the surface,

$$F_{ij} = \left(\frac{R}{R_o}\right) \delta_{ij} \quad (5.45)$$

to give

$$-T_{rr} = P^f \hat{J} + \frac{2\hat{T}}{R_o}. \quad (5.46)$$

The second variation appearing in the last line of Eq. (5.38), $\delta \mathcal{V}^{s'} = \mathcal{A}^{s'} \delta R_o$, concerns the change in energy when the reference volume of the crystal is altered by an accretion of lattice sites (unit cells) to the crystal under conditions of constant temperature and chemical potentials.

$$\omega_{v'}^s + \frac{2\gamma'}{R_o} = J\omega_v^f \qquad (5.47)$$

Equation (5.47) expresses equilibrium with respect to a phase transformation between the crystal and fluid phases. Equilibrium requires that the change in the grand canonical potential arising from accretion to the crystal ($\omega_{v'}^s$) and creation of new surface ($2\gamma'/R_o$) is exactly offset by the change in the grand canonical potential of the fluid when referred to the same reference state. Equation (5.47) can be expressed in terms of the actual state by multiplying by J^{-1} to give

$$\omega_v^s + \frac{2\gamma}{R} = \omega_v^f \qquad (5.48)$$

16. Applications

In this section, we present several examples illustrating the relative contributions of the surface stress and surface free energy to equilibrium phenomena. We also show that the two chemical potentials defined in the section on hydrostatically stressed crystals, which were shown to be equal at equilibrium in bulk systems, are no longer equal in small crystalline systems for which surface effects become important. Some of the examples considered have been treated by Gibbs[20] and Cahn.[53]

a. Diffusion Potential

The diffusion potential tracks the change in energy associated with an exchange of two components on a specific sublattice holding the number of lattice sites (unit cells) fixed. For example, M_{AV}^I is the change in energy associated with the placement of an A atom in a vacant site on sublattice I. The diffusion potentials are equivalent to the chemical potentials μ_i^c defined initially by Eq. (2.11).

In order to obtain an expression for the dependence of a given diffusion potential on particle size, we begin by considering the diffusion potential defined by the exchange of an A atom and a vacancy, M_{AV}, in a multicomponent particle surrounded by a fluid. We then obtain an expression for the dependence of the diffusion potential on the particle stress state and relate the particle stress state to the particle size and imposed fluid pressure. Limiting the development henceforth to small strains, Eq. (4.52) gives an expression for the dependence of the diffusion potential on temperature, composition, and stress for an isotropic system in which the elastic constants are independent of composition. Recognizing that

---

[53] J. W. Cahn, *Acta metall.* **28**, 1333 (1980).

for an isotropic crystal under hydrostatic pressure P^s, the stress field is given by $\sigma_{kk} = -3P^s$, Eq. (4.52) yields for the diffusion potential M_{AV}

$$M_{AV}(\theta, c, P^s) = M_{AV}(\theta, c) + \frac{3\epsilon_A^c P^s}{\rho_o'}. \tag{5.49}$$

The term $3\epsilon_A^c$ is the small strain approximation to the change in volume per unit reference volume accompanying the exchange process. Expressing the crystal pressure in terms of the external pressure of the fluid and the surface stress using Eq. (5.43) gives, for the size dependence of the diffusion potential,

$$M_{AV}(\theta, c, R) = M_{AV}(\theta, c) + \bar{V}_A P^f + \frac{2\bar{V}_A \hat{f}}{R}. \tag{5.50}$$

The magnitude of a typical surface stress is on the order of $10^2 - 10^4 \text{N/m}$[44]. For a particle with a radius of 100nm, the pressure induced in the crystal by the surface stress is of the order $2\hat{f}/R \approx 1 - 100\text{MPa}$. This surface stress contribution to crystal pressure is usually far in excess of the contribution owing to the fluid. Thus the term containing the fluid pressure in Eq. (5.20) can usually be neglected, giving

$$M_{AV}(\theta, c, R) = M_{AV}(\theta, c) + \frac{2\bar{V}_A \hat{f}}{R}. \tag{5.51}$$

This expression for the diffusion potential is equally valid for the diffusion potentials associated with all crystal sublattices, whether interstitial or substitutional. The diffusion potential M_{BA} can be obtained from Eq. (5.51) as

$$M_{BA}(\theta, c, R) = M_{BV} - M_{AV} = M_{BA}(\theta, c) + \frac{2(\bar{V}_B - \bar{V}_A)\hat{f}}{R}. \tag{5.52}$$

The exchange mechanism of the diffusion potentials creates no new surface area as viewed in the reference state, as there is no change in the number or position of the lattice sites. Consequently, no work is done in creating new surface and the diffusion potential does not depend on the surface energy γ'. The mechanical work performed in the exchange process stems from the change in volume of the crystal working against the pressure acting on the crystal. The change in volume of the crystal occurs when the partial molar volumes of A and B differ, or equivalently, when the lattice parameter depends on composition. Because the crystal pressure depends on the surface stress, it is the surface stress that appears in the expression for the diffusion potential along with the difference in partial molar volumes, $\bar{V}_B - \bar{V}_A$.

b. Chemical Potential

The chemical potential μ_i^v, defined initially by Eq. (2.16), concerns the addition of component i to the crystal while holding the number of other components and number of vacant lattice sites fixed. Addition of the atom thus entails the simultaneous addition of a lattice site or unit cell to the crystal. This process leads to an increase in surface area as viewed in the reference state and, thereby, energy must be expended in creating the new surface. The accretion of component i changes the volume of the crystal and work is also done against the mechanical forces. Consequently, the chemical potential μ_i^v should also reflect the mechanical work done against the crystal pressure.

First consider a single-component crystal composed of species A and vacancies. The diffusion potential and chemical potential of component A are related according to

$$\mu_A^v = M_{AV} + \mu_V^v = M_{AV} + \mu_o \qquad (5.53)$$

where μ_o is the energy associated with the addition of a primitive unit cell holding the number of atoms of component A constant. Equation (5.53) is just an accounting of an energy balance at equilibrium: The energy associated with increasing the number of lattice sites in the crystal by addition of an A atom (μ_A^v) must equal the sum of the energies associated with increasing the number of lattice sites in the crystal by addition of a vacancy (μ_V^v) and the change in energy when a vacancy is replaced by an A atom holding the number of lattice sites fixed (M_{AV}). μ_o is equivalent to the chemical potential of the vacancy, μ_V^v.

An expression for μ_o can be obtained from Eq. (5.53) and the thermodynamic equilibrium conditions. From Eq. (5.23), μ_o is

$$\rho_o \mu_o = \omega_v^s + P^s. \qquad (5.54)$$

Substituting Eq. (5.44) for P^s and Eq. (5.48) for ω_v^s into Eq. (5.54) gives

$$\mu_o = \frac{2(\hat{f} - \gamma)}{\rho_o R} \qquad (5.55)$$

Substituting Eq. (5.43) along with Eq. (5.51) for the diffusion potential into Eq. (5.53) yields the following expression for the chemical potential μ_A^v:

$$\mu_A^v(\theta, \mathbf{c}, \sigma_{ij}) = M_{AV}(\theta, \mathbf{c}) + \frac{3\eta_A}{\rho_o}\left(P^f + \frac{2\hat{f}}{R}\right) + \frac{2(\hat{f} - \gamma)}{\rho_o R} \qquad (5.56)$$

where η_A is a measure of the difference in partial molar volumes of the vacancies and species A. Because chemical equilibrium requires $M_{AV} = \mu_A^f$,

$$\mu_A^v(\theta, \mathbf{c}, \sigma_{ij}) = \mu_A^f - \frac{2\gamma}{\rho_o R} + \frac{2\hat{f}}{\rho_o R}(1 + 3\eta_A) + \frac{3\eta_A P^f}{\rho_o}. \quad (5.57)$$

The chemical potentials are evaluated at the respective pressures of the crystal and fluid. These pressures differ when $\hat{f} \neq 0$. If $\hat{f} = 0$, the pressures of the two phases are identical and the chemical potentials of the crystal and fluid are related by

$$\mu_A^v(\theta, \mathbf{c}, P^f) = \mu_A^f - \frac{2\gamma}{\rho_o R}, \quad (5.58)$$

assuming the phases are incompressible. In general, the interfacial stress and surface energy of the crystal are not equal and, neglecting the elasticity correction, the chemical potentials of the crystal and fluid in equilibrium differ by

$$\mu_A^v(\theta, \mathbf{c}, \sigma_{ij}) = \mu_A^f - \frac{2(\hat{f} - \gamma)}{\rho_o R}. \quad (5.59)$$

Following Gibbs and Cahn,[53] we define the chemical potential of component i for a small particle as μ_i^v (see Eq. (2.16)). This chemical potential corresponds to the physical process by which an atom of component i is added to the crystal by the simultaneous creation of a new lattice site on the surface. The procedure used to establish the chemical potential of the small particle for the single component system is extendable directly to a multicomponent system. Of course, other definitions of chemical potential can be invoked that may lead to different relationships between the defined chemical potential and the corresponding chemical potential in the fluid,[16,54] but the definitions used here, we believe, are consistent with well-defined physical processes and the use of chemical potential in bulk systems. Cahn[53] gives other examples of how the surface stress affects vapor pressures and melting temperatures with references to differences between various treatments of small crystals.

VI. Crystal–Fluid Equilibrium

17. INTRODUCTION

In the previous two sections, we examined the thermodynamics of an isolated crystal subjected to nonhydrostatic stresses and the thermodynamics of a small,

---

[54] M. Hillert and J. Ågren, *Acta mater.* **50**, 2429 (2002).

hydrostatically stressed (isotropic) crystal in equilibrium with a fluid accounting for the excess properties associated with the crystal–fluid interface. In this section, we extend this thermodynamic formalism to the more general case of a nonhydrostatically stressed crystal in contact with a fluid. The interface is allowed to be nonplanar and to possess superficial properties associated with the crystal–fluid interface. The conditions for thermodynamic equilibrium in the two-phase system are determined first. Then, in order to illustrate the manner in which these conditions can be applied to physical problems of interest, we examine the thermodynamic and morphological stability of a uniaxially stressed crystal in contact with a fluid.

The model used to incorporate the interfacial energy into the thermodynamic description of the two-phase system is the same as that employed for the small, hydrostatically stressed crystal. However, there are a couple of important caveats that render this approach less precise than when applied to fluid systems or to the hydrostatically stressed crystal of the previous section. Interfacial properties for the nonhydrostatically stressed system are again defined using the dividing surface construction of Gibbs, wherein the interfacial energy is given by the difference between the energy of the actual system with the interface and that of an imagined system in which the two phases exhibit their bulk properties up to the two-dimensional dividing surface.[20,48] For the small, hydrostatically stressed spherical crystal, the thermodynamic state of the volume elements located within the central region of the crystal are equivalent at equilibrium. Consequently, the thermodynamic properties of the crystal (and fluid phase) were assumed to be uniform up to the dividing surface and the excess properties were assigned to the dividing surface. For the nonhydrostatically stressed crystal with a nonspherical surface, the thermodynamic state of each volume element will, in general, be different at equilibrium, and the thermodynamic state of the crystal cannot be considered to be uniform. Consequently, for systems containing nonplanar interfaces and/or systems in which the deformation state of the crystal phases are nonuniform, the dividing surface construction of Gibbs is an approximation.

Crystals are different from fluids in that the bulk crystalline phases can sustain nonhydrostatic stress at equilibrium. Similarly, the interfacial properties of a crystal are different from those of a fluid in that, for the crystal, the interfacial stress is distinct from the interfacial energy.[15,20,53,55–57] In the previous section for a spherical crystal, we showed that this difference is manifested in the form of the equilibrium conditions. The interfacial stress of a crystal can be either

---

[55] J. W. Cahn, *Acta metall.* **37**, 773 (1989).
[56] C. Rottman, *Phys. Rev. B* **38**, 12031 (1988).
[57] W. C. Johnson, *Acta. Mater.* **48**, 433 (2000).

positive or negative, unlike the interfacial energy, which is always positive.[15] In contrast, the values of the superficial stress and superficial energy density of a fluid are identical, and are commonly referred to as the interfacial tension. The presence of a nonzero interfacial stress in an anisotropic crystal with a nonspherical geometry greatly complicates the description of the mechanical state of the crystal and interface.[58] These treatments necessitate that the surface stress and deformation tensors be expressed in general curvilinear coordinates and as projections of three-dimensional tensors on to the local two-dimensional tangent plane to the crystal. Consequently, we focus in this chapter on those cases where the surface stress in the solid can be ignored or, more precisely, is zero. This allows us to derive the equilibrium conditions and to illustrate their use without the complications of treating the surface tensors. We therefore concentrate on the central effects of the bulk stress components on thermodynamic equilibrium and on morphological evolution of the interface in crystal–fluid systems. Finally, although the interfacial energy density in crystals can be a function of the interface normal (i.e., the interface is anisotropic), we assume the interfacial energy to be isotropic and independent of the interface normal. Thermodynamic treatments that include the effects of anisotropic interfacial energies can be found in References [59, 60].

As in the previous sections, we commit to a defect structure in the crystal and assume that the crystal is composed of substitutional species A, B, and vacancies. For reasons of convenience, when these conditions are applied later to examine the stability of a crystal–liquid interface, we choose species A and B as the independent composition variables. Because the nonhydrostatically stressed crystal is frequently heterogeneous in its thermodynamic state, we assume that the only location where lattice sites can be added to, or removed from, the crystal is at the crystal–fluid interface. Thermodynamic densities of the crystal are, therefore, expressed in terms of the uniform reference state and it is assumed that there are no sources of vacancies within the crystal.

18. Energy Functional

In keeping with Gibbs' dividing surface approach of surface thermodynamics, the total energy of the two-phase system, $\mathcal{E}$, is assumed to be partitioned among the crystal, fluid, and interface. Because the thermodynamic state of the nonhydrostatically stressed system is a function of position, the energy of each phase must be expressed as an integral of the appropriate energy density over the volume of the

---

[58] For example, see Leo and Sekerka[59] or Gurtin[13] for a dynamical treatment of the interfacial stress.
[59] P. H. Leo and R. F. Sekerka, *Acta metall.* **37**, 3119 (1989).
[60] J. I. D. Alexander and W. C. Johnson, *J. Appl. Phys.* **58**, 816 (1985).

phase. Thus,

$$\mathcal{E} = \int_{V^{s'}} e_{v'}^s(s_{v'}, F_{ij}, \rho'_A, \rho'_B) dV + \int_{V^f} e_v^f(s_v^f, \rho_A, \rho_B) dV$$
$$+ \int_{\Sigma'} e_{a'}^{\Sigma}(s_{a'}^{\Sigma}, \Gamma'_A, \Gamma'_B, \kappa'_1, \kappa'_2) dA \qquad (6.1)$$

where Σ' denotes the crystal–fluid interface expressed in the reference state of the crystal, $e_{a'}^{\Sigma}$ and $s_{a'}^{\Sigma}$ are the excess internal energy and entropy densities, respectively, due to the presence of the interface, Γ'_i is the excess mass per unit area of the reference state of component i, and κ'_1, κ'_2 are the principal radii of curvature of the interface in the reference state and are, in general, functions of position. Owing to the assumption of a vanishing surface stress, the superficial energy density does not depend on the surface deformation tensor as in the previous section. Because the thermodynamic state of the interface can vary with position along the interface, the interfacial energy must also be expressed as a surface integral of the superficial energy density in Eq. (6.1).

The other extensive properties of the system are also expressed as integrals of the appropriate densities. The system entropy, $\mathcal{S}$, is

$$\mathcal{S} = \int_{V^{s'}} s_{v'}^s dV + \int_{V^f} s_v^f dV \int_{\Sigma'} s_{a'}^{\Sigma} dA. \qquad (6.2)$$

The total number of atoms of components A and B, $\mathcal{N}_A$ and $\mathcal{N}_B$, are

$$\mathcal{N}_A = \int_{V^{s'}} \rho_A^{s'} dV + \int_{V^f} \rho_A^f dV + \int_{\Sigma'} \Gamma'_A dA \qquad (6.3)$$

and

$$\mathcal{N}_B = \int_{V^{s'}} \rho_B^{s'} dV + \int_{V^f} \rho_B^f dV + \int_{\Sigma'} \Gamma'_B dA. \qquad (6.4)$$

19. An Extremum in the Energy

In order to obtain the conditions for thermodynamic equilibrium, Gibbs' variational approach is employed once again. An arbitrary region of the entire system containing both the crystal and fluid phases is identified and imagined to be isolated from its surroundings. The energy of the isolated system is calculated for a given set of thermodynamic fields as in Eq. (6.1). The thermodynamic fields are then perturbed under the conditions of constant system entropy and number of each component species assuming that the outer surface remains rigid, assuring

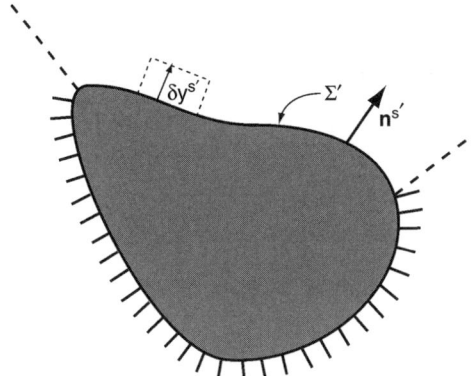

FIG. 23. Solid–liquid system with crystal (grey) having imposed displacements along part of its surface and in contact with a fluid (white) along other part. $\delta y'_s$ represents the accretion of lattice sites to the crystal by a displacement of the crystal–fluid interface. The displacement, represented by the lines, is applied along the interface Σ' over the region between the two dashed lines.

that there is no interaction with the surroundings. The change in energy owing to the perturbation in the fields is then calculated and the equilibrium conditions are obtained by noting the conditions under which the change in system energy vanishes.

The variation in a thermodynamic variable must account for changes in that variable at each point within the bulk phases and along the interface, as well as the change that accompanies a phase transformation at the interface. As for the case of the spherical particle, the extent of the phase transformation is tracked with respect to the reference state of the crystal, because this description allows the two physical processes that can give rise to the motion of the crystal–fluid interface, accretion, and mechanical deformation, to be clearly differentiated. The change in volume of a phase gives rise to an increase or decrease in a quantity ϕ determined by the product of the density $\phi_{v'}$ and the volume added to the reference state of the crystal. Accretion of material (lattice sites or unit cells) to the crystal reference state occurs at the crystal surface. An accretion vector, $\mathbf{y}^{s'}$, is a measure of the extent of the phase transformation (accretion) and refers to the change in position of the crystal interface as measured in the reference state (see Fig. (23)). Accretion is here defined to occur perpendicular to the crystal interface so that $\delta \mathbf{y}^{s'} = \delta y^{s'} \mathbf{n}'$ where $\delta y^{s'}$ gives the perturbation in the position of the reference interface owing to a small accretion to the crystal at the interface. The change in ϕ associated with the crystal owing to the phase transformation at the interface is, in the limit of infinitesimally small variations,

$$\delta \int_{V^{s'}} \phi^s_{v'} dV = \int_{V^{s'}} \delta \phi^s_{v'} dV + \int_{\Sigma'} \phi^s_{v'} \delta y^{s'} dA \qquad (6.5)$$

where $\delta y^{s'}$ is the magnitude of the displacement along the outward-pointing normal of the crystal in the reference state. The same analysis can be applied to the fluid. Because quantities in the fluid are referred to the actual state of the fluid,

$$\delta \int_{V^f} \phi_v^f dV = \int_{V^f} \delta\phi_v^f dV + \int_{\Sigma} \phi_v^f \delta y^f dA \qquad (6.6)$$

where δy^f is a measure of the accretion to the fluid in the direction normal to the fluid surface in the actual state. $\delta y^{s'}$ and δy^f are not independent, as they refer to the same variation in the position of the interface. The relationship between these two quantities, which depends on the local deformation state, is given in Appendix C.

Accretion to the crystal also changes the surface area of the crystal as measured in the reference state when the crystal interface is curved. For the spherical precipitate, this change in reference area was shown to be proportional to $2\delta R_o/R_o$, where δR_o has the same meaning as the more general accretion term $\delta y^{s'}$ introduced here. For the general case, a displacement $\delta y^{s'}$ of an element of area dA' along the crystal normal results in the creation of new surface of area $\kappa' \delta y^{s'} dA'$. $\kappa' = (\kappa_1' + \kappa_2')/2$ is the mean curvature of the interface at a point where κ_1' and κ_2' are the principal curvatures in the reference state (see Appendix A). Because we have chosen to displace the surface along the normal to the crystal, κ is taken to be positive for a spherical crystalline particle in a fluid. Thus, a positive displacement along the normal at a point with positive curvature increases the interfacial area. The change in some extensive quantity ϕ at the interface owing to accretion to the crystal is expressed as

$$\delta \int_{\Sigma'} \phi_{a'}^\Sigma dA = \int_{\Sigma'} \delta\phi_{a'}^\Sigma dA + \int_{\Sigma'} \phi_{a'}^\Sigma 2\kappa' \delta y^{s'} dA \qquad (6.7)$$

where $\phi_{a'}^\Sigma$ is the superficial density of ϕ as measured in the reference state.

Using Eqs. (6.5)–(6.7) with the internal energy in place of ϕ, the first variation in the energy becomes

$$\delta \mathcal{E} = \int_{V^{s'}} \delta e_{v'}^s dV + \int_{V^f} \delta e_v^f dV + \int_{\Sigma'} \delta e_{a'}^\Sigma dA + \int_{\Sigma'} e_{v'}^s \delta y^{s'} dA$$
$$+ \int_{\Sigma} e_v^f \delta y^f dA + \int_{\Sigma'} e_{a'}^\Sigma \kappa' \delta y^{s'} dA. \qquad (6.8)$$

The perturbation in the energy density of the crystal is given by the Gibbs equation for the crystal, Eq. (4.8), as

THE THERMODYNAMICS OF ELASTICALLY STRESSED CRYSTALS   115

$$\delta e^s_{v'} = \theta^s \delta s^s_{v'} + T_{ji}\delta F_{ij} + M_{AV}\delta\rho^{s'}_A + M_{BV}\delta\rho^{s'}_B. \quad (6.9)$$

Similarly, the variation in the energy density of the fluid is

$$\delta e^f_v = \theta^f \delta s^f_v + \mu^f_A \delta\rho^f_A + \mu^f_B \delta\rho^f_B. \quad (6.10)$$

For the interface, we have assumed $e^\Sigma_{a'} = e^\Sigma_{a'}(s^\Sigma_{a'}, \Gamma'_A, \Gamma'_B, \kappa'_1, \kappa'_2)$ so that the formal variation in the superficial energy density for the interface becomes

$$\delta e^\Sigma_{a'} = \frac{\partial e^\Sigma_{a'}}{\partial s^\Sigma_{a'}}\delta s^\Sigma_{a'} + \frac{\partial e^\Sigma_{a'}}{\partial \Gamma'_A}\delta\Gamma'_A + \frac{\partial e^\Sigma_{a'}}{\partial \Gamma'_B}\delta\Gamma'_B + \frac{\partial e^\Sigma_{a'}}{\partial \kappa'_1}\delta\kappa'_1 + \frac{\partial e^\Sigma_{a'}}{\partial \kappa'_2}\delta\kappa'_2. \quad (6.11)$$

With analogy to the definitions used for the bulk phases, the partial derivatives of the superficial energy density are identified as follows:

$$\theta^\Sigma = \frac{\partial e^\Sigma_{a'}}{\partial s^\Sigma_{a'}} \quad (6.12)$$

where θ^Σ is the temperature at the interface. There is no lattice constraint for the crystal–fluid interface and the superficial densities of each component can be varied independently. Thus the derivatives with respect to the superficial densities of components A and B give rise to chemical potentials at the interface, μ^Σ_A and μ^Σ_B, rather than to diffusion potentials that involve an exchange of components.

$$\mu^\Sigma_A = \frac{\partial e^\Sigma_{a'}}{\partial \Gamma'_A} \quad \text{and} \quad \mu^\Sigma_B = \frac{\partial e^\Sigma_{a'}}{\partial \Gamma'_B} \quad (6.13)$$

Finally, the terms representing the curvature dependence of the interfacial energy are replaced by coefficients K_1 and K_2, defined by

$$K_1 = \frac{\partial e^\Sigma_{a'}}{\partial \kappa'_1} \quad (6.14)$$

and

$$K_2 = \frac{\partial e^\Sigma_{a'}}{\partial \kappa'_2}. \quad (6.15)$$

Thus, the variation in the superficial energy density becomes

$$\delta e^\Sigma_{a'} = \theta^\Sigma \delta s^\Sigma_{a'} + \mu^\Sigma_A \delta\Gamma'_A + \mu^\Sigma_B \delta\Gamma'_B + K_1 \delta\kappa'_1 + K_2 \delta\kappa'_2. \quad (6.16)$$

The first variation of the system energy is obtained by substituting Eqs. (6.9), (6.10), and (6.15) into Eq. (6.8) to yield

$$\delta \mathcal{E} = \int_{V^{s'}} \{\theta^s \delta s_{v'}^s + T_{ji} \delta F_{ij} + M_{AV} \delta \rho_A^{s'} + M_{BV} \delta \rho_B^{s'}\} dV$$

$$+ \int_{V^f} \{\theta^f \delta s_v^f + \mu_A^f \delta \rho_A^f + \mu_B^f \delta \rho_B^f\} dV$$

$$+ \int_{\Sigma'} \{\theta^\Sigma \delta s_{a'}^\Sigma + \mu_A^\Sigma \delta \Gamma_A' + \mu_B^\Sigma \delta \Gamma_B' + K_1 \delta \kappa_1' + K_2 \delta \kappa_2'\} dA$$

$$+ \int_{\Sigma'} e_{v'}^s \delta y^{s'} dA + \int_{\Sigma} e_v^f \delta y^f dA + \int_{\Sigma'} e_{a'}^\Sigma 2\kappa' \delta y^{s'} dA. \quad (6.17)$$

Similarly, the first variation in the system entropy is, from Eqs. (6.2) and (6.7),

$$\delta \mathcal{S} = \int_{V^{s'}} \delta s_{v'}^s dV + \int_{V^f} \delta s_v^f dV + \int_{\Sigma'} \delta s_{a'}^\Sigma dA + \int_{\Sigma'} s_{v'}^s \delta y^{s'} dA$$

$$+ \int_{\Sigma} s_v^f \delta y^f dA + \int_{\Sigma'} s_{a'}^\Sigma 2\kappa' \delta y^{s'} dA. \quad (6.18)$$

The first variations of each of the system mass constraints are

$$\delta \mathcal{N}_A = \int_{V^{s'}} \delta \rho_A^{s'} dV + \int_{V^f} \delta \rho_A^f dV + \int_{\Sigma'} \delta \Gamma_A' dA + \int_{\Sigma'} \rho_A^{s'} \delta y^{s'} dA$$

$$+ \int_{\Sigma} \rho_A^f \delta y^f dA + \int_{\Sigma'} \Gamma_A' 2\kappa' \delta y^{s'} dA \quad (6.19)$$

and

$$\delta \mathcal{N}_B = \int_{V^{s'}} \delta \rho_B^{s'} dV + \int_{V^f} \delta \rho_B^f dV + \int_{\Sigma'} \delta \Gamma_B' dA + \int_{\Sigma'} \rho_B^{s'} \delta y^{s'} dA$$

$$+ \int_{\Sigma} \rho_B^f \delta y^f dA + \int_{\Sigma'} \Gamma_B' 2\kappa' \delta y^{s'} dA. \quad (6.20)$$

The constraints on the total entropy and total mass of each component are included in the variation of the energy $\mathcal{E}$ by defining a new energy $\mathcal{E}^*$,

$$\mathcal{E}^* = \mathcal{E} - \theta_L \mathcal{S} - \lambda_A \mathcal{N}_A - \lambda_B \mathcal{N}_B \quad (6.21)$$

where θ_L, λ_A, and λ_B are the respective Lagrange multipliers for the entropy and mass constraints. The first variation of $\mathcal{E}^*$ is

$$\delta \mathcal{E}^* = \delta \mathcal{E} - \lambda_s \delta \mathcal{S} - \lambda_A \delta \mathcal{N}_A - \lambda_B \delta \mathcal{N}_B. \quad (6.22)$$

Substituting Eqs. (6.17) and (6.18)–(6.20) in Eq. (6.22) and setting $\delta \mathcal{E}^* = 0$, yields

$$0 = \int_{V^{s'}} \{(\theta^s - \theta_L)\delta s^s_{v'} + (M_{AV} - \lambda_A)\delta\rho^{s'}_A + (M_{BV} - \lambda_B)\delta\rho^{s'}_B + T_{ji}\delta F_{ij}\}dV$$

$$+ \int_{V^f} \{(\theta^f - \theta_L)\delta s^f_v + (\mu^f_A - \lambda_A)\delta\rho^f_A + (\mu^f_B - \lambda_B)\delta\rho^f_B\}dV$$

$$+ \int_{\Sigma'} \{(\theta^\Sigma - \theta_L)\delta s^\Sigma_{a'} + (\mu^\Sigma_A - \lambda_A)\delta\Gamma'_A + (\mu^\Sigma_B - \lambda_B)\delta\Gamma'_B$$

$$+ K_1\delta\kappa'_1 + K_2\delta\kappa'_2\}dA$$

$$+ \int_{\Sigma'} \left(e^s_{v'} - \theta_L s^s_{v'} - \lambda_A \rho^{s'}_A - \lambda_B \rho^{s'}_B\right)\delta y^{s'} dA$$

$$+ \int_{\Sigma} \left(e^f_v - \theta_L s^f_v - \lambda_A \rho^f_A - \lambda_B \rho^f_B\right)\delta y^f dA$$

$$+ \int_{\Sigma'} \left(e^\Sigma_{a'} - \theta_L s^\Sigma_{a'} - \lambda_A \Gamma'_A - \lambda_B \Gamma'_B\right) 2\kappa' \delta y^{s'} dA. \quad (6.23)$$

Not all the variations appearing in Eq. (6.23) are independent. The term involving the first variation in the deformation gradient tensor can be written as the sum of a surface and volume integral according to Eq. (4.26). Using Eq. (4.26) in Eq. (6.23) gives

$$0 = \int_{V^{s'}} \{(\theta^s - \theta_L)\delta s^s_{v'} + (M_{AV} - \lambda_A)\delta\rho^{s'}_A + (M_{BV} - \lambda_B)\delta\rho^{s'}_B - T_{ji,j}\delta u_i\}dV$$

$$+ \int_{V^f} \{(\theta^f - \theta_L)\delta s^f_v + (\mu^f_A - \lambda_A)\delta\rho^f_A + (\mu^f_B - \lambda_B)\delta\rho^f_B\}dV$$

$$+ \int_{\Sigma'} \{(\theta^\Sigma - \theta_L)\delta s^\Sigma_{a'} + (\mu^\Sigma_A - \lambda_A)\delta\Gamma'_A + (\mu^\Sigma_B - \lambda_B)\delta\Gamma'_B$$

$$+ K_1\delta\kappa'_1 + K_2\delta\kappa'_2\}dA$$

$$+ \int_{\Sigma'} \omega^s_{v'}\delta y^{s'} dA + \int_{\Sigma} \omega^f_v \delta y^f dA + \int_{\Sigma'} 2\gamma' \kappa' \delta y^{s'} dA + \int_{\Sigma'} T_{ji} n^{s'}_j \delta u_i dA$$

$$(6.24)$$

where we have used the grand canonical free energy densities defined for the crystal, fluid, and interface as

$$\omega^s_{v'} = e^s_{v'} - \theta s^s_{v'} - M_{AV}\rho^{s'}_A - M_{BV}\rho^{s'}_B \quad (6.25)$$

$$\omega^f_v = e^f_v - \theta s^f_v - \mu^f_A \rho^f_A - \mu^f_B \rho^f_B \quad (6.26)$$

and

$$\gamma' = e_{a'}^\Sigma - \theta s_{a'}^\Sigma - \mu_A^\Sigma \Gamma'_A - \mu_B^\Sigma \Gamma'_B \tag{6.27}$$

where γ' is the excess grand canonical free energy associated with the interface.

The terms in Eq. (6.23) containing the variations of the curvatures can be rewritten as

$$K_1 \delta\kappa'_1 + K_2 \delta\kappa'_2 = \frac{1}{2}(K_1 + K_2)(\delta\kappa'_1 + \delta\kappa'_2) + \frac{1}{2}(K_1 - K_2)(\delta\kappa'_1 - \delta\kappa'_2). \tag{6.28}$$

Gibbs proved that when the dividing surface is "sensibly coincident with the surface of discontinuity," each of the coefficients K_1 and K_2 can be made to assume either positive or negative values. He then showed that the term containing the difference $(K_1 - K_2)$ becomes small with respect to the term containing the sum $(K_1 + K_2)$ and can be neglected so long as the thickness of the interface is small relative to the radii of curvature. As in the previous section for a spherical crystal, we follow Gibbs and choose the location of the dividing surface such that $K_1 + K_2 = 0$. This means that the location of the dividing surface is chosen so as to give an extremum in the superficial density γ'. With this understanding, the variations containing $\delta\kappa'_1$ and $\delta\kappa'_2$ are eliminated from Eq. (6.23).

The variations in the displacement of the interface are related by the condition that the solid and fluid must remain in contact in the varied state. As shown in Appendix C,

$$-\delta y^f = \left(F_{ij} n_j^{s'} \delta y^{s'} + \delta u_i\right) n_i^s. \tag{6.29}$$

Equation (6.29) states that the displacement of the fluid interface is a result of two contributions: δu_i, the mechanical displacement of the crystal, and the term containing $\delta y^{s'}$, which is a result of the phase transformation at the surface.

Equation (6.29) can be used to eliminate δy^f in favor of $\delta y^{s'}$ in Eq. (6.24). Using Nanson's formula, Eq. (3.52), the integral over the interface in the actual state containing the term δy^f in the integrand can be transformed to one over the interface in the reference state as follows:

$$\begin{aligned}
\int_\Sigma \omega_v^f \delta y^f dA &= -\int_\Sigma \omega_v^f \left[F_{ij} n_j^{s'} \delta y^{s'} + \delta u_i\right] n_i^s dA \\
&= -\int_{\Sigma'} \omega_v^f \left[F_{ij} n_j^{s'} \delta y^{s'} + \delta u_i\right] J n_k^{s'} F_{ki}^{-1} dA \\
&= -\int_{\Sigma'} \omega_v^f F_{ij} F_{ki}^{-1} n_j^{s'} n_k^{s'} \delta y^{s'} dA - \int_{\Sigma'} \omega_v^f J F_{ki}^{-1} n_k^{s'} \delta u_i dA \\
&= -\int_{\Sigma'} J \omega_v^f \delta y^{s'} dA - \int_{\Sigma'} \omega_v^f J F_{ki}^{-1} n_k^{s'} \delta u_i dA, \tag{6.30}
\end{aligned}$$

where we have used the identities $F_{ki}^{-1} F_{ij} = \delta_{jk}$ and $n_j^{s'} n_k^{s'} \delta_{jk} = 1$. Using Eqs. (6.28) and (6.30) gives for Eq. (6.24)

$$0 = \int_{V^{s'}} \{(\theta^s - \theta_L)\delta s_{v'}^s + (M_{AV} - \lambda_A)\delta \rho_A^{s'} + (M_{BV} - \lambda_B)\delta \rho_B^{s'} - T_{ji,j}\delta u_i\}dV$$

$$+ \int_{V^f} \{(\theta^f - \theta_L)\delta s_v^f + (\mu_A^f - \lambda_A)\delta \rho_A^f + (\mu_B^f - \lambda_B)\delta \rho_B^f\}dV$$

$$+ \int_{\Sigma'} \{(\theta^\Sigma - \theta_L)\delta s_{a'}^\Sigma + (\mu_A^\Sigma - \lambda_A)\delta \Gamma_A' + (\mu_B^\Sigma - \lambda_B)\delta \Gamma_B'\}dA$$

$$+ \int_{\Sigma'} (\omega_v^{s'} - J\omega_v^f + 2\gamma'\kappa')\delta y^{s'}dA + \int_{\Sigma'} (T_{ji}n_j^{s'} - \omega_v^f J F_{ji}^{-1} n_j^{s'})\delta u_i dA. \quad (6.31)$$

The individual perturbations appearing in Eq. (6.31) are now independent, and an extremum in the energy $\delta \mathcal{E}^*$ is assured when each of the coefficients multiplying the individual variations vanishes. Thus,

$$\theta^s = \theta^\Sigma = \theta^f = \theta_L, \quad (6.32)$$

$$M_{AV} = \mu_A^\Sigma = \mu_A^f = \lambda_A$$
$$M_{BV} = \mu_B^\Sigma = \mu_A^f = \lambda_B, \quad (6.33)$$

$$T_{ji,j} = 0, \quad (6.34)$$

$$\omega_v^{s'} - J\omega_v^f + 2\gamma'\kappa' = 0, \quad (6.35)$$

and

$$T_{ji}n_j^{s'} - \omega_v^f J F_{ji}^{-1} n_j^{s'} = 0. \quad (6.36)$$

The first equation is the condition for thermal equilibrium and requires the temperature to be constant throughout the system. Equation (6.33) states that the diffusion potentials are constant within the solid and equal to the chemical potential for that component at the interface and equal to the chemical potential of the component in the fluid. Equation (6.34) is the condition for mechanical equilibrium and, as shown previously, requires that the divergence of the Piola-Kirchhoff stress tensor equal zero.

The remaining two equilibrium conditions pertain to the interface. Equation (6.35) is a balance of configurational forces, those forces that operate on the atoms in the system, whereas Eq. (6.36) is a balance of mechanical forces at the interface. These equations illustrate some of the important differences in the thermodynamic description between a crystal and a fluid. Because we have assumed that the interfacial stress is zero, the mechanical equilibrium condition involves only the bulk forces, and the term containing the interfacial stress derived in the previous chapter, does not appear. The interfacial energy density is nonzero, however, and this energy contributes to the jump in the grand canonical free energies at the surface in Eq. (6.35).

20. SMALL-STRAIN LIMIT

a. Equilibrium Conditions

In this section, the equilibrium conditions given by Eqs. (6.34)–(6.36) are expressed in the limit of small strains. The use of the resulting equations is then illustrated in the following section by examining the influence of stress on the morphological stability of an initially planar interface.

The small strain limit corresponding to Eq. (6.34) was obtained previously and is given in terms of the Cauchy stress tensor by Eq. (3.63) as $\sigma_{ij,i} = 0$. The condition for mechanical equilibrium at the interface can also be expressed in terms of the Cauchy stress tensor. Substituting Eq. (3.58) into Eq. (6.36) gives

$$J\sigma_{pi} F_{jp}^{-1} n_j^{s'} - \omega_v^f J F_{ji}^{-1} n_j^{s'} = 0. \tag{6.37}$$

Changing the dummy indices and factoring allows Eq. (6.37) to be rewritten as

$$\left[\sigma_{ji} - \omega_v^f \delta_{ij}\right] J F_{kj}^{-1} n_k^{s'} = 0. \tag{6.38}$$

Multiplying Eq. (6.38) by the element of surface area dS' and using Eq. (3.51) yields

$$\left[\sigma_{ji} - \omega_v^f \delta_{ij}\right] n_j^s dS = 0. \tag{6.39}$$

Recognizing that $\omega_v^f = -P^f$, Eq. (6.39) for the local mechanical equilibrium condition at the surface becomes

$$\sigma_{ji} n_j^s = \omega_v^f n_i^s \tag{6.40}$$

or, using Eq. (3.53),

$$t_i = -P^f n_i^s. \tag{6.41}$$

The lefthand side of Eq. (6.41) is the traction vector. This equilibrium condition, which holds for both small strain and finite deformations, requires that the traction vector be normal to the surface at every point along the surface for a crystal in mechanical equilibrium with a fluid.

In order to express Eq. (6.35) in the limit of small strains, we recall that $J = 1 + \epsilon_{kk}$. Substituting Eq. (4.123) for $\omega_{v'}$ into Eq. (6.35) gives

$$\mu_V^v \rho_o^{s'} + \frac{1}{2} S_{ijkl} \sigma_{ij} \sigma_{kl} + \left(M_{AV}^o - M_{AV}\right)\rho_A^{s'} + \left(M_{BV}^o - M_{BV}\right)\rho_B^{s'}$$
$$+ P^f(1 + \epsilon_{kk}) + 2\gamma'\kappa' = 0 \qquad (6.42)$$

where M_{AV}^o is the diffusion potential evaluated at temperature θ, composition $\rho_A^{s'}, \rho_B^{s'}$, and zero stress. μ_V^v is the chemical potential of a vacancy. If the standard state is chosen as the homogeneous crystal at pressure P^o, instead of the zero stress state as in Eq. (6.42), we have from Eq. (4.124)

$$0 = \mu_V^v \rho_o^{s'} + \frac{1}{2} S_{ijkl} \sigma_{ij} \sigma_{kl} - \frac{1}{2} S_{iikk}(P^o)^2 - P^o(1 + \epsilon_{kk})$$
$$+ \left(M_{AV}^p - M_{AV}\right)\rho_A^{s'} + \left(M_{BV}^p - M_{BV}\right)\rho_B^{s'} + P^f(1 + \epsilon_{kk}) + 2\gamma'\kappa'. \qquad (6.43)$$

21. APPLICATIONS

We shall now apply the equilibrium conditions to two cases: the equilibrium between a solid and fluid in a binary alloy, and the morphological stability of a solid–vacuum interface.

a. Morphological Stability of a Solid Surface

We shall consider the stability of a planar solid–vacuum interface where the solid is under uniaxial tension. As was found by Asaro and Tiller,[61] Grinfeld[62] (ATG), and others,[63-65] the surface of a uniaxially stressed solid is morphologiclly unstable. As mentioned in the introduction, this ATG instability can play a central role in the formation of islands during heteroepitaxy. The underlying cause of the instability was illustrated in Figure 5. When the surface of the solid is planar, the solid is uniformly stressed. Thus the elastic energy density along the surface is also uniform. In contrast, when the surface is perturbed by a sinusoidal perturbation, the elastic energy density along the surface is nonuniform, as is illustrated by the variation in separation between the lattice planes in Figure 5 for a film initially in a state of compression. In this case, the lattice planes are further compressed in the volume elements beneath the troughs, in a manner similar to the stress concentration at a crack tip. The compressive stresses in the film are relaxed near

---

[61] R. J. Asaro and W. A. Tiller, *Metall. Trans. A* **3**, 1789 (1972).
[62] M. A. Grinfeld, *Sov. Phys. Dokl.* **31**, 831 (1986).
[63] D. J. Srolovitz, *Acta. Mater.* **37**, 621 (1989).
[64] B. J. Spencer, P. W. Voorhees, and S. H. Davis, *Phys. Rev. Lett.* **67**, 3696 (1991).
[65] A. Onuki, *J. Phys. Soc. Japan* **60**, 345 (1991).

the surface peaks, because these regions are shielded partly from the compressive stress engendered by the substrate. Consequently, the elastic energy density for a volume element in a trough exceeds that of a volume element near a peak. Because the diffusion potential of atoms in a region of high elastic energy density is greater than that in a region of low elastic energy density, atoms tend to flow from the trough to the peak with the result that the troughs deepen and the peaks grow. Opposing the development of a corrugated surface is surface energy, which would be minimized if the surface were planar. The surface energy, therefore, tends to stabilize the planar surface. In order to determine which of these two mechanisms is dominant, we shall develop an expression for the growth rate of small-amplitude morphological perturbations to an initially planer, solid–vacuum interface. As is the case in many thin-film deposition processes, surface diffusion is taken to be the dominant mass transport process. Because mass flow along this nonplanar surface is in response to gradients in the diffusion potential, we also determine the dependence of the diffusion potential at the surface on strain energy and surface energy.

b. *Diffusion Potential of a Vacancy at a Solid–Vacuum Surface*

We consider a single component crystal in contact with a vacuum. There are no defects in the crystal other than vacancies. In this isothermal case, the diffusion potential at the surface is defined by the surface equilibrium conditions Eq. (6.41) and Eq. (6.42), and the bulk condition for mechanical equilibrium $\sigma_{ij,j} = 0$.

In order to examine the stability of a crystalline surface in contact with a vacuum and to focus on the effects of the elastic energy density at the surface, we take the elastic constants to be independent of vacancy concentration and we neglect the dependence of the lattice parameter of the crystal on vacancy concentration.[66] This implies that $M^o_{AV} = M_{AV}$. We take the reference state for the measurement of strain to be the crystal at zero pressure. Because the solid is in contact with a vacuum, $P^f = 0$ and Eq. (6.42) becomes

$$\rho_o^{s'} \mu_V^v + \frac{1}{2} S_{ijkl} \sigma_{ij} \sigma_{kl} + 2\gamma' \kappa' = 0. \tag{6.44}$$

Thus the chemical potential of a vacancy is set by the elastic strain energy density at the surface and the curvature of the surface.

c. *Surface Diffusion and Morphological Instability*

The surface evolves by the diffusion of atoms along the surface. There is no lattice constraint for surface atoms because, for example, atoms can jump from

---

[66] The conditions for stability when the lattice depends on vacancy concentration have been examined by Spencer *et al.*[72]

one terrace to another across a step edge. Thus the flux of atoms along the surface, J_Σ, is assumed to be given by the gradient in the chemical potential of component A, μ_A^v, as[67]

$$\mathbf{J}_\Sigma = -b\nabla_\Sigma \mu_A \qquad (6.45)$$

where $b = D_s \Gamma_o'/k\theta$, D_s is the surface mobility of A, Γ_o' is the surface density of lattice sites, k is Boltzmann's constant, θ is the temperature, and ∇_Σ is the gradient along the surface assumed to be taken with respect to the referential state. For a planar film with normal parallel to the x_3-axis of a Cartesian coordinate system, $\nabla_\Sigma = \partial_{x_1}\hat{i} + \partial_{x_2}\hat{j}$. If the surface is nonplanar, the surface gradient involves the shape of the surface.[68] If the surface atoms are in equilibrium with the underlying crystal, then $\mu_A^v = M_{AV}$, giving for the surface flux

$$\mathbf{J}_\Sigma = -b\nabla_\Sigma M_{AV}. \qquad (6.46)$$

The diffusion potential for a given state of stress is equal to its value under hydrostatic stress, because we are neglecting the dependence of the lattice parameter on vacancy concentration and the dependence of the elastic constants on vacancy concentration. Thus,

$$M_{AV} = M_{AV}^o = \mu_A^v - \mu_V^v. \qquad (6.47)$$

The chemical potential of component A for an ideal solution of vacancies and A atoms is

$$\mu_A^v = \tilde{\mu}_A + k\theta \ln(C_A) \qquad (6.48)$$

where $\tilde{\mu}_A$ is the chemical potential of a crystal of pure A and C_A is the mole fraction of A. Because $C_A = 1 - C_V$ and $C_V \ll 1$,

$$\mu_A^v \approx \tilde{\mu}_A - k\theta C_v. \qquad (6.49)$$

Because the temperature at which surface diffusion is dominant is on the order of one half the melting point of the solid, the vacancy concentration is extremely small and the second term is negligible compared to the first. We thus approximate

$$M_{AV} = \tilde{\mu}_A^v - \mu_V^v. \qquad (6.50)$$

---

[67] W. W. Mullins, *J. Appl. Phys.* **28**, 333 (1957).
[68] C. E. Weatherburn, *Differential Geometry of Three Dimensions*, Cambridge University Press, New York (1927).

Substituting Eq. (6.50) into Eq. (6.46) and using Eq. (6.44),

$$\mathbf{J}_\Sigma = -b \nabla_\Sigma \left(\frac{1}{2} S_{ijkl} \sigma_{ij} \sigma_{kl} + 2\gamma' \kappa' \right) \quad (6.51)$$

because $\tilde{\mu}_A$ is a constant. Thus atoms flow from regions of high elastic energy density to regions of low elastic energy density in the absence of capillary effects, as mentioned previously. In the limit where the crystal is unstressed, this reduces to the expression for the flux of atoms along a surface as given by Mullins.[67]

The mass conservation condition at the surface of a crystal in the limit of a small vacancy concentration is found by considering the flow of atoms into and out of a small region of interface. There are two flows of interest here: The atoms can move along the surface by surface diffusion and they can move to and from the surface as the crystal grows or shrinks. Considering each of these flows the mass conservation condition is[69]

$$\left(2\Gamma'_A \kappa' + \rho'_A \right) v + \frac{\partial \Gamma'_A}{\partial t} = -\nabla_\Sigma \cdot \mathbf{J}_\Sigma \quad (6.52)$$

where v is the velocity of the surface along the normal directed into the vacuum and Γ'_A is the density of A atoms at the surface. In the limit $v = 0$, Eq. (6.52) reduces to the usual mass conservation condition for a diffusive flow of atoms. In this limiting case, the atoms are those on the surface. If the interface is moving, however, atoms are lost from or added to the surface and the first term in Eq. (6.52) can become important. The first term in the brackets in Eq. (6.52) that involves κ accounts for the extra atoms that flow into the surface as the area of a nonplanar interface changes during evolution. For most surface velocities and interfacial curvatures, the terms involving Γ'_A are negligible and thus,

$$v = -\frac{1}{\rho'_o} \nabla_\Sigma \cdot \mathbf{J}_\Sigma, \quad (6.53)$$

where we note that the density of lattice sites is essentially the same as the density of A atoms, $\rho'_o = \rho'_A$. Substituting Eq. (6.51) into Eq. (6.53) yields the normal velocity for a surface element of a stressed crystal in contact with a vacuum,

$$v = B \nabla^2 \left(\frac{1}{2} S_{ijkl} \sigma_{ij} \sigma_{kl} + 2\gamma' \kappa' \right), \quad (6.54)$$

where $B = b/\rho'_o$ and we have assumed that B is constant. This is not true in general. For example, B would be expected to depend on stress and composition. The

---

[69] M. E. Gurtin and P. W. Voorhees, *J. Appl. Math. Phys.* **41**, 782 (1990).

dependence of the mobility on stress can also lead to a morphological instability of a planar surface.[70,71]

Equation (6.54) can be used to determine the stability of a planar surface. We shall follow closely the approach given by Voorhees and Aziz in their work on the stability of the amorphous-crystalline interface in silicon[71] by examining the stability of the surface in response to infinitesimally small perturbations. As is clear from Eq. (6.54), the elastic energy density is required to determine the interfacial velocity.

The elastic energy density is a function of the stress state. The stress field is found by requiring mechanical equilibrium, which in the small strain approximation is determined by

$$\sigma_{ij,j} = 0. \tag{6.55}$$

Assuming an isotropic crystal, the stress and strain are related by

$$\sigma_{ij} = \frac{E_y}{1+\nu} \left(\frac{\nu}{1-2\nu} \epsilon_{kk} \delta_{ij} + \epsilon_{ij} \right) \tag{6.56}$$

where E_y is Young's modulus and ν is Poisson's ratio. From Eq. (3.19), the strain is related to gradients in the displacement u_j by

$$\epsilon_{ij} = (u_{i,j} + u_{j,i})/2. \tag{6.57}$$

Substituting Eqs. (6.56) and (6.57) into Eq. (6.55) yields Navier's equations for the displacement field in the crystal,

$$(1 - 2\nu) u_{i,kk} + u_{k,ki} = 0. \tag{6.58}$$

The displacement field is determined by solving Eq. (6.58) subject to boundary conditions. Because the solid is in contact with a vacuum, there is no applied traction normal to the surface and

$$\sigma_{ij} n_j = 0 \tag{6.59}$$

where n_j is the normal to the interface pointing into the vacuum. However, the crystal surface is biaxially stressed. If the normal to the planar interface is along the x_3-direction, as in Figure 24, then the stress field of the crystal requires

$$\sigma_{11} = \sigma_{22} = \sigma_a. \tag{6.60}$$

---

[70] W. Barvosa-Carter, M. J. Aziz, L. J. Gray, and T. Kaplan, *Phys. Rev. Lett.* **81**, 1445 (1998).
[71] P. W. Voorhees, in *Interfaces in the 21st Century*, eds. M. K. S. et al., Imperial College Press (2002), 167.

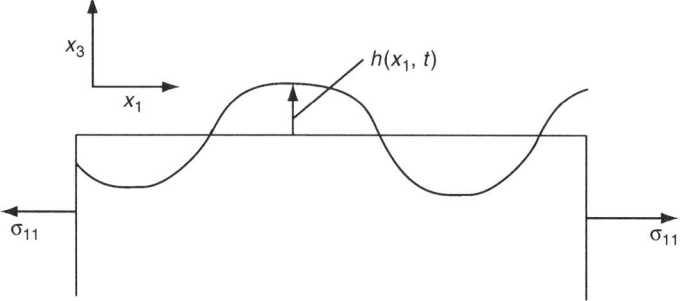

FIG. 24. Perturbation of the planar solid–vacuum interface. The perturbation along the x_2 direction is not shown.

Once the displacements are determined, the elastic strain energy density, W, follows from

$$W = \frac{1}{2} S_{ijkl} \sigma_{ij} \sigma_{kl}. \tag{6.61}$$

The linear stability of the planar surface located initially at $x_3 = 0$ (see Fig. 24) is examined by perturbing the height of the surface in the x_3-direction, $h = h(x_1, x_2)$, and all other quantities in normal modes. In vector form, the perturbed quantities are given as

$$\begin{pmatrix} h \\ v \\ u_i \\ W \\ \kappa \end{pmatrix} = \begin{pmatrix} 0 \\ 0 \\ u_i^0 \\ W^0 \\ 0 \end{pmatrix} + \delta \begin{pmatrix} \hat{h} \\ v^1 \\ u_i^1 \\ W^1 \\ \kappa^1 \end{pmatrix} \Phi \tag{6.62}$$

where $\Phi = \exp(i(q_1 x_1 + q_2 x_2) + rt)$, q_1 and q_2 are the x_1 and x_2 components of the wave vector, respectively, δ is the magnitude of the perturbation assumed to be small, $\delta \ll 1$, and t is time. The growth rate of a perturbation with a certain wave vector is r. If $r > 0$, the perturbation grows and, if $r < 0$, the perturbation decays. If r is imaginary, then oscillatory modes are possible. One would expect that, if the wave number of the perturbation is large or the wavelength is small, surface energy would stabilize this perturbation and $r < 0$. However, for smaller wave number disturbances, those with large wavelengths, elastic stress may cause these modes to be unstable and $r > 0$.

The perturbation analysis is performed as follows. Equation (6.62) is substituted into Eq. (6.54) and like-powers of δ are equated. This process separates the basic state, where $\delta = 0$, from the perturbed state. In the unperturbed, or basic state, the interface is planar and the stress state is uniform. There is a nonzero displacement along x_3, because this surface is traction free.

To illustrate the manner in which the perturbed quantities are computed, consider the interfacial curvature. The curvature is given by

$$\kappa = \frac{h_{,11} + h_{,22}}{\left[1 + \left(h_{,1}^2 + h_{,2}^2\right)^{1/2}\right]^{3/2}}. \tag{6.63}$$

Substituting the expression given in Eq. (6.62) for h into Eq. (6.63) gives the curvature to first order in δ as

$$\kappa = \delta[(\hat{h}\Phi)_{,11} + (\hat{h}\Phi)_{,22}]. \tag{6.64}$$

There is no zeroth order term because the unperturbed interface is planar. Substituting for Φ in Eq. (6.64) gives, for the perturbed curvature, κ_1,

$$\kappa^1 = -q^2 \hat{h} \tag{6.65}$$

where $q = \sqrt{(q_1^2 + q_2^2)}$. This implies that perturbations with large wavenumber have large curvatures and the magnitude of the curvature scales with the amplitude of the perturbation.

The normal velocity of the interface is

$$v = \frac{h_{,t}}{\left(1 + h_{,1}^2 + h_{,2}^2\right)^{1/2}} \tag{6.66}$$

where $h_{,t} = \partial h/\partial t$. Using the normal mode expansion as before yields

$$v^1 = r\hat{h}. \tag{6.67}$$

Thus r measures the growth rate of the perturbation.

The elastic strain energy density in the basic state follows from the stresses given above,

$$W^0 = \sigma_a^2/Y \tag{6.68}$$

where $Y = E_y/(1 - \nu)$. The strain energy density in the perturbed state is determined by solving Eq. (6.58) with the boundary condition Eq. (6.59) and the condition that the $u_i^1(x_3) \to 0$ as $x_3 \to -\infty$.[72] Using these displacements in Eq. (6.57) and substituting into Eq. (6.56) and using Eq. (6.62) yields the magnitude of the

---

[72] B. J. Spencer, P. W. Voorhees, and S. H. Davis, *J. Appl. Phys.* **73**, 4955 (1993).

perturbed elastic strain energy density along the interface[73],

$$W^1 = 2(1+v)q\sigma_a^2 \hat{h}/Y. \tag{6.69}$$

The perturbation in the elastic energy has a weaker wavenumber dependence than that of the curvature. In addition, the elastic energy increases for both applied compressive and tensile stresses, since $W^{(1)} \sim \sigma_a^2$.

Substituting Eq. (6.69), Eq. (6.65), Eq. (6.67), Eq. (6.66) into Eq. (6.62) and using this result in Eq. (6.54) we find to first order in δ,

$$r = 2(1+v)q^3 \sigma_a^2/Y - \gamma' q^4. \tag{6.70}$$

This equation gives the growth rate of a perturbation of wavenumber q. If σ_a is nonzero, elastic stress is destabilizing, because the first term of Eq. (6.70) is always positive. In contrast, surface energy is always stabilizing because the last term on the righthand side of Eq. (6.70) is always negative. However, if σ_a is nonzero because the elastic stress and surface energy contributions scale differently with wave vector, r must be greater than zero for sufficiently small q, or large wavenumber. The critical wavenumber, q_c, where $r = 0$ is given by

$$q_c = \frac{2(1+v)\sigma_a^2}{Y\gamma'}. \tag{6.71}$$

This provides a reasonable order of magnitude estimate for the length scale at which one would expect the instability to be observed. Because q_c is positive for any nonzero σ_a, a planar interface will always be unstable to a stress-driven instability. As expected, the ATG instability is independent of the sign of the applied stress. This follows from the dependence of the elastic energy on the square of the stress. Thus a planar interface under compression will form ripples along the surface as well as one under tension. In the thin-film applications these surface perturbations grow, touch the substrate, and create coherent islands.

d. Equilibrium Condition at a Crystal–Fluid Interface

An adequate description of the kinetics and morphological evolution of diffusional phase transformations in crystal–fluid alloys frequently requires an expression for the crystal and fluid concentrations along their common interface as a function of the local interfacial curvature and stress state. These concentrations often provide the boundary condition for the mass flow problem in each phase and for the motion of the interface. In this example, we use the conditions for thermodynamic equilibrium previously developed to determine the mass concentrations in

---

[73] J. E. Guyer and P. W. Voorhees, *Phys. Rev. B* **54**, 11710 (1995–6).

the crystal and fluid phases evaluated at their interface when the crystal is nonhydrostatically stressed and the interface is nonplanar.

We assume an isothermal, binary alloy with vacancies for which local equilibrium exists at the crystal–fluid interface. The conditions for equilibrium are given by Eqs. (6.33)–(6.36). We use the small-strain approximation and take the stress-free crystal as the reference state for the measurement of strain. In most cases the fluid pressure is negligibly small compared to the elastic strain energy density, and the term containing the fluid pressure can be neglected. The surface equilibrium condition, Eq. (6.42), then becomes

$$\rho_o^{s'} \mu_V^v + \frac{1}{2} S_{ijkl} \sigma_{ij} \sigma_{kl} + \rho_A^{s'} \left(M_{AV}^o - M_{AV} \right) + \rho_B^{s'} \left(M_{BV}^o - M_{BV} \right) = -2\gamma'\kappa'. \quad (6.72)$$

The difference in the diffusion potential in the presence of stress, M_{AV}, and the diffusion potential in the zero-stress reference state, M_{AV}^o, arises from internal stresses generated within the crystal by nonuniform composition fields when the lattice parameter depends on the composition. This coupling can be important in many situations. For example, it can drive morphological instabilities of the surface[39,40,74–76] and alter the manner in which a system undergos spinodal decomposition.[77,78] For simplicity, we neglect these stresses here so that $M_{AV} = M_{AV}^o$, and write for Eq. (6.72)

$$\rho_o^{s'} \mu_V^v = -\frac{1}{2} S_{ijkl} \sigma_{ij} \sigma_{kl} - 2\gamma'\kappa'. \quad (6.73)$$

The diffusion potentials at the crystal–liquid interface can be rewritten in terms of the chemical potentials as

$$\begin{aligned} M_{AV} &= M_{AV}^o = \mu_A^v - \mu_V^v \\ M_{BV} &= M_{BV}^o = \mu_B^v - \mu_V^v. \end{aligned} \quad (6.74)$$

Using Eq. (6.74) in the equilibrium condition (Eq. (6.33)) yields

$$\begin{aligned} \mu_A^v - \mu_V^v &= \mu_A^f \\ \mu_B^v - \mu_V^v &= \mu_B^f. \end{aligned} \quad (6.75)$$

---

[74] F. C. Leonard and R. C. Desai, *Appl. Phys. Lett.* **73**, 208 (1998).
[75] B. J. Spencer, P. W. Voorhees, and J. Tersoff, *Phys. Rev. Lett.* **84**, 2449 (2000).
[76] B. J. Spencer, P. W. Voorhees, and J. Tersoff, *Phys. Rev. B* 235318 (2001).
[77] P. H. Leo and R. F. Sekerka, *Acta mater.* **49**, 1771 (2001).
[78] J. W. Cahn, *Acta metall.* **9**, 795 (1961).

Substituting Eq. (6.73) into Eq. (6.75) gives

$$\rho'_o \mu^v_A + \frac{1}{2} S_{ijkl} \sigma_{ij} \sigma_{kl} + 2\gamma' \kappa' = \rho'_o \mu^f_A$$
$$\rho'_o \mu^v_B + \frac{1}{2} S_{ijkl} \sigma_{ij} \sigma_{kl} + 2\gamma' \kappa' = \rho'_o \mu^f_B.$$
(6.76)

In general, the chemical potentials in the crystal are functions of both the mole fraction of component A and the vacancy concentration. However, as shown in the previous sections, this dependence can be neglected when the vacancy concentration is very small. Thus, the interfacial condition Eq. (6.76) provides two equations for the two unknown interfacial compositions in the crystal and fluid, C^S_A and C^f_A.

The two equations defining the equilibrium compositions at the interface are nonlinear functions of composition and cannot, in general, be solved analytically. Analytic approximations to the equilibrium compositions can be obtained, however, when the equations are linearized about a suitable reference state. If the shift in composition owing to the elastic stress and curvature are not too large, then a convenient reference state about which to linearize the equations is the state of two-phase equilibrium between unstressed crystal and fluid across a planar interface. (Because the fluid pressure is taken to be vanishingly small, the crystal can be considered to be stress free.) The compositions of the fluid and crystal in this state are those given by the equilibrium phase diagram at zero pressure, and are given in most reference books. All quantities in this unstressed equilibrium state are denoted by a superscript e. Expanding the chemical potentials in a Taylor series about $\mu^v_A = \mu^e_A$ corresponding to $c_A = c^e_A$ gives, for both the crystal and fluid phases,

$$\mu^v_A(c_A) = \mu^{ve}_A(c^e_A) + (c_A - c^e_A)\mu^{ve}_{A,c_A}(c^e_A)$$
$$\mu^v_B(c_A) = \mu^{ve}_B(c^e_A) + (c_A - c^e_A)\mu^{ve}_{B,A}(c^e_A)$$
(6.77)

where the comma denotes differentiation with respect to the mole fraction c_A. The derivatives of the chemical potentials of the fluid can be expressed in terms of the compositional derivatives of the Gibbs free energy. This is also the case for the crystal, as the reference state for the crystal is the stress-free equilibrium state and the chemical potential for vacancies in the crystal is zero. Therefore, for both phases,

$$\mu^v_A = \mu^{ve}_A + (1 - c^e_A)G^e_{o,c_A c_A}(c_A - c^e_A)$$
$$\mu^v_B = \mu^{ve}_B - c^e_A G^e_{o,c_A c_A}(c_A - c^e_A)$$
(6.78)

where G^e_o is the Gibbs free energy per atom evaluated at the stress-free equilibrium composition. Substituting Eq. (6.78) into Eq. (6.75) and recognizing that, for the equilibrium state, $\mu^{vfe}_A = \mu^{vse}_A$ and $\mu^{vfe}_B = \mu^{vse}_B$, yields two linear equations for the equilibrium compositions. Solution yields the compositions at the interface of a stressed crystal in equilibrium with a fluid,

$$c_A^s = c_A^{se} + \frac{S_{ijkl}\sigma_{ij}\sigma_{kl}/2 + 2\gamma'\kappa'}{\rho_o'(c_A^{se} - c_A^{fe})G_{o,c_Ac_A}^{se}}$$

$$c_A^f = c_A^{fe} + \frac{S_{ijkl}\sigma_{ij}\sigma_{kl}/2 + 2\gamma'\kappa'}{\rho_o'(c_A^{se} - c_A^{fe})G_{o,c_Ac_A}^{fe}}.$$

(6.79)

In the limit where the stress is zero, we recover the usual Gibbs-Thompson equation for a solid immersed in a fluid in which both components are soluble. Thus the concentration at the interface of a crystal under stress is less than that of a stress-free solid if the concentration of solute in the fluid is greater than that in the solid. Thus, a stress-free crystal would be expected to grow at the expense of a crystal under stress.

VII. Two-Phase Crystalline Systems

22. INTRODUCTION

Building on the previous developments for the thermodynamics of crystals under hydrostatic stress, isolated crystals under nonhydrostatic stress, and two-phase crystal–fluid systems, we now determine the conditions for thermodynamic equilibrium at a coherent interface between two nonhydrostatically stressed crystalline phases. After deriving the equilibrium conditions, we illustrate their use by considering three applications. In the first example, we derive the local Gibbs-Thompson equation for a system under nonhydrostatic stress. Secondly, we examine the effects of elastic stress on the characteristics of phase equilibria in a coherent two-phase crystal, emphasizing the construction of phase diagrams and the use of the common-tangent construction. Finally, we illustrate the influence of a misfit (transformation) strain and an applied stress on the equilibrium shape of a coherent second-phase particle within the context of bifurcation theory.

The interface between the two crystals is assumed to be nonplanar and to possess an energy density distinct from those of the contiguous phases. As was done in the previous section, we allow the interfacial energy to be nonzero; however, for the two-phase crystal, we assume that the interfacial stress vanishes. A nonzero interfacial stress greatly complicates the derivation of the equilibrium conditions in that it necessitates treating surface tensors in general curvilinear coordinates. By neglecting the interfacial stress, we can focus on the effects of the bulk stress on the equilibrium conditions in two-phase crystalline systems. See Leo and Sekerka[59] for a variational treatment of the interfacial stress and Gurtin[13] for a dynamical treatment. Finally, we take the interface between the phases to be coherent and the interfacial energy to be isotropic, although the interfacial energy in crystalline solids can be anisotropic.

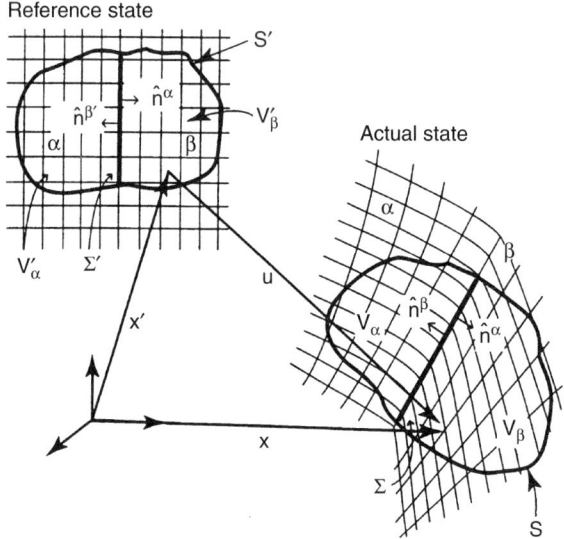

FIG. 25. A region V containing both the α and β phases is imagined to be thermodynamically isolated from the rest of the system in its actual state. For a coherent interface, all points in V can be mapped back to a continuous reference state.

The assumption of coherency implies that each of the crystalline phases, α and β, can be mapped back into an identical reference state, as shown in Figure 25. Physically, this means there is a continuity of the stacking sequence across the interface, Σ, from one phase into the other. The reference state is chosen as the unstressed lattice of one of the phases, at a given temperature and composition. Unlike the crystal–fluid interface, lattice points cannot be created at a coherent interface, although lattice points can be transferred between phases at the interface. There are no sources of vacancies in the crystal. Other models for the interface between the two phases have been employed. For example, the conditions at a incoherent interface have been derived.[18,59,79,80]

23. Equilibrium Conditions

a. Energy Functional

The total energy of the system, composed of two substitutional species A and B, in the reference state is partitioned into the energy of the α and β phases and

---

[79] F. C. Larché and J. W. Cahn, *Acta metall.* **26**, 1579 (1978).
[80] P. Cermelli and M. E. Gurtin, *Arch. Rat. Mech. Anal.* **127**, 41 (1994).

interface,

$$\mathcal{E} = \int_{V^{\alpha'}} e_{v'}^{\alpha}(s_{v'}, F_{ij}, \rho_A', \rho_B',) dV + \int_{V^{\beta'}} e_{v'}^{\beta}(s_{v'}, F_{ij}, \rho_A', \rho_B',) dV$$
$$+ \int_{\Sigma'} e_{a'}^{\Sigma}(s_{a'}^{\Sigma}, \Gamma_A', \Gamma_B', \kappa_1', \kappa_2') dA \qquad (7.1)$$

where Σ' is the interface in the reference state, $e_{a'}^{\Sigma}$ and $s_{a'}^{\Sigma}$ are the excess internal energy and entropy, respectively, due to the presence of the interface, Γ_i' is the excess mass per unit area of the reference state of component i, and κ_1', κ_2' are the principal curvatures of the interface in the reference state. We also allow for the possibility that the independent thermodynamic variables describing the interface can vary with position along the interface, thus the integral of the interfacial energy per unit area appears in Eq. (7.1). Because the system is coherent, there is a common reference lattice to which both crystals can be referred. Consequently, there is only one line of points in the referential state that identifies the interface. This differs from an incoherent interface where the presence of a reference state for each crystal implies that there are two interfaces, one referred to a reference state for each phase.

b. An Extremum in the Energy

In addition to variations in the independent thermodynamic variables of the bulk phases and interface, we must allow for phase changes that occur at the crystalline interface. This variation in the position of the interface owing to a change in the number of lattice sites associated with each phase is taken in the direction normal to the interface and is measured in the referential state. The change in number of lattice sites associated with a phase changes the reference volume of each phase and the value of each extensive property ϕ. The change in ϕ is calculated using a surface integral of the thermodynamic density $\phi_{v'}$ over the interface in the unvaried state multiplied by the magnitude of the variation in interface position. This approach is strictly valid only in the limit of infinitesimally small variations. The change in the extensive property ϕ associated with the α phase is

$$\delta \phi^{\alpha} = \delta \int_{V^{\alpha'}} \phi_{v'}^{\alpha} dV = \int_{V^{\alpha'}} \delta \phi_{v'}^{\alpha} dV + \int_{\Sigma'} \phi_{v'}^{\alpha} \delta y^{\alpha'} dA \qquad (7.2)$$

where $\delta y^{i'}$ is the displacement of the interface owing to phase transformation directed along the outward pointing normal of phase i in the reference state. Similarly, for the β phase,

$$\delta \phi^{\beta} = \delta \int_{V^{\beta'}} \phi_{v'}^{\beta} dV = \int_{V^{\beta'}} \delta \phi_{v'}^{\beta} dV + \int_{\Sigma'} \phi_{v'}^{\alpha} \delta y^{\beta'} dA. \qquad (7.3)$$

The relationship between $\delta y^{\beta'}$ and $\delta y^{\alpha'}$ at a coherent interface is given in Appendix B.

The interface is nonplanar and, when a phase transformation occurs at the interface, the referential shape and area of the interface will also change. The transformation results in a change in the value of ϕ associated with the interface, which can be expressed as

$$\delta \phi^{\Sigma} = \delta \int_{\Sigma'} \phi_{a'}^{\Sigma} dA = \int_{\Sigma'} \delta \phi_{a'}^{\Sigma} dA + \int_{\Sigma'} \phi_{a'}^{\Sigma} 2\kappa' \delta y^{\beta'} dA \qquad (7.4)$$

where $\phi_{a'}^{\Sigma}$ is the area density of ϕ in the reference state, κ' is the mean of the principal interfacial curvatures in the reference state at a point, $\kappa' = (\kappa_1' + \kappa_2')/2$ where κ_1' and κ_2' are the principal curvatures at a point on the interface. For the change in area to be consistent with a displacement along the outward pointing normal to the β phase, κ' is taken to be positive for a spherical β phase particle.

Using Eqs. (7.2)–(7.4) with the internal energy in place of ϕ, yields the first variation in the system energy,

$$\delta \mathcal{E} = \int_{V^{\alpha'}} \delta e_{v'}^{\alpha} dV + \int_{V^{\beta'}} \delta e_{v'}^{\beta} dV + \int_{\Sigma'} \delta e_{a'}^{\Sigma} dA + \int_{\Sigma'} e_{v'}^{\alpha} \delta y^{\alpha'} dA$$
$$+ \int_{\Sigma'} e_{v'}^{\beta} \delta y^{\beta'} dA + \int_{\Sigma'} 2\kappa' e_{a'}^{\Sigma} \delta y^{\beta'} dA, \qquad (7.5)$$

where for the α and β phases,

$$\delta e_{v'}^{\alpha} = \frac{\partial e_{v'}^{\alpha}}{\partial s_{v'}^{\alpha}} \delta s_{v'}^{\alpha} + \frac{\partial e_{v'}^{\alpha}}{\partial F_{ij}} \delta F_{ij}^{\alpha} + \frac{\partial e_{v'}^{\alpha}}{\partial \rho_A^{\alpha'}} \delta \rho_A^{\alpha'} + \frac{\partial e_{v'}^{\alpha}}{\partial \rho_B^{\alpha'}} \delta \rho_B^{\alpha'} \qquad (7.6)$$

$$\delta e_{v'}^{\beta} = \frac{\partial e_{v'}^{\beta}}{\partial s_{v'}^{\beta}} \delta s_{v'}^{\beta} + \frac{\partial e_{v'}^{\beta}}{\partial F_{ij}} \delta F_{ij}^{\beta} + \frac{\partial e_{v'}^{\beta}}{\partial \rho_A^{\beta'}} \delta \rho_A^{\beta'} + \frac{\partial e_{v'}^{\beta}}{\partial \rho_B^{\beta'}} \delta \rho_B^{\beta'} \qquad (7.7)$$

and for the interface,

$$\delta e_{a'}^{\Sigma} = \frac{\partial e_{a'}^{\Sigma}}{\partial s_{a'}^{\Sigma}} \delta s_{a'}^{\Sigma} + \frac{\partial e_{a'}^{\Sigma}}{\partial \Gamma_A'} \delta \Gamma_A' + \frac{\partial e_{a'}^{\Sigma}}{\partial \Gamma_B'} \delta \Gamma_B' + \frac{\partial e_{a'}^{\Sigma}}{\partial \kappa_1'} \delta \kappa_1' + \frac{\partial e_{a'}^{\Sigma}}{\partial \kappa_2'} \delta \kappa_2'. \qquad (7.8)$$

The partial derivatives appearing in Eqs. (7.6)–(7.8) have been defined in previous sections. The first variations in the energy densities for the α and β phases become, respectively,

$$\delta e_{v'}^{\alpha} = \theta^{\alpha} \delta s_{v'}^{\alpha} + T_{ji}^{\alpha} \delta F_{ij}^{\alpha} + M_{AV}^{\alpha} \delta \rho_A^{\alpha'} + M_{BV}^{\alpha} \delta \rho_B^{\alpha'} \qquad (7.9)$$

$$\delta e_{v'}^{\beta} = \theta^{\beta} \delta s_{v'}^{\beta} + T_{ji}^{\beta} \delta F_{ij}^{\beta} + M_{AV}^{\beta} \delta \rho_A^{\beta'} + M_{BV}^{\beta} \delta \rho_B^{\beta'}. \qquad (7.10)$$

In an analogy to the definitions used for the bulk phases,

$$\theta^\Sigma = \frac{\partial e_{a'}^\Sigma}{\partial s_{a'}^\Sigma} \tag{7.11}$$

where θ^Σ is the temperature of the interface. Because the interface is coherent, the lattice is continuous across the interface. The derivatives of the interfacial energy with respect to the excess mass densities must occur via an exchange with a vacancy, as there is no change in the total number of lattice sites. Thus, the derivatives with respect to number density of components A and B are related to the diffusion potentials by

$$M_{AV}^\Sigma = \frac{\partial e_{a'}^\Sigma}{\partial \Gamma_A'}$$

and

$$M_{BV}^\Sigma = \frac{\partial e_{a'}^\Sigma}{\partial \Gamma_B'}. \tag{7.12}$$

Finally, the terms representing the curvature dependence of the interfacial energy are replaced by the coefficients,

$$K_1 = \frac{\partial e_{a'}^\Sigma}{\partial \kappa_1'}$$

and

$$K_2 = \frac{\partial e_{a'}^\Sigma}{\partial \kappa_2'}. \tag{7.13}$$

Substituting Eqs. (7.9)–(7.13) into Eq. (7.5) yields, for the first variation of the system energy,

$$\begin{aligned}\delta \mathcal{E} = & \int_{V_{\alpha'}} \{\theta^\alpha \delta s_{v'}^\alpha + T_{ji}^\alpha \delta F_{ij}^\alpha + M_{AV}^\alpha \delta \rho_A^{\alpha'} + M_{BV}^\alpha \delta \rho_B^{\alpha'}\} dV \\ & + \int_{V_{\beta'}} \{\theta^\beta \delta s_{v'}^\beta + T_{ji}^\beta \delta F_{ij}^\beta + M_{AV}^\beta \delta \rho_A^{\beta'} + M_{BV}^\beta \delta \rho_B^{\beta'}\} dV \\ & + \int_{\Sigma'} \{\theta^\Sigma \delta s_{a'}^\Sigma + M_{AV}^\Sigma \delta \Gamma_A' + M_{BV}^\Sigma \delta \Gamma_B' + K_1 \delta \kappa_1' + K_2 \delta \kappa_2'\} dA \\ & + \int_{\Sigma'} e_{v'}^\alpha \delta y^{\alpha'} dA + \int_{\Sigma'} e_{v'}^\beta \delta y^{\beta'} dA + \int_{\Sigma'} 2 e_{a'}^\Sigma \kappa' \delta y^{\beta'} dA. \end{aligned} \tag{7.14}$$

Following Gibbs' variational approach for determining the thermodynamic equilibrium conditions, the two-phase system is imagined to be isolated and the energy minimized subject to the constraints that the number of atoms of A and B are fixed, the total entropy is constant, and the surface enclosing the system is nondeformable. Thus, the first variation in the energy given by Eq. (7.14) must be set to zero subject to the constraints of constant entropy $\mathcal{S}$,

$$\mathcal{S} = \int_{V^{\alpha'}} s_v^{\alpha} dV + \int_{V^{\beta'}} s_v^{\beta} dV + \int_{\Sigma'} s_a^{\Sigma} dA \tag{7.15}$$

and constant number of components A and B,

$$\begin{aligned} \mathcal{N}_A &= \int_{V^{\alpha'}} \rho_A^{\alpha'} dV + \int_{V^{\beta'}} \rho_A^{\beta'} dV + \int_{\Sigma'} \Gamma_A' dA \\ \mathcal{N}_B &= \int_{V^{\alpha'}} \rho_B^{\alpha'} dV + \int_{V^{\beta'}} \rho_B^{\beta'} dV + \int_{\Sigma'} \Gamma_B' dA. \end{aligned} \tag{7.16}$$

The first variations of Eqs. (7.15) and (7.16) are, using Eqs. (7.2)–(7.4),

$$\begin{aligned} \delta\mathcal{S} = &\int_{V^{\alpha'}} \delta s_v^{\alpha} dV + \int_{V^{\beta'}} \delta s_v^{\beta} dV + \int_{\Sigma'} \delta s_a^{\Sigma} dA + \int_{\Sigma'} s_v^{\alpha} \delta y^{\alpha'} dA \\ &+ \int_{\Sigma'} s_v^{\beta} \delta y^{\beta'} dA + \int_{\Sigma'} s_a^{\Sigma} \kappa' \delta y^{\beta'} dA \end{aligned} \tag{7.17}$$

$$\begin{aligned} \delta\mathcal{N}_A = &\int_{V^{\alpha'}} \delta\rho_A^{\alpha'} dV + \int_{V^{\beta'}} \delta\rho_A^{\beta'} dV + \int_{\Sigma'} \delta\Gamma_A' dA + \int_{\Sigma'} \rho_A^{\alpha'} \delta y^{\alpha'} dA \\ &+ \int_{\Sigma'} \rho_A^{\beta'} \delta y^{\beta'} dA + \int_{\Sigma'} \Gamma_A' \kappa' \delta y^{\beta'} dA \end{aligned} \tag{7.18}$$

$$\begin{aligned} \delta\mathcal{N}_B = &\int_{V^{\alpha'}} \delta\rho_B^{\alpha'} dV + \int_{V^{\beta'}} \delta\rho_B^{\beta'} dV + \int_{\Sigma'} \delta\Gamma_B' dA + \int_{\Sigma'} \rho_B^{\alpha'} \delta y^{\alpha'} dA \\ &+ \int_{\Sigma'} \rho_B^{\beta'} \delta y^{\beta'} dA + \int_{\Sigma'} \Gamma_B' \kappa' \delta y^{\beta'} dA. \end{aligned} \tag{7.19}$$

These constraints are included in the variation by minimizing a new energy, $\mathcal{E}^*$, such that

$$\mathcal{E}^* = \mathcal{E} - \theta_L \mathcal{S} - \lambda_A \mathcal{N}_A - \lambda_B \mathcal{N}_B \tag{7.20}$$

where θ_L, λ_A, and λ_B are Lagrange multipliers for the entropy constraint and for the mass constraints for components A and B, respectively. The first variation of $\mathcal{E}^*$ is

$$\delta\mathcal{E}^* = \delta\mathcal{E} - \theta_L \delta\mathcal{S} - \lambda_A \delta\mathcal{N}_A - \lambda_B \delta\mathcal{N}_B. \tag{7.21}$$

THE THERMODYNAMICS OF ELASTICALLY STRESSED CRYSTALS

Using Eq. (7.14) and Eqs. (7.17)–(7.19) in Eq. (7.21), and setting $\delta \mathcal{E}^* = 0$, yields

$$0 = \int_{V^{\alpha'}} \{(\theta^\alpha - \theta_L)\delta s^\alpha_{v'} + (M^\alpha_{AV} - \lambda_A)\delta \rho^{\alpha'}_A + (M^\alpha_{BV} - \lambda_B)\delta \rho^{\alpha'}_B + T^\alpha_{ji}\delta F^\alpha_{ij}\}dV$$

$$+ \int_{V^{\beta'}} \{(\theta^\beta - \theta_L)\delta s^\beta_{v'} + (M^\beta_{AV} - \lambda_A)\delta \rho^{\beta'}_A + (M^\beta_{BV} - \lambda_B)\delta \rho^{\beta'}_B$$

$$+ T^\beta_{ji}\delta F^\beta_{ij}\}dV$$

$$+ \int_{\Sigma'} \{(\theta^\Sigma - \theta_L)\delta s^\Sigma_{a'} + (M^\Sigma_{AV} - \lambda_A)\delta \Gamma'_A + (M^\Sigma_{BV} - \lambda_B)\delta \Gamma'_B + K_1 \delta \kappa'_1$$

$$+ K_2 \delta \kappa'_2\}dA$$

$$+ \int_{\Sigma'} \left(e^\alpha_{v'} - \theta_L s^\alpha_{v'} - \lambda_A \rho^{\alpha'}_A - \lambda_B \rho^{\alpha'}_B\right)\delta y^{\alpha'} dA$$

$$+ \int_{\Sigma'} \left(e^\beta_{v'} - \theta_L s^\beta_{v'} - \lambda_A \rho^{\beta'}_A - \lambda_B \rho^{\beta'}_B\right)\delta y^{\beta'} dA$$

$$+ \int_{\Sigma'} \left(e^\Sigma_{a'} - \theta_L s^\Sigma_{a'} - \lambda_A \Gamma'_A - \lambda_B \Gamma'_B\right) 2\kappa' \delta y^{\beta'} dA. \tag{7.22}$$

Not all variations appearing in Eq. (7.22) are independent. The term involving the first variation in the deformation gradient tensor can be written as the sum of a surface integral and a volume integral (see Eq. (4.26)). Substituting Eq. (4.26) into Eq. (7.22) gives

$$0 = \int_{V^{\alpha'}} \{(\theta^\alpha - \theta_L)\delta s^\alpha_{v'} + (M^\alpha_{AV} - \lambda_A)\delta \rho^{\alpha'}_A + (M^\alpha_{BV} - \lambda_B)\delta \rho^{\alpha'}_B - T^\alpha_{ji,j}\delta u^\alpha_i\}dV$$

$$+ \int_{V^{\beta'}} \{(\theta^\beta - \theta_L)\delta s^\beta_{v'} + (M^\beta_{AV} - \lambda_A)\delta \rho^{\beta'}_A + (M^\beta_{BV} - \lambda_B)\delta \rho^{\beta'}_B$$

$$- T^\beta_{ji,j}\delta u^\beta_i\}dV$$

$$+ \int_{\Sigma'} \{(\theta^\Sigma - \theta_L)\delta s^\Sigma_{a'} + (M^\Sigma_{AV} - \lambda_A)\delta \Gamma'_A + (M^\Sigma_{BV} - \lambda_B)\delta \Gamma'_B$$

$$+ K_1 \delta \kappa'_1 + K_2 \delta \kappa'_2\}dA$$

$$+ \int_{\Sigma'} \left(e^\alpha_{v'} - \theta_L s^\alpha_{v'} - \lambda_A \rho^{\alpha'}_A - \lambda_B \rho^{\alpha'}_B\right)\delta y^{\alpha'} dA$$

$$+ \int_{\Sigma'} \left(e^\beta_{v'} - \theta_L s^\beta_{v'} - \lambda_A \rho^{\beta'}_A - \lambda_B \rho^{\beta'}_B\right)\delta y^{\beta'} dA$$

$$+ \int_{\Sigma'} \left(e^\Sigma_{a'} - \theta_L s^\Sigma_{a'} - \lambda_A \Gamma'_A - \lambda_B \Gamma'_B\right) 2\kappa' \delta y^{\beta'} dA$$

$$+ \int_{\Sigma'} \left(T^\alpha_{ji} n^{\alpha'}_j \delta u^\alpha_i + T^\beta_{ji} n^{\beta'}_j \delta u^\beta_i\right)dA. \tag{7.23}$$

The variations appearing in the first three integrals of Eq. (7.23) are now independent. An extremum in the energy exists when

$$\theta^\alpha = \theta^\Sigma = \theta^\beta = \theta_L$$
$$M^\alpha_{AV} = M^\Sigma_{AV} = M^\beta_{AV} = \lambda_A$$
$$M^\alpha_{BV} = M^\Sigma_{BV} = M^\beta_{AV} = \lambda_B \quad (7.24)$$
$$T^\alpha_{ji,j} = 0$$
$$T^\beta_{ji,j} = 0.$$

The first equation for constant temperature is the condition for thermal equilibrium in the system. The diffusion potentials are constant within both phases and equal to the chemical potential for that component at the interface. Finally, as was found previously, mechanical equilibrium requires the divergence of the stress tensor in both the α and β phases to be zero in the absence of body forces.

Eliminating the remaining Lagrange multipliers using Eq. (7.24), and choosing the local position of the dividing surface such that the terms involving the curvature dependence of the interfacial energy are of second order, as was done for the crystal–fluid system, the variations that involve the crystal interface become

$$0 = \int_{\Sigma'} \omega^\alpha_{v'} \delta y^{\alpha'} dA + \int_{\Sigma'} \omega^\beta_{v'} \delta y^{\beta'} dA + \int_{\Sigma'} 2\gamma' \kappa' \delta y^{\beta'} dA$$
$$+ \int_{\Sigma'} \left(T^\alpha_{ji} n^{\alpha'}_j \delta u^\alpha_i + T^\beta_{ji} n^{\beta'}_j \delta u^\beta_i \right) dA \quad (7.25)$$

where $\omega_{v'}$ is the grand canonical free energy per volume of the noted phase in the reference state, and

$$\omega_{v'} = e_{v'} - \theta s_{v'} - M_{AV} \rho'_A - M_{BV} \rho'_B. \quad (7.26)$$

The excess grand canonical free energy density associated with the interface, the interfacial energy density γ', is defined as

$$\gamma' = e^\Sigma_{a'} - \theta^\sigma s^\Sigma_{a'} - M^\Sigma_{AV} \Gamma'_A - M^\Sigma_{BV} \Gamma'_B. \quad (7.27)$$

The variations in the displacement of the interface are related by the condition that the two-phase crystal must remain coherent during the variation. Thus, as shown in Appendix B,

$$\delta u^\alpha_i = \delta u^\beta_i + \left(F^\beta_{ij} - F^\alpha_{ij} \right) n^{\beta'}_j \delta y^{\beta'}. \quad (7.28)$$

Substituting Eq. (7.28) into Eq. (7.25), and using the referential accretion condition for a coherent interface, $\delta y^{\alpha'} = -\delta y^{\beta'}$, gives

$$0 = \int_{\Sigma'} \left(\omega_{v'}^{\beta} - \omega_{v'}^{\alpha} + T_{ji}^{\alpha} n_j^{\alpha'} \left(F_{ik}^{\beta} - F_{ik}^{\alpha} \right) n_k^{\beta'} + 2\gamma' \kappa' \right) \delta y^{\beta'} dA$$
$$+ \int_{\Sigma'} \left(T_{ji}^{\alpha} n_j^{\alpha'} + T_{ji}^{\beta} n_j^{\beta} \right) \delta u_i^{\alpha} dA. \tag{7.29}$$

The variations are now independent and an extremum in the energy requires the following interfacial conditions to be satisfied:

$$T_{ji}^{\alpha} n_j^{\alpha'} + T_{ji}^{\beta} n_j^{\beta} = 0 \quad \text{or} \quad \left(T_{ji}^{\alpha} - T_{ji}^{\beta} \right) n_j^{\alpha'} = 0 \tag{7.30}$$

and

$$\omega_{v'}^{\beta} - \omega_{v'}^{\alpha} - T_{ji}^{\beta} n_j^{\beta'} \left(F_{ik}^{\beta} - F_{ik}^{\alpha} \right) n_k^{\beta'} + 2\gamma' \kappa' = 0 \tag{7.31}$$

where we have used Eq. (7.30) in Eq. (7.29) to yield Eq. (7.31).

Unlike the crystal-fluid case, the chemical equilibrium condition, Eq. (7.31), involves a jump in the deformation gradient at the interface. This reflects the energy required to keep the interface coherent upon accretion. When the interface is incoherent, this term also does not appear.[59,79] Equation (7.30) shows that the jump in the normal stress at the interface is zero, for the assumption that the interfacial stress vanishes. If the interfacial stress is nonzero, a term involving the surface divergence of the interfacial stress appears in Eq. (7.31).

c. Equilibrium Conditions in the Small Strain Limit

These interfacial conditions can be written in a more familiar form in the limit of small strain. In the small-strain limit, Eqs. (7.30) and (7.31) become, respectively,

$$\left(\sigma_{ij}^{\alpha} - \sigma_{ij}^{\beta} \right) n_j^{\alpha} = 0 \tag{7.32}$$

and

$$\omega_v^{\beta} - \omega_v^{\alpha} - \sigma_{ij}^{\beta} \left(\epsilon_{ij}^{\beta} - \epsilon_{ij}^{\alpha} \right) + 2\gamma \kappa = 0 \tag{7.33}$$

where the prime superscripts denoting the referential state have been dropped, as the difference in the magnitudes of the primed and unprimed densities are second order in the strain. These equations are supplemented by the conditions that follow from the bulk equilibrium conditions:

$$M_{AV}^{\alpha} = M_{AV}^{\Sigma} = M_{AV}^{\beta}$$
$$M_{BV}^{\alpha} = M_{BV}^{\Sigma} = M_{BV}^{\beta} \tag{7.34}$$
$$\theta^{\alpha} = \theta^{\Sigma} = \theta^{\beta}.$$

If the energy of the system is unaffected by vacancies, but they are present to mediate mass flow in the crystal, the interfacial conditions as given by Eq. (7.32) and Eq. (7.33) are unchanged, but the bulk equilibrium conditions for the diffusion potential are given by

$$M_{AB}^{\alpha} = M_{AB}^{\Sigma} = M_{AB}^{\beta} \tag{7.35}$$

and the definition of ω_v becomes

$$\omega_v = e_v - \theta s_v - M_{AB}\rho_A. \tag{7.36}$$

This alternate formulation is particularly useful in systems where the energy of vacancies is negligible, there is no need to determine the diffusion potential of a vacancy, or the intrinsic diffusivities of the mass components are similar.

In order to apply Eq. (7.33), it is necessary to express ω_v in terms of the stress state, the diffusion potentials, and the pressure. First consider the case of an alloy with substitutional components A, B, and vacancies and assume that the energy of the system is a function of the vacancy concentration. Taking the reference state for strain to be hydrostatically stressed α and β phases at a pressure P^o, and using Eq. (6.42) in Eq. (7.33), yields

$$0 = \rho_o\left(\mu_V^{v\beta} - \mu_V^{v\alpha}\right) + \frac{1}{2}[\![S_{ijkl}\sigma_{ij}\sigma_{kl}]\!] + \frac{1}{2}[\![S_{iikk}]\!](P^o)^2 + [\![\rho_A(M_{AV}^o - M_{AV})]\!]$$
$$+ [\![\rho_B(M_{BV}^o - M_{BV})]\!] - \sigma_{ij}^{\beta}[\![\epsilon_{ij}]\!] + 2\gamma\kappa \tag{7.37}$$

where, for a quantity ϕ at the interface, $[\![\phi]\!] = \phi^{\beta} - \phi^{\alpha}$, M_{AV}^o is the diffusion potential of A in the referential state, and we have used the definition of the chemical potential μ_V^v. In most cases the compressibility of the crystal is sufficiently large that, for pressures of interest in many materials applications, on the order of atmospheric pressure or less, the term $S_{iikk}(P^o)^2$ is small compared to the other terms and Eq. (7.37) simplifies to

$$0 = \rho_o\left(\mu_V^{v\beta} - \mu_V^{v\alpha}\right) + \frac{1}{2}[\![S_{ijkl}\sigma_{ij}\sigma_{kl}]\!] + [\![\rho_A(M_{AV}^o - M_{AV})]\!]$$
$$+ [\![\rho_B(M_{BV}^o - M_{BV})]\!] - \sigma_{ij}^{\beta}[\![\epsilon_{ij}]\!] + 2\gamma\kappa. \tag{7.38}$$

At the crystal–fluid interface, the condition on the grand canonical free energies sets the value of the chemical potential at the interface. For the crystal–crystal interface, however, Eq. (7.38) shows that the equilibrium condition on the grand canonical free energies only sets the jump in the chemical potential of vacancies at the interface.

When vacancies are absent from the system, or when they are present but do not affect the thermodynamic description of the crystal, the equilibrium condition containing the grand canonical free energy density is not given by Eq. (7.38).

Following the derivation leading to Eq. (6.42) without allowing for vacancies yields

$$\omega_v = -P^o + \rho_o \mu_B^v + \frac{1}{2} S_{ijkl} \sigma_{ij} \sigma_{kl} + \frac{1}{2} S_{iikk}(P^o)^2 + \rho_A(M_{AB}^o - M_{AB}). \quad (7.39)$$

In the limit where the pressure in the hydrostatic state is the same for both phases and the elastic strain energy due to this pressure is small, substitution of Eq. (7.39) into Eq. (7.38) yields

$$\rho_o(\mu_B^{v\beta} - \mu_B^{v\alpha}) + \frac{1}{2}[\![S_{ijkl}\sigma_{ij}\sigma_{kl}]\!] + [\![\rho_A(M_{AB}^o - M_{AB})]\!] - \sigma_{ij}^\beta[\![\epsilon_{ij}]\!] + 2\gamma\kappa = 0. \quad (7.40)$$

In the small strain limit without vacancies, the condition for the jump in the grand canonical free energies at the interface defines the jump in the chemical potentials of species B rather than the jump in the chemical potential of vacancies.

24. APPLICATIONS OF EQUILIBRIUM CONDITIONS

We now illustrate the implications of the thermodynamic equilibrium conditions obtained for two-phase crystals by considering three examples. The first example details the derivation of the Gibbs-Thompson equation for determining the phase compositions at a curved, coherent interface in a system under stress. As a second example, we discuss some of the characteristics of thermodynamic equilibrium in two-phase coherent crystals under stress, including the construction of phase and field diagrams, the applicability of the common-tangent construction, and the existence of a phase rule. Finally, we show how the surface and elastic energies of a coherent, misfitting particle influence the equilibrium shape of the particle and give rise to elastically induced shape bifurcations with increasing particle size.

a. The Gibbs-Thompson Equation

In the absence of stress, the equilibrium composition of a phase at an interface is a function of the local interfacial curvature. This relationship, known as the Gibbs-Thompson equation, plays an important role in defining the boundary conditions for many models of phase transformation processes when local thermodynamic equilibrium at the interface can be assumed. In this section, we use the conditions for thermodynamic equilibrium in a two-phase coherent crystal just developed to derive the Gibbs-Thompson equation for a stressed, binary system containing vacancies. We treat two situations. The first is a system for which the vacancies must be considered explicitly, while the second is the often useful limiting case in which the vacancies can be ignored.

We assume a coherent crystal composed of substitutional species A and B and vacancies V. The stresses can result from externally applied tractions or internal misfit strains. We assume that the lattice parameters and elastic constants of both phases are independent of composition. This implies that concentration gradients in the two phases do not generate stress. If this were not the case, then the concentrations at the interface become a function of the stress generated by the nonuniform composition fields in both phases and thus a functional of the composition field itself. This long-range coupling between the compositions at an interface and the entire composition field in the crystal can be quite important and has been investigated in other contexts (see[77,81–85]).

There are two unknown compositions for each phase at the interface, and four equations are required to determine uniquely the four equilibrium compositions. However, there are only three equilibrium conditions applicable to the interface. Two equations are obtained from the chemical equilibrium conditions corresponding to the equality of the diffusion potentials between phases at the interface and one equation is obtained from the condition on thermodynamic equilibrium relating to the phase transformation. This set of equations is under determined, and we expect only to be able to establish relationships between the interfacial compositions, and not their precise values, without additional information on the system being supplied. In order to establish expressions between the equilibrium compositions, we begin with the interfacial condition on the equality of the diffusion potentials at the interface.

$$M_{AV}^\alpha = M_{AV}^\beta \quad \text{and} \quad M_{BV}^\alpha = M_{BV}^\beta. \tag{7.41}$$

Each diffusion potential can be expressed in terms of the appropriate chemical potentials at the interface, because lattice sites can be added to either phase by a transfer process at the interface. Thus,

$$M_{AV} = \mu_A^v - \mu_V^v \quad \text{and} \quad M_{BV} = \mu_B^v - \mu_V^v. \tag{7.42}$$

Substituting Eqs. (7.42) into Eqs. (7.41) and rearranging yields the following relationship between the chemical potentials at the interface:

$$\begin{aligned}\mu_A^{v\alpha} &= \mu_A^{v\beta} + \mu_V^{v\alpha} - \mu_V^{v\beta} \\ \mu_B^{v\alpha} &= \mu_B^{v\beta} + \mu_V^{v\alpha} - \mu_V^{v\beta}\end{aligned} \tag{7.43}$$

where Eqs. (7.43) must be satisfied point-to-point along the interface.

---

[81] P. W. Voorhees and W. C. Johnson, *J. Chem. Phys.* **84**, 5108 (1986).
[82] W. C. Johnson and P. W. Voorhees, *Metall. Trans.* **16A**, 337 (1985).
[83] F. C. Larché and J. W. Cahn, *Acta metall.* **30**, 1835 (1982).
[84] F. C. Larché and J. W. Cahn, *Acta metall.* **40**, 947 (1992).
[85] J. W. Cahn and R. Kobayashi, *Acta Metall. Mater.* **43**, 931 (1995).

The third equation for determining the phase compositions at the interface is obtained from the condition for phase equilibrium, Eq. (7.38). Because the lattice parameters and elastic constants are taken to be independent of the composition, the diffusion potentials in the actual state are equal to those in the referential state (see Eq. (3.111)). Thus $M_{AV}^o = M_{AV}$ and $M_{BV}^o = M_{BV}$. Solving Eq. (7.38) for the jump in the chemical potential of vacancies, $\mu_V^{v\alpha} - \mu_V^{v\beta}$, using the condition on the equality of the reference state and actual diffusion potentials, and substituting the result into Eqs. (7.43) yields three equations for the four unknown compositions at the interface, as follows:

$$\mu_A^{v\alpha} = \mu_A^{v\beta} + \frac{1}{\rho_o}\left(\frac{1}{2}[\![S_{ijkl}\sigma_{ij}\sigma_{kl}]\!] - \sigma_{ij}^{\beta}[\![\epsilon_{ij}]\!] + 2\gamma\kappa\right)$$

$$\mu_B^{v\alpha} = \mu_B^{v\beta} + \frac{1}{\rho_o}\left(\frac{1}{2}[\![S_{ijkl}\sigma_{ij}\sigma_{kl}]\!] - \sigma_{ij}^{\beta}[\![\epsilon_{ij}]\!] + 2\gamma\kappa\right) \qquad (7.44)$$

$$\mu_V^{v\alpha} = \mu_V^{v\beta} + \frac{1}{\rho_o}\left(\frac{1}{2}[\![S_{ijkl}\sigma_{ij}\sigma_{kl}]\!] - \sigma_{ij}^{\beta}[\![\epsilon_{ij}]\!] + 2\gamma\kappa\right).$$

Equations (7.44) define the jump in each of the chemical potentials across the interface. The chemical potentials of a phase are functions of any two of the three phase compositions, C_A, C_B, and C_V. There are four unknown compositions, two compositions in each phase, to determine using the three equilibrium equations. Thus the equilibrium conditions of Eq. (7.44) are insufficient to set the interfacial compositions. The fourth equation necessary to determine the four unknown compositions is supplied by a mass conservation condition for systems in global equilibrium or, for a system undergoing a phase transformation, by a boundary condition that follows from the description of diffusion in the local equilibrium approximation. This result is identical to that encountered in establishing the equilibrium compositions at a two-phase interface in a ternary fluid system where there are only three equilibrium conditions on the equality of the chemical potentials across the interface that can be used to establish the four interfacial compositions.[86]

In conditions where the presence of vacancies can be neglected, analytic approximations to the composition at the interface can be derived. A similar situation holds when the composition of vacancies in each phase is very small and, to a good approximation, the chemical potential of the majority components can be taken to be independent of the vacancy composition. Under these conditions, the chemical potentials of the mass components are functions of only one composition (C_A or C_B) and only two equilibrium conditions are needed to establish the equilibrium interfacial compositions.

---

[86] J. S. Kirkaldy and D. J. Young, *Diffusion in the Condensed State*, Institute of Metals, London (1987).

When the chemical potentials are assumed independent of the vacancy composition, the first two equilibrium conditions of Eq. (7.44) can be employed to determine the equilibrium interfacial compositions. We first show this assumption is equivalent to ignoring vacancies altogether by using the interfacial equilibrium conditions given by Eqs. (7.35) and (7.40) to determine the interfacial compositions, C_A^α and C_A^β. Retaining the assumption that the lattice parameter of each phase is independent of composition, Eq. (7.40) can be rewritten as

$$\mu_B^{v\alpha} = \mu_B^{v\beta} + \frac{1}{\rho_o}\left(\frac{1}{2}[\![S_{ijkl}\sigma_{ij}\sigma_{kl}]\!] - \sigma_{ij}^\beta[\![\epsilon_{ij}]\!] + 2\gamma\kappa\right). \tag{7.45}$$

The condition on the equality of the diffusion potential at the interface can be written in terms of the chemical potentials as

$$\mu_A^{v\beta} - \mu_B^{v\beta} = \mu_A^{v\alpha} - \mu_B^{v\alpha}. \tag{7.46}$$

Substituting Eq. (7.45) into Eq. (7.46) yields

$$\mu_A^{v\alpha} = \mu_A^{v\beta} + \frac{1}{\rho_o}\left(\frac{1}{2}[\![S_{ijkl}\sigma_{ij}\sigma_{kl}]\!] - \sigma_{ij}^\beta[\![\epsilon_{ij}]\!] + 2\gamma\kappa\right). \tag{7.47}$$

Eq. (7.45) and Eq. (7.47) are identical to the first two of Eqs. (7.44), showing that the two approaches are consistent in the limit where the vacancy composition is extremely small.

In order to obtain analytic approximations to C_A^α and C_A^β, the chemical potentials are expanded to first order in the deviation of the composition from a prescribed equilibrium composition. Because we are interested in the effects of stress and curvature on the interfacial compositions, we take a stress-free, two-phase system with a planar interface to be the reference state. The chemical potential in this state is designated $\mu_A^{v\alpha e}$ and corresponds to a composition $C_A^{\alpha e}$. In this reference state, $\mu_A^{v\alpha e} = \mu_A^{v\beta e}$. Using Eq. (6.77) to expand the chemical potentials in terms of the composition C_A to first order about C_A^e under conditions of zero stress yields

$$\begin{aligned}\mu_A^v &= \mu_A^{ve} + (1 - C_A^e)f_{C_AC_A}^o(C_A - C_A^e) \\ \mu_B^v &= \mu_B^{ve} - C_A^e f_{C_AC_A}^o(C_A - C_A^e)\end{aligned} \tag{7.48}$$

where $f_{C_AC_A}^o$ is the second derivative of the Helmholtz free energy per atom with respect to composition C_A. In the reference state there is an equality of the chemical potentials at the planar interface giving $\mu_A^{v\alpha e} = \mu_A^{v\beta e}$ and $\mu_B^{v\alpha e} = \mu_B^{v\beta e}$. To obtain the equilibrium compositions, Eq. (7.48) is first used to linearize the chemical potentials appearing in the first two expressions of Eq. (7.44). The resulting two linear equations are then solved simultaneously to obtain the following approximations for the equilibrium compositions at a coherent interface in terms of the stress state and local curvature:

$$C_A^\alpha - C_A^{\alpha e} = \frac{[\![S_{ijkl}\sigma_{ij}\sigma_{kl}]\!]/2 - \sigma_{ij}^\beta [\![\epsilon_{ij}]\!] + 2\gamma\kappa}{\rho_o(C_A^{\beta e} - C_A^{\alpha e})f_{C_A C_A}^{o\alpha e}}$$

$$C_A^\beta - C_A^{\beta e} = \frac{[\![S_{ijkl}\sigma_{ij}\sigma_{kl}]\!]/2 - \sigma_{ij}^\beta [\![\epsilon_{ij}]\!] + 2\gamma\kappa}{\rho_o(C_A^{\beta e} - C_A^{\alpha e})f_{C_A C_A}^{o\beta e}}$$

(7.49)

where the free energy derivatives are evaluated for a stress-free phase at the reference state composition. The interfacial compositions are a function of both the local stress state and the interfacial energy. The compositions are usually a function of position along the interface for an evolving system. These conditions have been employed to determine the interfacial compositions in a wide array of problems ranging from Ostwald ripening in stressed solids[5,87–91] to morphological stability of interfaces.[92]

In order to obtain a better qualitative understanding of the effects of stress and interfacial energy on the equilibrium composition at an interface, we nondimensionalize Eq. (7.49). The stress is scaled by a characteristic stress σ_o where σ_o is an estimate of the magnitude of the stress in the system. For systems under applied stress it would be the applied stress, for a system with misfitting particles $\sigma_o = C\epsilon^T$ where ϵ^T is the magnitude of the misfit strain and C is an appropriate elastic constant. The jump in strains at the interface appearing in Eq. (7.49) can also be expressed in terms of stress as $[\![\epsilon_{ij}]\!] = [\![S_{ijkl}\sigma_{kl}]\!]$. The curvature is scaled by a characteristic length ℓ. ℓ is chosen such that the scaled curvature is $O(1)$. The appropriate choice of ℓ depends on the problem studied. For a particle growing into a supersaturate matrix, one can use $\ell = V^{1/3}$, where V is the volume of a particle. For an interfacial stability problem, ℓ can be chosen as the wavelength of a perturbation with zero growth rate. Using these scalings in Eq. (7.49) yields for the α phase,

$$C_A^\alpha - C_A^{\alpha e} = \frac{L\{[\![\tilde{S}_{ijkl}\tilde{\sigma}_{ij}\tilde{\sigma}_{kl}]\!]/2 - \tilde{\sigma}_{ij}^\beta [\![\tilde{S}_{ijkl}\tilde{\sigma}_{ij}]\!]\} + 2\tilde{\kappa}}{(C_A^{\beta e} - C_A^{\alpha e})\tilde{f}_{C_A C_A}^{o\alpha e}}$$

(7.50)

where the tilde denotes a dimensionless quantity, $\tilde{S}_{ijkl} = S_{ijkl}/S$, $S = 1/C$, and $\tilde{f}_{C_A C_A}^{o\alpha e} = f_{C_A C_A}^{o\alpha e}\rho_o\ell/\sigma_o$. Most importantly when writing the Gibbs-Thompson

---

[87] P. W. Voorhees, G. B. McFadden, and W. C. Johnson, *Acta metall. mater.* **40**, 2979 (1992).
[88] C. H. Su and P. W. Voorhees, *Acta mater.* **44**, 2001 (1996).
[89] N. Akaiwa, K. Thornton, and P. W. Voorhees, *J. Comput. Phys.* **173**, 61 (2001).
[90] K. Thornton, N. Akaiwa, and P. W. Voorhees, *Phys. Rev. Lett.* **86**, 1259 (2001).
[91] P. H. Leo, W. W. Mullins, R. F. Sekerka, and J. Vinals, *Acta metall. et mater.* **38**, 1573 (1990).
[92] P. H. Leo and R. F. Sekerka, *Acta metall.* **37**, 3139 (1989).

equation in this dimensionless form, the dimensionless parameter L appears as,[93]

$$L = \frac{S\sigma_o^2 \ell}{\gamma}. \tag{7.51}$$

When the numerator and denominator of Eq. (7.51) are multiplied by ℓ^2, it can be seen that L is the ratio of the characteristic elastic and interfacial energies. The parameter L sets the magnitude of the elastic energy compared to that of the surface energy. If L is $O(1)$ then the two energies are comparable and the elastic stress can play a major role in setting the concentrations at the interface and, thus, plays a strong role in the evolution of the interface or the shape of a particle. In systems with coherent interfaces, γ can be small and it is not uncommon for L to be greater than 1. A second important insight to be gained from Eq. (7.51) is that, since $L \sim \ell$, the influence of stress on the equilibrium composition scales with the characteristic size. For example, consider a system with misfitting precipitates in a matrix. When the precipitates are sufficiently small, $L \ll 1$, stress would be expected to have little effect on the morphology, growth, and/or coarsening behavior of the precipitates. However, as the precipitates grow, L eventually exceeds one, and stresses can be expected to play a more dominant role in setting particle morphologies and growth kinetics. This size-dependent interplay between interfacial energy and elastic stress is common to a vast array of problems, and a parameter similar to L appears in many contexts.

b. Phase Equilibria: Two-phase Coherent Systems

During many diffusional phase transformations, the atoms comprising the crystal rearrange themselves on the lattice to form new phases of different composition. During this process, the lattice often remains intact, even though it is elastically distorted when the partial molar volumes (strains) of the component species differ. If the lattice remains intact and the individual lattice sites can be associated with the sites in a perfect lattice (or reference state) then the phase transformation is coherent and the equilibrium that is established in this case is termed coherent equilibrium.[94] Coherent equilibrium is not restricted to systems in which the two phases possess the same crystal structure. For example, a diffusional transformation can result in a coherent tetragonal or orthorhombic precipitate in a cubic matrix.

Just as a change in pressure will alter the relative stability of two phases and thereby affect the equilibrium compositions and volume fraction in a multiphase system, the application of an applied stress or the presence of misfit strains will

---

[93] M. E. Thompson and P. W. Voorhees, *Acta metall. mater.* **47**, 983 (1999).
[94] J. W. Cahn, *Acta metall.* **10**, 179 (1962).

influence phase equilibria in multiphase crystals.[95–97] Sometimes the changes induced by elastic deformation are qualitatively similar to those induced by a pressure, such as when the phases are constrained to be epitaxial thin films on a thick or rigid substrate.[94,98,99] In other cases, qualitatively different phenomena are observed, including the existence of more than one thermodynamically stable state and discontinuous jumps in the equilibrium phase fraction with a continuous change in temperature or bulk composition.[100–106]

The difference in the predicted behavior between elastically stressed coherent systems and incoherent (or fluid) systems is thought to be a result of the long-range elastic interaction between phases in a coherent system. This elastic interaction renders the state of deformation and, therefore, the thermodynamic state of one phase dependent on material properties and volume fraction of the other phases present in the system.[107] In addition, the thermodynamic state of a phase will depend on the morphology and spatial distribution of the coexisting phases (see Figure 26). This is in contrast to a fluid system subjected to a pressure in that the pressure in each of the phases and individual phase domains at equilibrium, in the absence of capillarity effects, is equal to the applied pressure and is independent of the domain morphology.[20]

In this section, the conditions for thermodynamic equilibrium will be used to illustrate some of the qualitative changes in phase equilibria and phase diagrams engendered by elastic stress. In order to do so, it is important first to define carefully what is meant by a phase and to classify appropriately the various thermodynamic variables.

a. Thermodynamic Relationships

Thermodynamic Fields and Densities. Following Griffiths and Wheeler,[108] thermodynamic quantities can be divided into two categories, thermodynamic

---

[95] A. L. Roytburd, *J. Appl. Phys.* **83**, 228 (1998).
[96] A. L. Roytburd, *J. Appl. Phys.* **83**, 239 (1998).
[97] S. P. Alpay and A. L. Roytburd, *J. Appl. Phys.* **83**, 4714 (1998).
[98] W. C. Johnson and C.-S. Chiang, *J. Appl. Phys.* **64**, 1155 (1988).
[99] A. A. Mbaye, D. M. Wood, and A. Zunger, *Phys. Rev. B* **37**, 3008 (1988).
[100] R. O. Williams, *Metall. Trans.* **11A**, 247 (1980).
[101] R. O. Williams, *CALPHAD* **8**, 1 (1984).
[102] A. L. Roitburd, *Sov. Phys. Sol. State* **25**, 17 (1983).
[103] A. L. Roitburd, *Sov. Phys. Sol. St.* **26**, 1229 (1984).
[104] W. C. Johnson and P. W. Voorhees, *Metall. Mater. Trans.* **18A**, 1213 (1987).
[105] W. C. Johnson, in *Mater. Res. Soc. Symp. Proc.*, vol. 103, eds. F. S. T W Barbee and L. Greer (Materials Research Society, 1987), 61.
[106] A. L. Roytburd and J. Slutsker, *Mater. Sci. Engr.* **A238**, 23 (1997).
[107] W. C. Johnson and W. H. Mueller, *Acta metall. mater.* **89**, 1991 (1991).
[108] R. B. Griffiths and J. W. Wheeler, *Phys. Rev. A* **2**, 1047 (1970).

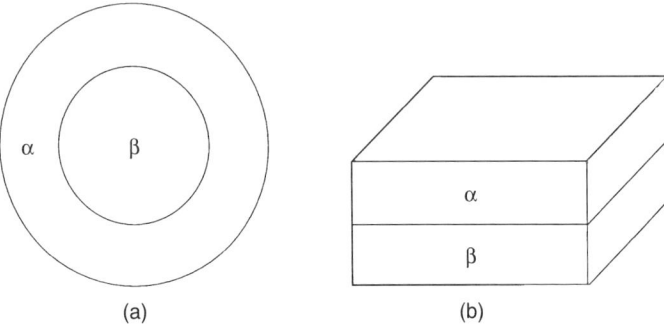

FIG. 26. Two idealized system geometries are depicted that allow the conditions for thermodynamic equilibrium to be satisfied simultaneously. In (a) the α and β phases appear as concentric spheres with the stress state of the α phase being a function of position. The elastic constants of both phases must be isotropic in this configuration. In (b) the phases appear as parallel plates. If the system is not allowed to bend, the stress state will be uniform in each phase. If no external stresses are applied and the system (alloy) composition and temperatures are the same, the equilibrium compositions and volume fractions could differ between the geometries because of the different stress states.

fields and *densities*. Thermodynamic fields have the property that they assume identical values within and between all homogeneous phases at equilibrium. As such, they are equivalent to thermodynamic intensive variables. Common examples of thermodynamic fields in unstressed systems include the temperature and chemical potentials. Thermodynamic densities are those thermodynamic variables that, at equilibrium, are constant throughout a given homogeneous phase but do not assume identical values in the different phases at equilibrium. In addition, thermodynamic densities must remain unchanged by the combination of two identical systems. The entropy per unit volume or the phase compositions are examples of thermodynamic densities.

In addition to the thermodynamic fields and densities, we consider externally controlled variables. These variables are problem dependent and are fixed by experiment. The externally controlled variables are usually a combination of thermodynamic fields and thermodynamic densities. Because thermodynamic densities are not equal between phases at equilibrium, we will refer to thermodynamic densities that relate to the system in its entirety as system densities. A system density commonly employed as an externally controlled variable is the alloy composition. For example, in a single-component fluid system that reaches equilibrium under conditions of imposed constant temperature and pressure, the externally controlled variables are the two thermodynamic fields, temperature and pressure. For a binary system at constant temperature, pressure, and composition, the externally controlled variables would correspond to two thermodynamic fields (temperature and pressure) and one system density (bulk composition). The correct recognition of the appropriate thermodynamic fields, thermodynamic densities, and externally controlled variables is necessary to the construction and understanding of phase diagrams in coherent solids.

A system with n independent variables can be characterized by $n + 1$ thermodynamic fields, h_i, $i = 0, 1, \ldots, n$.[20] One thermodynamic field can be treated as a function of the other n thermodynamic fields[109] and assumes equal values in coexisting, homogeneous phases at equilibrium. A particularly simple example is a one-component fluid for which the temperature and pressure can be varied. The chemical potential or molar Gibbs free energy assumes the role of the dependent thermodynamic field as it is equal within and between phases at equilibrium and is a function of the temperature and pressure. The dependent thermodynamic field is designated as h_0 or ϕ.[110] Any of the thermodynamic fields can be chosen as the dependent thermodynamic field ϕ. However, a particularly convenient choice can be identified from the condition of thermodynamic equilibrium that applies at a two-phase interface; i.e., that condition relating to the transformation of one phase into the other. In an unstressed or fluid system, this equilibrium condition is the equality of the pressures, $P^\alpha = P^\beta$, across the interface. This is equivalent to the equality of the density of the grand canonical or thermodynamic potential across the interface, $\omega_v^\alpha = \omega_v^\beta$, an energy per unit volume. The free energy and thermodynamic field, ω_v, is a function of the three independent thermodynamic fields θ, μ_A, and μ_B in the binary fluid, because $\omega_v = f_v - \mu_A \rho_A - \mu_v \rho_B$ and using the definition of f_v for a fluid, we have

$$-dP = d\omega_v = -s_v d\theta - \rho_A d\mu_A - \rho_B d\mu_B. \tag{7.52}$$

The thermodynamic densities are obtained by differentiation of ω_v with respect to the appropriate thermodynamic field.

$$s_v = -\left(\frac{\partial \omega_v}{\partial \theta}\right)_{\mu_A, \mu_B} \quad \rho_A = -\left(\frac{\partial \omega_v}{\partial \mu_A}\right)_{\theta, \mu_B} \quad \rho_B = -\left(\frac{\partial \omega_v}{\partial \mu_B}\right)_{\theta, \mu_A}. \tag{7.53}$$

In general, the n thermodynamic densities are given by

$$d_i = \frac{\partial \phi}{\partial h_i} \quad \text{for} \quad i = 1, \ldots, n \tag{7.54}$$

where the thermodynamic fields other than h_i are held constant in the differentiation of Eq. (7.54). Equation (7.52) is a fundamental equation and the thermodynamic densities are the conjugate thermodynamic variables to the thermodynamic fields.

---

[109] Because only n thermodynamic fields are independent, one of the thermodynamic fields must be dependent.
[110] As will be shown, the dependent thermodynamic field ϕ is often a density, but not a thermodynamic density, in that it has units of energy per unit volume. In such cases it will be identified by ϕ_v.

This choice of thermodynamic fields is particularly useful as ω is the free energy that is minimized at equilibrium for the fixed set of fields θ, μ_A, and μ_B.

Field Diagrams and Phase Diagrams. The n independent thermodynamic fields, h_i, can be considered to form an n-dimensional thermodynamic space, which shall be referred to as the thermodynamic field space. Two systems that occupy the same point in the thermodynamic field space are the same phase if the corresponding thermodynamic densities of the phases are equal and are different phases when at least one of the thermodynamic densities is different.[108] Two phases, α and β, cannot coexist unless they occupy the same point in the thermodynamic field space because, if they are in equilibrium, the following conditions for the independent thermodynamic fields must be satisfied:

$$h_i^\alpha = h_i^\beta. \tag{7.55}$$

The n equalities of Eq. (7.55), along with the equilibrium condition $\phi^\alpha = \phi^\beta$, define a hypersurface of dimension $n-1$ in the n-dimensional field space.[111] This hyper- or coexistence surface represents a first-order phase transformation between the two phases when at least one of the thermodynamic densities is discontinuous between the two phases.[108] Three-phase coexistence occurs along a hypersurface of dimension $n-2$ (the intersection of two hypersurfaces of dimension $n-1$). Hypersurfaces of dimension $n-2$ also result from the termination of a hypersurface of dimension $n-1$ and give a surface of critical points. The depiction of phase equilibria in the thermodynamic field space is referred to as a field-space diagram or, simply, a field diagram. The depiction of the phase boundaries in a thermodynamic space that is spanned by at least one of the thermodynamic densities is referred to as a phase diagram.

Phase Rule. The Gibbs phase rule relates the number of homogeneous phases present in a system to the number of thermodynamic fields that can be varied independently without changing the number of phases in the system. In the previous paragraph it was argued that, for a system with n independent thermodynamic fields, single-phase equilibria exists in a region of dimension n in the thermodynamic field space, two-phase equilibria is found along a hypersurface of dimension $n-1$ and three-phase equilibria exists along a hypersurface of dimension $n-2$. In general, the coexistence of p phases in an n-dimensional field space corresponds to a hypersurface of dimension $n-p+1$. Recognizing that a hypersurface of

---

[111] As an example, consider the temperature-pressure phase diagram of a single-component system. There are two independent thermodynamic fields (temperature and pressure) and the dependent field is the molar Gibbs free energy. In the two-dimensional space of temperature and pressure, single phases are found in areas (two dimensions), regions of two-phase coexistence along lines (one dimension), and three-phase coexistence at points.

dimension $n - p + 1$ allows $n - p + 1$ thermodynamic fields to be independently varied while still remaining on the hypersurface, the Gibbs phase rule follows directly as

$$f = n - p + 1 \qquad (7.56)$$

where f is the number of thermodynamic fields that can be changed independently leaving the number of phases in the system constant, and is referred to as the number of degrees of freedom. (In fluids, the thermodynamic fields are equivalent to the potentials.[112]) This development applies equally well to nonhydrostatically stressed coherent systems so long as the thermodynamic fields are correctly identified and the phases are homogeneously deformed. It is important to remember that the number of degrees of freedom in the system is different from the number of quantities that must be specified in order to define a system uniquely.[113]

If all the externally controlled thermodynamic variables are thermodynamic fields, then f is also the number of those externally controlled thermodynamic variables that can be changed independently, keeping the number of phases in the system constant. However, if one or more of the externally controlled variables is a thermodynamic density, the degrees of freedom in the system (given by Eq. (7.56)), might no longer be equal to the number of externally controlled thermodynamic variables that can be changed while leaving the number of phases present in the system fixed. This is not to say that the Gibbs phase rule (which applies to the fields) is no longer applicable, as it is valid regardless of the choice of the externally controlled variables, but rather that it is possible to change the value of a system density that is an externally controlled variable, holding the other fields fixed, without changing the number of phases present in the system. This is accomplished by a change in the phase fraction of the system as discussed following.

A simple example illustrating this point is a binary fluid system (assume the independent thermodynamic fields are θ, μ_B, and P, and the dependent thermodynamic field is μ_A) in which the externally controlled variables are chosen as θ, P, and c_o where c_o is the alloy (bulk) composition measured in terms of the mole fraction of component B. Within a single-phase field of a temperature-pressure-composition phase diagram, θ, P, and c_0 can all be changed independently without changing the number of phases (the number of degrees of freedom is equal to the number of independent thermodynamic fields). However, at a point within a two-phase field of the same phase diagram that is not contiguous to a phase boundary, the externally controlled variables θ, P, and c_0 can again all be changed independently leaving the two phases in equilibrium. If one of the thermodynamic field variables (θ or P) is changed, all the thermodynamic densities change, as

---

[112] Z.-K. Liu and J. Agren, *Acta Metall.* **38**, 561 (1990).
[113] A unique description of the system not only requires knowledge of which phases are present but also the relative amounts of each phase.

the phases would now occupy a new point in the field space. (In other words, a change in θ or P results in a change in the equilibrium composition, entropy density, and molar volume of each phase.) In contrast, if the alloy composition (a system density) is changed at constant θ and P and the two-phase system remains in equilibrium, the thermodynamic densities remain constant; i.e., the equilibrium composition, molar volume, and entropy density of each phase does not change. If this were not the case, the system would occupy a new point in the thermodynamic field space which, in general, would not permit two-phase equilibrium. Although the number of externally controlled variables that can be changed independently in the two-phase system, without changing the number of phases present in the system, is again equal to three, the same number as for the single-phase field, there are only two degrees of freedom for the two-phase system. These arguments are also applicable to systems with more than two phases and more than one density as externally controlled variables. The correct identification of the thermodynamic fields and externally controlled variables is important in discussing the applicability of the Gibbs phase rule to coherent systems.

Common-tangent Construction. The common-tangent construction[20,114] is a graphical means of satisfying the thermodynamic equilibrium conditions of systems in which one or more of the externally controlled variables is a system density. Implementing the common-tangent construction for coherent systems requires the appropriate thermodynamic free energy to be identified and plotted as a function of those densities that are experimentally controlled. Phase equilibria between two or more phases are possible when hyperplanes tangent to these surfaces coincide. The densities of the phases at equilibrium are established by the point of tangency of the hyperplane and the free energy curve. The common-tangent construction of coherent systems follows directly from the thermodynamic equilibrium conditions as shown following.

Assume that the externally controlled thermodynamic variables of a two-phase homogeneous system consist of $n - r$ thermodynamic fields h_i ($i = r+1, \ldots, n$) and r system densities D_i ($i = 1, \ldots, r$). The system densities are the weighted sums of the thermodynamic densities of the individual phases,

$$D_i = (1-z)d_i^\alpha + zd_i^\beta \quad i = 1, \ldots, n, \tag{7.57}$$

where z is the volume fraction of the β phase. If ϕ is the free energy that is extremized when all the thermodynamic fields are specified experimentally, the free energy extremized at equilibrium under the preceding experimental conditions,

---

[114] M. Hillert, *Int. Metal. Rev.* **30**, 45 (1985).

K, is

$$K = [(1-z)k_v^\alpha + zk_v^\beta]V \tag{7.58}$$

with

$$k_v = \phi_v - \sum_{i=1}^{r} d_i h_i. \tag{7.59}$$

V is the volume of the system, referred to an appropriate reference state, and k_v and ϕ_v are the volume density of the respective free energies.

When the $n - r$ thermodynamic fields h_i ($i = r+1, \ldots, n$) are held constant, the free energy density of each phase, k_v, can be plotted as a function of the r thermodynamic densities d_i ($i = 1, \ldots, r$). For simplicity, we denote a point in the space spanned by the r thermodynamic densities as $\mathbf{d}$. The components of the vector $\mathbf{d}$ are the r thermodynamic densities d_i ($i = 1, \ldots, r$). Thus $\mathbf{d} = (d_1, d_2, \ldots, d_r)$. If a similar vector $\mathbf{h}$ is introduced for the thermodynamic fields considered to be established by experiment, that is $\mathbf{h} = (h_{r+1}, h_{r+2}, \ldots, h_r)$, then the free energy density, k_v, is a function of $\mathbf{h}$ and $\mathbf{d}$: $k_v = k_v(\mathbf{d}, \mathbf{h})$. The free energy density, k_v, will be different for the α and β phases.

The equation of the hyperplane tangent to the surface of $k_v^\alpha(\mathbf{d}, \mathbf{h})$ at the specific point $\mathbf{d}^\alpha$ in the thermodynamic space spanned by the free energy k_v and the thermodynamic densities d_i ($i = 1, \ldots, r$) is

$$k_T^\alpha = k_v^\alpha(\mathbf{d}^\alpha, \mathbf{h}) + \sum_{i=1}^{r}(d_i - d_i^\alpha)\left(\frac{\partial k_v^\alpha}{\partial d_i}\right)_{\mathbf{d}=\mathbf{d}^\alpha} \tag{7.60}$$

where k_T^α designates the tangent plane to the α phase. The partial derivatives are evaluated at the point of tangency $\mathbf{d} = \mathbf{d}^\alpha$.

If a common-tangent construction exists for identifying the equilibrium thermodynamic densities of the phases in equilibrium, then the tangent plane to the α phase at point $\mathbf{d}^\alpha$ must coincide with the tangent plane to the β phase at the point of tangency $\mathbf{d}^\beta$. This requires

$$k_T^\alpha = k_T^\beta. \tag{7.61}$$

Using Eq. (7.60), Eq. (7.61) can be written

$$k_v^\alpha(\mathbf{d}^\alpha, \mathbf{h}) + \sum_{i=1}^{r}(d_i - d_i^\alpha)\left(\frac{\partial k_v^\alpha}{\partial d_i}\right)_{\mathbf{d}=\mathbf{d}^\alpha}$$
$$= k_v^\beta(\mathbf{d}^\beta, \mathbf{h}) + \sum_{i=1}^{r}(d_i - d_i^\beta)\left(\frac{\partial k_v^\beta}{\partial d_i}\right)_{\mathbf{d}=\mathbf{d}^\beta}. \tag{7.62}$$

Rearranging Eq. (7.62) gives

$$0 = k_v^\alpha(\mathbf{d}, \mathbf{h}) - \sum_{i=1}^{r} d_i^\alpha \left(\frac{\partial k_v^\alpha}{\partial d_i}\right)_{\mathbf{d}=\mathbf{d}^\alpha} - k_v^\beta(\mathbf{d}, \mathbf{h}) + \sum_{i=1}^{r} d_i^\beta \left(\frac{\partial k_v^\beta}{\partial d_i}\right)_{\mathbf{d}=\mathbf{d}^\beta}$$
$$+ \sum_{i=1}^{r} d_i \left[\left(\frac{\partial k_v^\alpha}{\partial d_i}\right)_{\mathbf{d}=\mathbf{d}^\alpha} - \left(\frac{\partial k_v^\beta}{\partial d_i}\right)_{\mathbf{d}=\mathbf{d}^\beta}\right]. \quad (7.63)$$

Equation (7.63) is satisfied identically for all $\mathbf{d}$ (that is, arbitrary d_i) only when each term appearing in square brackets is identically zero,

$$\left(\frac{\partial k_v^\alpha}{\partial d_i}\right)_{\mathbf{d}=\mathbf{d}^\alpha} = \left(\frac{\partial k_v^\beta}{\partial d_i}\right)_{\mathbf{d}=\mathbf{d}^\beta} \quad \text{for} \quad i = 1, 2, \ldots, r, \quad (7.64)$$

and when the constant term vanishes

$$k_v^\alpha(\mathbf{d}^\alpha, \mathbf{h}) - \sum_{i=1}^{r} d_i^\alpha \left(\frac{\partial k_v^\alpha}{\partial d_i}\right)_{\mathbf{d}=\mathbf{d}^\alpha} = k_v^\beta(\mathbf{d}^\beta, \mathbf{h}) - \sum_{i=1}^{r} d_i^\beta \left(\frac{\partial k_v^\beta}{\partial d_i}\right)_{\mathbf{d}=\mathbf{d}^\beta}. \quad (7.65)$$

The set of conditions given by Eq. (7.64) assures that the tangent planes are parallel. These conditions are equivalent to the equilibrium conditions given by Eq. (7.55) on the equality of the fields $h_i^\alpha = h_i^\beta$ ($i = 1, \ldots, r$) because $\partial k_v/\partial d_i = -h_i$ (see Eq. (7.59)). Equation (7.65) assures that the parallel tangent planes coincide by intersecting at the origin and is equivalent to the equilibrium condition on the equality of the dependent field in the two phases, $\phi_v^\alpha = \phi_v^\beta$. Thus the common-tangent construction is equivalent to satisfying the conditions for phase equilibria in systems comprised of homogeneous phases. These results are directly applicable to nonhydrostatically stressed coherent systems, provided the phases are homogeneously deformed at equilibrium.[107]

The points of tangency to the free energy curves, $\mathbf{d}^\alpha$ and $\mathbf{d}^\beta$, correspond to the equilibrium thermodynamic densities of the phases. It is only when the respective phases possess the thermodynamic densities $\mathbf{d}^\alpha$ and $\mathbf{d}^\beta$ that the tangent planes coincide and all conditions for thermodynamic equilibrium are satisfied. A line drawn connecting the points $\mathbf{d}^\alpha$ and $\mathbf{d}^\beta$ is termed a tie line. The ends of the tie line, by definition, give the equilibrium thermodynamic densities of the phases. (Note that not all thermodynamic densities are given by $\mathbf{d}^\alpha$ and $\mathbf{d}^\beta$, but just those defined by the thermodynamic space $\mathbf{d}$; i.e., those system densities that were experimentally controlled. The other system densities must be determined from constitutive relations.) The tie lines clearly span the space of the r thermodynamic densities and, when a phase diagram is viewed in any thermodynamic space that does not include all r thermodynamic densities corresponding to those densities controlled experimentally, it will appear as if the common-tangent construction is invalid. In certain stressed systems, the points of tangency are not necessarily equivalent to the phase boundaries but form a set of distinct lines (planes) called *density* lines.[107] The preceding analysis can be extended directly to systems with more than two phases.

b. Example: Parallel Plates. In this subsection, we use the conditions for thermodynamic equilibrium and the results of the previous subsection to examine the influence of elastic stress on two-phase equilibria and the construction of phase diagrams for coherent systems where crystals are binary substitutional alloys of species A and B. We neglect the dependence of the energy on vacancy concentration and thus use the equilibrium conditions as expressed by Eqs. (7.33) and (7.35), with $\omega_{v'}$ given by Eq. (7.36).

The elastic state of a phase depends on the geometry of the system. Consequently, investigations of phase equilibria in stressed systems usually assume a specific system geometry *a priori*. The conditions for thermodynamic equilibrium have been applied to at least two different system geometries, concentric spheres and parallel plates (see Fig. 26), in order to examine the characteristics of equilibrium in coherent systems.[98,104,115] A significant difference between these two cases is that in the concentric sphere geometry, the exterior phase is nonuniformly stressed. Thus, for example, the previous discussion on phase rules does not hold strictly in this case. We use the parallel plate geometry to illustrate how elastic stress can influence phase diagram construction. In this case, the phases are homogeneously deformed at equilibrium and the results of the previous section can be applied. Unless stated otherwise, the temperature is held fixed, the binary system is closed with respect to mass, and only systems with isotropic or cubic symmetry are considered. For an analysis of more complicated systems, see Roytburd.[116,117]

As we saw in the previous subsection, before phase diagrams can be constructed, it is necessary to establish which thermodynamic variables behave as thermodynamic fields, which as thermodynamic densities, and which variables are the externally controlled variables. The importance of these concepts is conveyed by considering the two types of mechanical loading conditions applied to the elastically stressed system of parallel plates shown in Figure 27. In Fig. 27(a), a stress, σ_{33}^o, is applied in the x_3 (vertical) direction. In the x_1 and x_2 (horizontal) directions, the positions of the surfaces have been fixed. By fixing the position of the surface, the strain in the plane of the plates, ϵ_{11} and ϵ_{22}, is established experimentally. In Figure 27(b), stresses are applied to all surfaces. In each case, the deformation in the x_1 and x_2 directions is constrained to remain homogeneous by the presence of rigid clamps along the surface. This configuration results in the stress and strain within each phase being constant and independent of position. It

---

[115] W. C. Johnson and C.-S. Chiang, *J. Mater. Res.* **4**, 678 (1989).
[116] A. L. Roytburd and Y. Yu, *Ferroelectrics* **144**, 137 (1993).
[117] A. L. Roytburd, *Phase Trans.* **45**, 1 (1993).

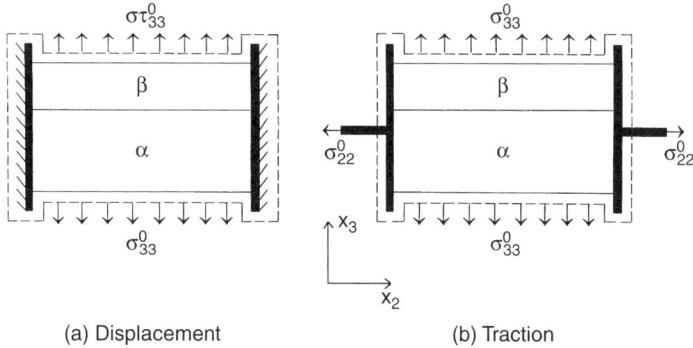

FIG. 27. Two thermodynamic systems for the parallel plate geometry are shown. In (a) displacement boundary conditions have been imposed to hold the dimensions of the system in the $x_1 - x_2$ plane constant. When tractions are imposed as in (b), the dimensions in the $x_1 - x_2$ plane can change, but the stresses must balance. In each case, the stress and strain within each phase are constant and independent of position.

also means that the components of the strain tensor (or deformation gradient tensor F_{ij}) lying within the plane of the interface are continuous across the interface: For an isotropic system or a cubic system with its $\langle 100 \rangle$ crystallographic axes lying parallel to the coordinate axes means that $\epsilon_{11}^\alpha = \epsilon_{11}^\beta = \epsilon_{11}$, $\epsilon_{22}^\alpha = \epsilon_{22}^\beta = \epsilon_{22}$ and all off-diagonal components of the strain tensor are zero.

As a result of the imposed system geometry, the strain components ϵ_{11} and ϵ_{22} behave as thermodynamic fields: They are constant within each phase and equal between phases at equilibrium. The strain component ϵ_{33}, however, will differ in each of the phases, even when the traction in the x_3 direction vanishes, because the material is free to deform in the x_3 direction in order to lower its elastic energy. The extent of deformation depends on the elastic constants (Poisson's ratio in the isotropic case) and the stress-free lattice parameter of each phase. The conjugate stress term, σ_{33}, is constant within each phase and equal between phases (see Eq. (7.32)) and is thus classified as a thermodynamic field. Thus the strain components ϵ_{11} and ϵ_{22} and the stress component σ_{33} behave as thermodynamic fields while their conjugate variables, σ_{11}, σ_{22}, and ϵ_{33} behave as thermodynamic densities. In order to simplify the following presentation, it is assumed that $\epsilon_{11} = \epsilon_{22}$, from which it follows that $\sigma_{11}^\alpha = \sigma_{22}^\alpha$ and $\sigma_{11}^\beta = \sigma_{22}^\beta$ ($\sigma_{11}^\alpha \neq \sigma_{11}^\beta$). The thermodynamic fields for this problem are temperature (θ), diffusion potential (M_{BA}), two strain components ($\epsilon_{11} = \epsilon_{22}$) and one stress component (σ_{33}).

Finally, for the purpose of constructing phase diagrams and free energy curves for stressed systems, it is necessary to identify the dependent thermodynamic field ϕ_v of the previous subsection. This is accomplished by considering the interfacial equilibrium condition in the small strain approximation, Eq. (7.33), and using the parallel plate condition for which $\epsilon_{11}^\alpha = \epsilon_{11}^\beta$, $\epsilon_{22}^\alpha = \epsilon_{22}^\beta$, and because the interfaces are planar, $\kappa' = 0$. (The normal to the interface lies in the x_3 direction.) Because the off-diagonal components of the strain tensor vanish in this configuration, the only non-zero strain component contributing to Eq. (7.33) is ϵ_{33}. Equation (7.33) can, therefore, be written for the assumed parallel plate geometry as

$$\omega_v^\alpha - \epsilon_{33}^\alpha \sigma_{33}^\alpha = \omega_v^\beta - \epsilon_{33}^\beta \sigma_{33}^\beta. \tag{7.66}$$

where the prime subscript denoting the referential state is dropped in keeping with the small strain approximation. For the present geometry, $\sigma_{33}^\alpha = \sigma_{33}^\beta = \sigma_{33}^o = \sigma_{33}$, and we can define the free energy density, Ξ_v, as

$$\Xi_v^\alpha = \omega^\alpha - \epsilon_{33}^\alpha \sigma_{33}. \tag{7.67}$$

Using the expression for ω_v given by Eq. (7.36) yields

$$\Xi_v^\alpha = e_v^\alpha - \theta s_v^\alpha - \rho_A^\alpha M_{AB} - \epsilon_{33}^\alpha \sigma_{33}. \tag{7.68}$$

Likewise, for the β phase,

$$\Xi_v^\beta = \omega_v^\beta - \epsilon_{33}^\beta \sigma_{33}^\beta = e_v^\beta - \theta s_v^\beta - \rho_A^\beta M_{AB} - \epsilon_{33}^\beta \sigma_{33}. \tag{7.69}$$

Differentiation of Ξ_v for either phase gives

$$d\Xi_v = -s_v d\theta - \rho_A dM_{AB} + \sigma_{11} d\epsilon_{11} + \sigma_{22} d\epsilon_{22} - \epsilon_{33} d\sigma_{33}. \tag{7.70}$$

Ξ_v is a fundamental equation with $\Xi_v = \Xi_v(\theta, M_{AB}, \epsilon_{11}, \epsilon_{22}, \sigma_{33})$. Thus Ξ_v is a thermodynamic field that is a function of the other (independent) thermodynamic fields θ, M_{AB}, ϵ_{11}, ϵ_{22}, and σ_{33}. As such, we can identify Ξ_v with the dependent thermodynamic field ϕ_v of the previous subsection.

Displacement Boundary Conditions. Specifying the displacement along the edges of the plate (directions perpendicular to the x_3 axis) establishes the strain components ϵ_{11} and ϵ_{22} in both phases. The externally controlled variables in this case are the thermodynamic fields θ, σ_{33}, and ϵ_{11} and the thermodynamic (system) density ρ_B^o, where ρ_B^o is the average alloy concentration of component B. In keeping with standard practice, we use the bulk composition $c_o = \rho_B^o/\rho_o$ as the appropriate system "density." If z is the phase fraction of β, then the equilibrium phase compositions, c^α and c^β, must satisfy the mass conservation equation (lever rule)

$$c_o = (1-z)c^\alpha + zc^\beta \tag{7.71}$$

where c is the mole fraction of component B in the phase of interest. Equation (7.71) is an example of the general relationship for homogeneous phases given by Eq. (7.57).

In this example, the alloy composition is the only system density that is an externally controlled variable. Consequently, the free energy extremized at equilibrium can be determined from the free energy density k_v using Eq. (7.59) as (dropping the prime superscript in keeping with the small-strain approximation)

$$k_v = \Xi_v - (-\rho_A)M_{AB} = e_v - \theta s_v - \sigma_{33}\epsilon_{33}. \qquad (7.72)$$

If the traction in the x_3 direction is furthermore taken to be zero, $\sigma_{33} = 0$, then the free energy extremized at equilibrium is $k_v = e_v - \theta s_v = f_v$, the Helmholtz free energy. The common-tangent construction is valid when the Helmholtz free energy, accounting for the elastic energy of each phase, is plotted as a function of composition. Because tie lines span the thermodynamic space defined by those externally controlled variables that are system densities, tie lines, in this case, are found to lie in any thermodynamic space containing the composition (a thermodynamic density).

A system that is well-modeled by the displacement boundary conditions and the parallel plate geometry is that of a thin film deposited epitaxially on a thick or rigid substrate. We assume that the two phases composing the film are of uniform thickness and lie parallel to the substrate surface and that no stress is applied in the x_3 direction perpendicular to the substrate. The substrate is assumed to fix the lattice parameter of each phase in the $x_1 - x_2$ plane of the film to the lattice parameter of the substrate throughout the thickness of the film, a good approximation away from the edges. Physically, this corresponds precisely to the displacement boundary conditions and system geometry of Figure 27(a).

The Helmholtz free energy density of each phase is given by the sum of the chemical and elastic energy density (W) contributions as Eq. (4.112),

$$f_v = \rho_o c \mu_B^v(\theta, c) + (1-c)\rho_o \mu_A^v(\theta, c) + W \qquad (7.73)$$

where the chemical potentials are evaluated in the stress-free condition and W is given by Eq. (6.61). In order to determine the strain energy density, a reference state for the measurement of strain is first defined. Although different reference states are possible, it is simplest here to choose the unstressed lattice parameter of the phase of interest as the reference state. Because the lattice parameter in the plane of the film is constrained by the epitaxy to be equal to the lattice parameter of the substrate, a_s, the strain components $\epsilon_{11} = \epsilon_{22}$ are given by

$$\epsilon_{11} = \epsilon_{22} = \epsilon^T = (a_s - a)/a \qquad (7.74)$$

where a is the unstressed lattice parameter of the phase of interest and ϵ^T is the transformation strain. The stress in the plane of the film $\sigma_{11} = \sigma_{22}$ and the strain component ϵ_{33} must be determined using the stress-strain constitutive relations. For phases with cubic symmetry and the crystallographic directions aligned parallel to the coordinate axes as in Figure 27, we have from Eq. (3.84)

$$\sigma_{ij} = \{C_{12}\delta_{ij}\delta_{kl} + C_{44}(\delta_{ik}\delta_{jl} + \delta_{il}\delta_{jk}) + (C_{11} - C_{12} - 2C_{44})\delta_{ijkl}\}\epsilon_{kl}. \quad (7.75)$$

The first equation for the two unknowns (σ_{11} and ϵ_{33}) is obtained by setting $i = j = 3$ with $\sigma_{33} = 0$.

$$\sigma_{33} = 0 = C_{12}\epsilon_{kk} + 2C_{44}\epsilon_{33} + (C_{11} - C_{12} - 2C_{44})\epsilon_{33} \quad (7.76)$$

Because $\epsilon_{kk} = \epsilon_{11} + \epsilon_{22} + \epsilon_{33} = 2\epsilon + \epsilon_{33}$, Eq. (7.76) can be solved directly for ϵ_{33}.

$$\epsilon_{33} = -2C_{12}\epsilon^T/C_{11}. \quad (7.77)$$

The stress component $\sigma_{11} = \sigma_{22}$ is obtained by setting $i = j = 1$ in Eq. (7.75) to give

$$\sigma_{11} = C_{12}\epsilon_{kk} + 2C_{44}\epsilon_{11} + (C_{11} - C_{12} - 2C_{44})\epsilon_{11}. \quad (7.78)$$

Using Eq. (7.77), Eq. (7.78) becomes

$$\sigma_{11} = \sigma_{22} = Y_{100}\epsilon^T \quad (7.79)$$

where Y is an effective elastic modulus defined by

$$Y_{100} = (C_{11} - C_{12})(C_{11} + 2C_{12})/C_{11}. \quad (7.80)$$

In general, the effective elastic modulus depends on the crystallographic orientation of the phases.[118] Of course, both the misfit strain and elastic moduli depend on the phase of interest.

The elastic energy density of a phase in the film is thus

$$W = \frac{1}{2}\sigma_{ij}\epsilon_{ij} = \frac{1}{2}(\sigma_{11}\epsilon_{11} + \sigma_{22}\epsilon_{22} + \sigma_{33}\epsilon_{33}) = (\epsilon^T)^2 Y. \quad (7.81)$$

The Helmholtz free energy density of each phase is thus

$$f_v = \rho_o c \mu_B^v(\theta, c) + (1-c)\rho_o \mu_A^v(\theta, c) + (\epsilon^T)^2 Y. \quad (7.82)$$

Because the phases are homogeneous at equilibrium, the total Helmholtz free energy, $\mathcal{F}$, becomes

$$\mathcal{F} = f_v^\alpha V_\alpha + f_v^\beta V_\beta \quad (7.83)$$

where V_α and V_β are the volumes of the α and β phases, respectively.

---

[118] J. E. Hilliard, *Phase Transformations*, American Society for Metals, Metals Park, OH (1970).

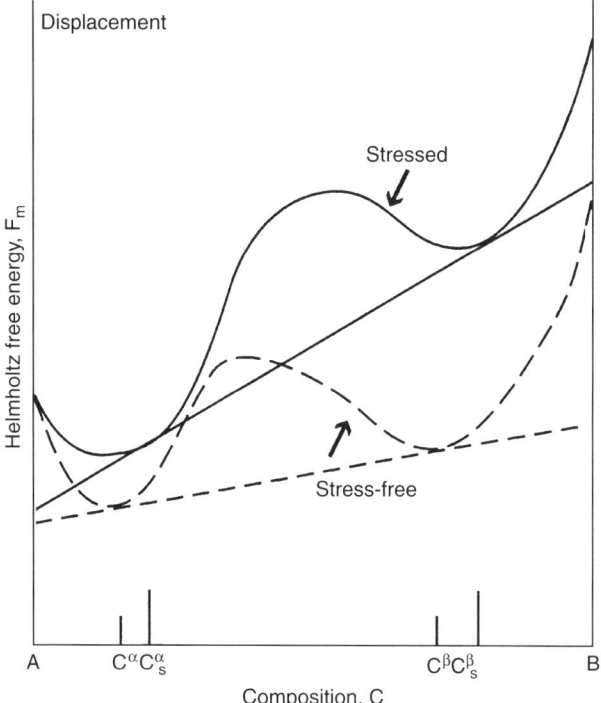

FIG. 28. The free energy of a phase in the absence of stress (dashed line) is compared to its free energy when the phase is constrained by displacement boundary conditions for the case where the alloy exhibits a miscibility gap. The common-tangent construction is applied to both sets of free energy curves and the equilibrium phase compositions for the unstressed (c^α and c^β) and the stressed (c_s^α and c_s^β) systems are shown. The imposed displacement is taken so that a phase composition of pure B does not engender stress.[98]

Figure 28 compares schematically the Helmholtz free energy of a system in the absence of stress (dashed line) with the Helmholtz free energy when the phase is epitaxial to a thick or rigid substrate (corresponding to displacement boundary conditions). The substrate lattice parameter is taken to be that of pure component A, so that no stresses are present for a bulk alloy composition $c = 0$. (This is why the two curves intersect at an alloy composition of pure A.) When the lattice parameter is a function of composition, changes in composition from pure A will induce a stress and the free energy curve will be shifted upward as shown. In keeping with the development of the previous section, the common-tangent construction can be used in the Helmholtz free energy-composition space in order to obtain the equilibrium compositions of the individual phases graphically. In Figure 28, the equilibrium phase compositions of the unstressed system are c^α and c^β whereas

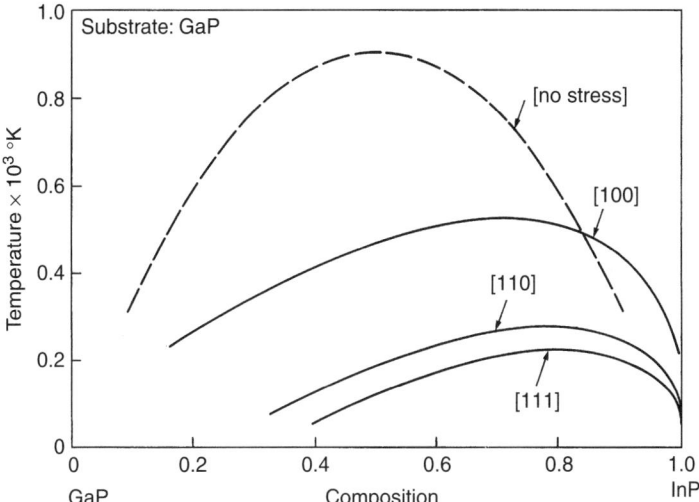

FIG. 29. Calculated miscibility gaps for the GaInP pseudobinary for three different substrate orientations; [100], [110], and [111] are shown. The substrate is taken to have the lattice parameter of GaP and is assumed thick or rigid enough to fix the displacement of the phases in the plane normal to the surface (displacement boundary condition.) Tie lines are contained within the plane of the figure.[98]

the equilibrium phase compositions of the stressed system have shifted to c_s^α and c_s^β.

Figure 29 depicts how the substrate-induced strains can affect the temperature-composition phase diagram for a ternary III-V semiconductor material.[98] In the absence of all stress effects, these pseudobinaries often exhibit regular solution behavior with a positive regular solution constant, Ω, yielding a free energy curve qualitatively similar to that depicted in Figure 28. The regular solution constant can be relatively small and the elastic energy can make a significant contribution to the system energy.[119] In Figure 29, the substrate is taken to be GaP. The broken line represents the miscibility gap in the absence of stress calculated for a regular solution constant, $\Omega = 15.1$kJ.[120] The solid lines depict the calculated miscibility gap for three substrate orientations when the films are epitaxial with the substrate. The curves can be obtained by using the common-tangent construction to the Helmholtz free energy curves as in Eq. (7.73). A compositional strain of $\epsilon^c = 0.077$[115] was used (see Eq. (3.111)). In addition, the elastic constants were assumed to be composition dependent and were obtained using a rule of mixtures between GaP and InP.[98,121]

---

[119] G. B. Stringfellow, *J. Electron. Mater.* **11**, 903 (1982).
[120] F. Glas, *J. Appl. Phys.* **62**, 15 (1987).
[121] M. Neuberger, *Handbook of Electronic Materials*, vol. 2, IFI/Plenum, New York (1971), 66.

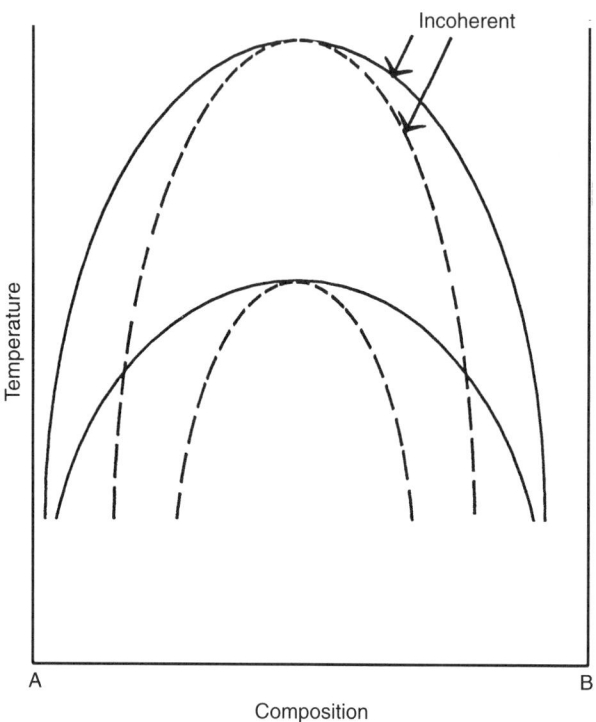

FIG. 30. The suppression of the miscibility gap owing to coherency strain is shown as calculated according to Cahn's seminal work.[94]

In the absence of stress, the critical temperature is $910°K$. Miscibility gaps appear for all three crystal orientations, although the critical temperature depends on the crystal orientation. The miscibility gap is smaller in the [110] and [111] directions than in [100] owing to the orientation dependence of the elastic energy: The effective elastic moduli for the [110] and [111] substrate orientations, Y_{110} and Y_{111}, are greater than Y_{100} defined by Eq. (7.80), which results in a greater elastic energy for planes with a [110] and [111] orientation.

The results presented here resemble those of Cahn's early calculations on the stress-induced suppression of the critical point.[94] Figure 30 depicts the change in the phase diagram predicted by Cahn owing to coherency strain and shows the coherent miscibility gap lying completely within the incoherent miscibility gap. This occurs when the displacement boundary conditions are applied to the unstressed matrix phase of the bulk composition. This is equivalent to fixing the substrate lattice parameter to that of the bulk composition before any phase separation has taken place. Thus, in comparison to Figure 29, Figure 30 is computed using a different substrate lattice parameter for each bulk alloy composition while Figure 29 uses the same substrate lattice parameter for all bulk alloy compositions.

Traction Boundary Conditions. When the tractions (stresses) acting in the x_1 and x_2 directions of the system with parallel plates are specified, as in Figure 27(b), the externally controlled variables consist of two thermodynamic fields, temperature and stress component (σ_{33}), and two system densities, stress component $\sigma_{11} = \sigma_{22}$ and alloy composition c_o. In addition to Eq. (7.71), the system densities must satisfy

$$\sigma_{11}^o = \sigma^o = (1-z)\sigma_{11}^\alpha + z\sigma_{11}^\beta \tag{7.84}$$

where σ_{11}^o is the average normal force per unit area (stress) applied to the surface with normal in the x_1 direction and z is the volume fraction of β. An identical expression connects the stress components (σ_{22}) in the x_2 direction.

The free energy extremized at equilibrium is obtained from the energy density, k_v, using Eq. (7.59), as

$$k_v = \Xi_v - (-\rho_A)M_{AB} - \sigma_{11}\epsilon_{11} - \sigma_{22}\epsilon_{22} = e_v - \theta s_v - \sigma_{11}\epsilon_{11}$$
$$- \sigma_{22}\epsilon_{22} - \sigma_{33}\epsilon_{33}. \tag{7.85}$$

For the case of traction boundary conditions, the tie lines lie in any space containing the composition and stress component $\sigma_{11} = \sigma_{22} = \sigma^o$.

Figure 31 illustrates how changing the mechanical loading from displacement to traction boundary conditions can lead to a qualitative change in phase diagram construction and the characteristics of phase equilibrium. In the first column, Figures 31(a–c), the temperature, strain ($\epsilon_{11} = \epsilon_{22}$) and composition have been chosen as the externally controlled variables. (This corresponds to the displacement boundary conditions of the previous subsection with two thermodynamic fields and one system density as externally controlled variables.) In the absence of all stress effects, it is assumed the alloy would exhibit a simple miscibility gap in the composition. The three-dimensional, temperature-strain-composition diagram is plotted in nondimensional units in part (a) for a case in which both the lattice parameter and elastic constants vary linearly with composition.[122] The phase boundary is depicted by the thick solid lines for several nondimensional temperatures (at constant strain) and strains (at constant temperature). The outlines of these two-dimensional cuts through the three-dimensional space are indicated by the dashed lines in Figure 31(a). The dotted line AB indicates the locus of critical points at which the (coherent) spinodal coincides with the phase boundary. Representative tie lines (thin solid lines) end on the phase boundary and are parallel to the axis of the only externally controlled density variable, the composition c. The material is taken to be cubic with a [100] orientation.

Figures 31(b) and (c) represent two-dimensional cuts through the three-dimensional phase diagram of Figure 31(a): Figure 31(b) is a temperature-

---

[122] J.-Y. Huh and W. C. Johnson, *Acta. Metall. Mater.* **43**, 1995 (1995).

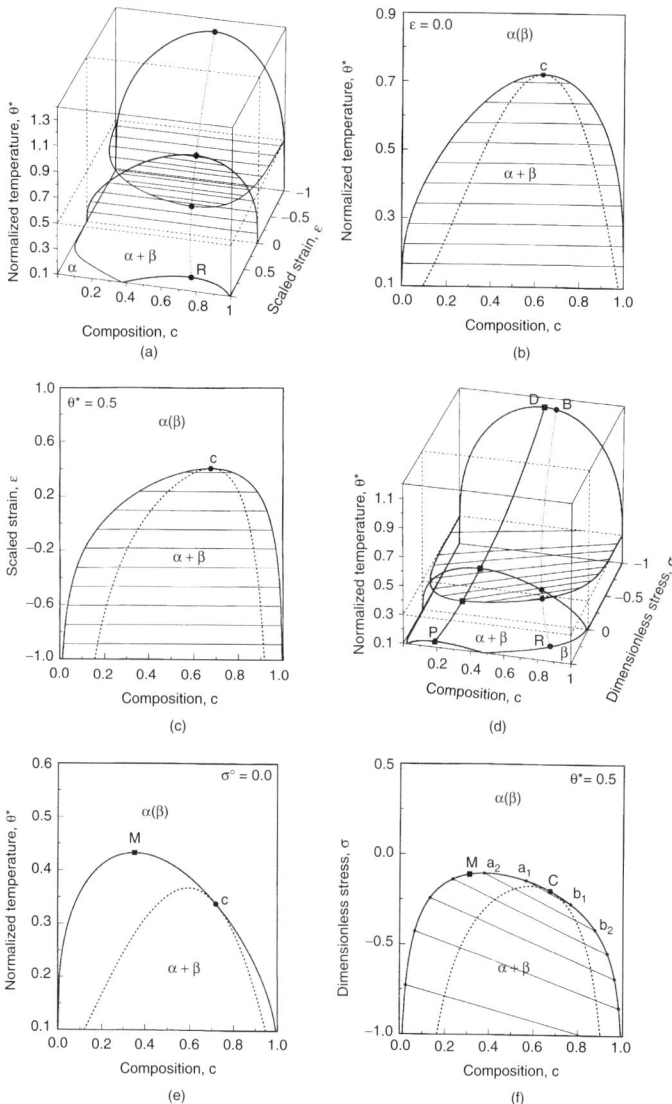

FIG. 31. Two sets of coherent phase diagrams are shown for a system with the parallel geometry.[122] In (a)–(c) the temperature, composition, and strain component $\epsilon_{11} = \epsilon_{22}$ are the externally controlled variables. The two-dimensional phase diagrams of (b) and (c) are slices through the three-dimensional phase diagram of (a). In (d)–(f), the temperature, composition, and stress component $\sigma^o_{11} = \sigma^o_{22} = \sigma^o$ are the externally controlled variables. (e) and (f) are two-dimensional slices through the three-dimensional phase diagram of (d). In all figures the thick solid lines denote the phase boundaries and the thin solid lines the tie lines. For (a)–(c) the tie lines give the equilibrium phase compositions and in (f) the equilibrium compositions and stresses. Dashed lines indicate the coherent spinodal.

composition phase diagram at constant strain, $\epsilon = 0$, and Figure 31(c) is a strain-composition phase diagram at constant temperature. Because both of these phase diagrams span the composition axis (the only externally controlled variable that is a system density) tie lines lie in the plane of the phase diagram in both cases. The dotted lines indicate the coherent spinodal. The critical point, which identifies a second-order transition in both the temperature-composition and strain-composition phase diagrams, is denoted with C.

Figures 31(d–f) represent the phase diagrams for the same material system as depicted in Figures 31(a–c) except that the applied stress in the x_1 and x_2 directions ($\sigma_{11}^o = \sigma_{22}^o = \sigma^o$) has been chosen as the externally controlled variable in place of ϵ_{11}. Once again the heavy solid lines denote the phase boundary. The dotted line AB and the solid line PQ are the line of consolute critical points and maximum temperatures of the miscibility gap, respectively. Two system densities, composition and stress component σ_{11}^o, are externally controlled variables. Tie lines lie in the $\sigma_{11} - c$ plane and are denoted by the thin solid lines.

Figure 31(e) is the temperature-composition phase diagram taken at constant applied stress ($\sigma_{11}^o = \sigma_{22}^o = \sigma^o = 0$). The phase diagram spans only one of the system densities, which is an externally controlled variable; i.e., the composition c. (The other system density is the stress component σ_{11}.) Therefore, tie lines do not lie in the plane of the phase diagram, even though the applied stress is zero. Unlike the constant *strain*, temperature-composition phase diagram of Figure 31(b), the critical temperature (point C) in the constant *stress* temperature-composition phase diagram does not coincide with the maximum temperature of the miscibility gap (point M).[123] Figure 31(f) is the constant temperature, stress-composition phase diagram. Tie lines lie in this thermodynamic space, because the phase diagram spans the space of the two externally controlled variables that are system densities. The tie lines are represented by the thin, downward-sloping lines that end on the phase boundaries, e.g., the line segment a_2b_2. The ends of the tie lines give the equilibrium composition and stress component σ_{11} for each of the phases. Like the tie lines of a ternary isotherm in an unstressed ternary alloy, there is no reason to expect the tie lines to be horizontal.

Interpretation of phase diagrams for homogeneously deformed stressed systems is analogous to unstressed systems.[107] Figures 32(a) and (b) depict two schematic phase diagrams for a system with the parallel plate geometry in which the temperature, stress component $\sigma_{11}^o = \sigma_{22}^o$, and composition have been chosen as the experimentally controlled variables. The phase boundaries and tie lines are denoted by the thick solid and dashed lines, respectively. In Figure 32(a), assume the applied stress in the x_1 and x_2 directions is zero and the bulk composition

---

[123] This behavior is completely analogous to that of unstressed ternary alloys. In a plot of temperature and the composition of just one of the alloy components, the critical temperature does not coincide with the maximum temperature of the miscibility gap.

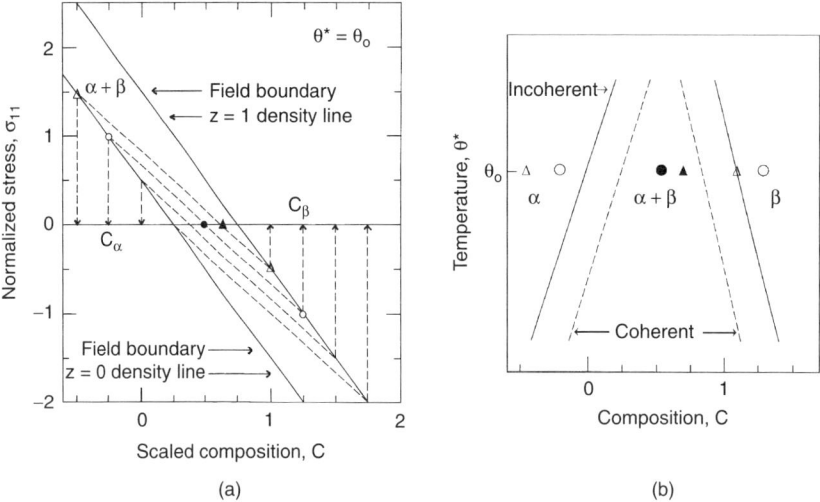

FIG. 32. A constant temperature, stress-composition phase diagram is shown in (a).[122] The tie lines (dashed lines) end on the phase boundaries and give the equilibrium phase composition and stress state of the phases. The arrows show the projection of the phase compositions onto the zero-stress line. The constant stress (zero applied stress), temperature-composition phase diagram corresponding to (a) is shown in (b). The equilibrium phase compositions do not lie on the phase boundaries in this projection as the tie lines do not lie in the temperature-composition space. (Tie lines lie in the $\sigma_{11} - c$ space.)

is 0.5. This point is given by the solid circle lying on the tie line ending at the points identified by the open circles. The open circles indicate the equilibrium stresses and compositions of the two phases in equilibrium: The composition of the α phase is $c^\alpha = 0.25$ and the nondimensional stress component σ_{11} is $\sigma_{11}^\alpha = 1$ while the corresponding equilibrium thermodynamic densities in the β phase are $c^\beta = 1.25$ and $\sigma_{11}^\beta = -1$. Now assume the bulk composition is increased from 0.5 to about 0.62 holding the applied stress at zero. The solid triangle identifies the point on the phase diagram of the new (imposed) system densities. This point lies on a different tie line from the tie line for the case in which the bulk composition was 0.5. The ends of the tie line are indicated by the open triangles. These open triangles identify the new equilibrium compositions and stresses of the phases in the absence of an applied stress. Because the phases have different lattice parameters in their unstressed states, the individual phases are stressed in the coherent condition.

Figure 32(b) is the corresponding temperature-composition phase diagram for the case where the applied stress vanishes. The solid lines denote the phase boundaries; i.e., they delineate the single-phase from two-phase regions. Tie lines do not lie in the plane of the temperature-composition phase diagram.[124] Indeed, if the equilibrium phase compositions of Figure 32(a) are projected onto the zero-stress plane, as indicated by the dotted arrows of Figure 32(a) and then plotted on the temperature-composition phase diagram of Figure 32(b), it appears that the compositions do not lie on the phase boundaries. (The open circles of Figure 32(b) give the equilibrium compositions of the phases when the bulk alloy composition is 0.5, denoted by the solid circle. The open triangles give the equilibrium compositions of the phases when the bulk composition is 0.62, denoted by the solid triangle.) If the bulk composition is changed at zero applied stress (for example, from 0.5 to 0.62) and the corresponding equilibrium phase compositions are plotted on the usual temperature-composition phase diagram, the compositions will shift in apparent violation[125] of the Gibbs phase rule.[100,104,112,126] Two-phase coherent systems will be under a state of internal stress, even when no external stress exists, owing to misfit strains. Changing the bulk composition changes the volume fraction of the phases and the equilibrium phase compositions. Therefore constructing phase diagrams by measuring phase compositions using, for example, analytical electron microscopy may not always be correct.[127]

Phase equilibria in two-phase coherent systems can be quite complex, with the existence of multiple equilibrium states for a given temperature, applied stress (or pressure), and alloy composition.[104]

Misfit strain can also affect equilibrium when a binary alloy can choose among three different phases. For example, consider the nondimensional, temperature-composition phase diagram of Figure 33 constructed in the absence of all stress, including misfit strain. For clarity, the equilibrium phase boundaries of the α and γ phases below the eutectoid temperature and in the absence of stress have been normalized to 0 and 1, respectively. Nondimensional compositions greater than one correspond to single-phase γ while nondimensional compositions less than zero lie within the single-phase α regime. The eutectoid temperature has been set to zero.

Figure 34(a) depicts the stress-composition phase diagram for a system with the parallel plate geometry with traction boundary conditions corresponding to the stress-free case of Figure 33. The temperature has been taken to be the eutectoid temperature ($\theta^* = 0$). The phase diagram, which is described in more detail next, was calculated assuming that the lattice parameters and elastic constants of the phases are independent of composition with $a_\beta > a_\gamma > a_\alpha$.

---

[124] The plane of the temperature-composition phase diagram corresponds to zero stress. But the individual phases are stressed, even when the applied stress vanishes, owing to the misfit strains.

[125] Of course, as proven in a previous subsection, a phase rule does exist for this system. The perception that the phase rule is violated arises when the thermodynamic fields and densities are not properly identified.

[126] J. W. Cahn and F. C. Larché, *Acta metall.* **32**, 1915 (1984).

[127] J.-Y. Huh, J. M. Howe, and W. C. Johnson, *Acta. Metall. Mater.* **41**, 2577 (1993).

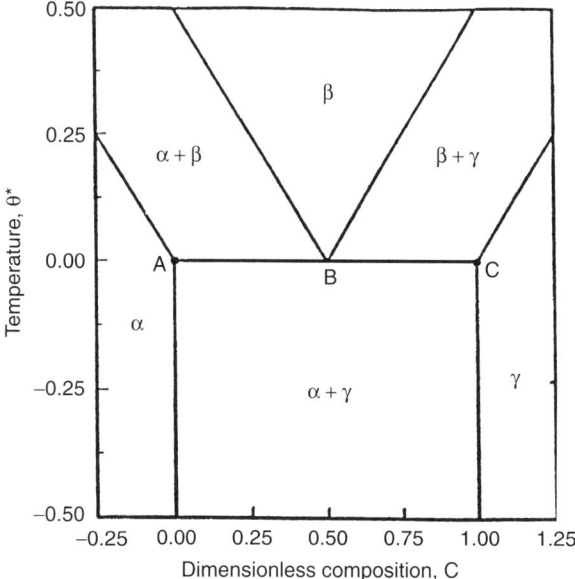

FIG. 33. A temperature-composition phase diagram for an incoherent system is shown in nondimensional units for a eutectoid system for which all stress effects vanish.[130] The equilibrium phase boundaries of α and γ below the eutectoid have been normalized to 0 and 1, respectively, and the eutectoid temperature has been set to zero. As expected for a stress-free system, there is only one equilibrium state for a given temperature and alloy composition and that state is always stable.

Several characteristics unique to coherent phase equilibria are found in Figure 34(a). The most apparent is the existence of several equilibrium states for a given composition and applied stress. For the material parameters employed, two regions can be identified: one for which only one equilibrium state exists and one for which three equilibrium states exist.[128] Those regions that give rise to three equilibrium states are contained within the triangle defined by vertices a, b, and c. The three-phase state of $\alpha + \beta + \gamma$ is possible everywhere within the triangle abc. The equilibrium stress states and compositions of the three phases are given by the vertices a, b, and c, respectively.[129] For example, if the dimensionless bulk composition and applied stress correspond to point p in Figure 34, the equilibrium composition and stress state of the α phase is given by point a for the three-phase system. Analysis of the *stability* of the three-phase equilibrium state with respect to changes in the relative phase fractions of the phases at constant

---

[128] An equilibrium state is one for which all conditions for thermodynamic equilibrium are satisfied. The state can correspond to an energy minimum, energy maximum, or a saddle point.

[129] The triangle abc is analogous to a tie-triangle observed for a three-phase region in an isothermal section of a ternary alloy.

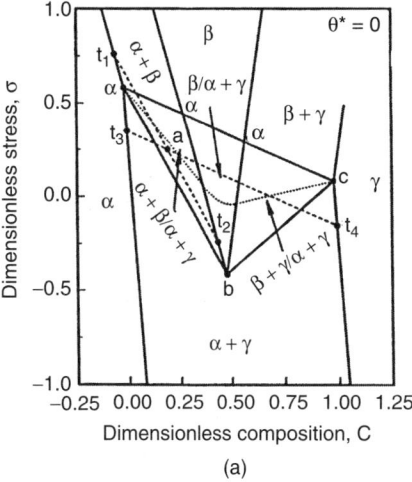 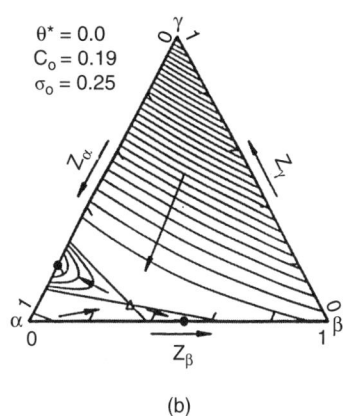

FIG. 34. The stress-composition phase diagram at the eutectoid temperature corresponding to the stress-free eutectoid system is shown in (a) when traction boundary conditions are imposed for the parallel plate geometry.[130] Multiple equilibrium states are possible within triangle abc. (b) Lines of equal free energy are plotted as a function of the volume fractions, z_i, for the alloy composition and applied stress given by point p. The arrows inside the phase-fraction triangle indicate the directions of decreasing system free energy. The stable equilibrium states are indicated by filled circles, whereas the unstable (saddle point) three-phase equilibrium state is denoted by the open triangle.

temperature, alloy composition, and applied stress shows that the three-phase state is thermodynamically unstable.

Two other equilibrium states are possible at point p, both of which are thermodynamically stable.[130] The first is an $\alpha + \gamma$ two-phase system. The equilibrium compositions and stress states of the α and γ phases are given by the intersection of the tie line passing through the point p (indicated by the dashed line) and the phase boundaries delineating the α and γ single-phase fields, points t_3 and t_4, respectively. The second thermodynamically stable equilibrium state corresponding to point p is a two-phase system of $\alpha + \beta$. The equilibrium compositions and stress states are given by the intersection of the tie line with the α and β single-phase fields, points t_1 and t_2. The regions in which two linearly stable equilibrium states can be found are indicated by arrows in Figure 34(a) and will now be delineated. Both an $\alpha + \gamma$ and an $\alpha + \beta$ two-phase system are stable within the triangle abe. The $\alpha + \gamma$ system is *absolutely* stable below the dotted curve that extends from point a to point c and the $\alpha + \beta$ two-phase system is linearly stable.[131] The

---

[130] J.-Y. Huh and W. C. Johnson vol. 311, Materials Research Society (1993), 119.
[131] By absolutely stable, it is meant that, for all permissible variations in the phase compositions and stress states, the system has the lowest energy. Linearly stable means the system has the lowest energy in some region around the equilibrium state. A linearly stable state sits in an energy well.

$\alpha + \gamma$ two-phase system is no longer an equilibrium solution for externally applied stresses that lie above the solid line ac. Within the triangle bde, both single-phase β and two-phase $\alpha + \gamma$ are thermodynamically stable, while within triangle bcd both two-phase systems $\beta + \gamma$ and $\alpha + \gamma$ are thermodynamically stable.

Lines of constant free energy of the entire system are plotted as a function of the volume fraction of the phases for the point p in Figure 34(b) assuming chemical, mechanical, and thermal equilibrium. The arrows within the phase-fraction triangle indicate the direction of decreasing free energy. The state of three-phase equilibrium ($\alpha + \beta + \gamma$) is indicated by the open triangle and corresponds to a saddle point in the free energy. It is unstable with respect to certain perturbations in the volume fractions and therefore would not be expected to be observed. The two stable states are given by the small, filled circles. These points are end-of-range minima in the free energy. Any physically realizable perturbation in the volume fractions about these points results in an increase in the system free energy. The stability of the two-phase equilibrium states does not arise from the presence of nucleation barriers for the formation of the equilibrium phase. In each of the thermodynamic states, formation of a nonzero volume fraction of the third phase results in an increase in the free energy.

The temperature-composition phase diagram (at constant applied stress, $\sigma = 0$) corresponding to the coherent system of Figure 34 is shown in Figure 35 assuming the applied stress vanishes. The phase diagram does not contain the tie lines and, hence, the equilibrium phase compositions cannot be obtained from this phase diagram. This phase diagram differs from its stress-free counterpart of Figure 33 in that an invariant temperature does not exist. The region in which three equilibrium states can exist is given by triangle ABC. The (unstable) $\alpha + \beta + \gamma$ three-phase equilibrium state is found within triangle ABC. The two-phase systems of $\alpha + \beta$ and $\alpha + \gamma$ are stable within triangle ABE; single-phase β and two-phase $\alpha + \gamma$ are stable equilibrium states within triangle EBD; and the two-phase systems of $\alpha + \gamma$ and $\beta + \gamma$ are stable within DBC. The $\alpha + \gamma$ two-phase system is absolutely stable for temperatures below the dotted curve connecting points A and C. Similar to an isopleth section of an incoherent ternary phase diagram, the tie lines cannot be drawn in the thermodynamic space of Figure 35 because the space is not spanned by all the density variables among the externally controlled variables. A projection of the equilibrium phase compositions onto the temperature-composition plane of Figure 35 would show that the equilibrium phase compositions would not correspond to the phase boundaries drawn in this plane.

The existence of more than one stable equilibrium state for a given temperature, alloy composition and applied stress can result in transformation hysteresis when cycling in temperature or applied stress. Suppose that the temperature of an alloy with a fixed composition ($c_o = 0.5$) and external stress ($\sigma_{11}^o = \sigma_{22}^o = 0$) is slowly decreased from a point in the β single-phase field to a point in the $\alpha + \gamma$ two-phase

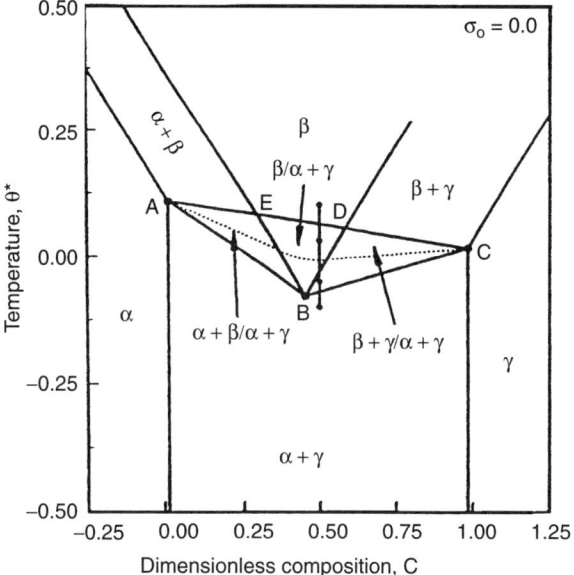

FIG. 35. The temperature-composition phase diagram for the eutectoid system with traction boundary conditions and no applied stress is shown.[130] Multiple equilibrium states are possible within triangle ABC.

field as shown by the vertical line in Figure 35. Four temperatures have been indicated, two of which lie in the region of multiple equilibrium states. The lines of constant free energy associated with these four points are plotted as a function of phase fraction in Figure 36. Figure 36(a) corresponds to the highest temperature and Figure 36(d) to the lowest. Chemical, mechanical, and thermal equilibrium has been assumed.

At the nondimensional temperature corresponding to $\theta^* = 0.1$, Figure 36(a), only one equilibrium state, identified by the solid circle and corresponding to single-phase β, exists. (The solid circle with an 'x' appearing through it located on the $\alpha + \gamma$ two-phase line at about $z_\alpha = 0.5$ gives a minimum in the free energy of the $\alpha + \gamma$ system but is unstable with respect to the formation of β. As the temperature is reduced to $\theta^* = 0.03$, Figure 36(b), two new equilibrium states appear: the unstable three-phase system indicated by the open triangle (saddle point) and the stable, $\alpha + \gamma$ two-phase system. Single-phase β is still the lowest energy state (absolute minimum). As the temperature is further reduced to that given by $\theta^* = -0.05$, Figure 36(c), two two-phase equilibrium states are stable: the $\alpha + \gamma$ system and a $\beta + \gamma$ system that is mostly β phase. The absolute minimum in the free energy has jumped discontinuously from single-phase β to the $\alpha + \gamma$

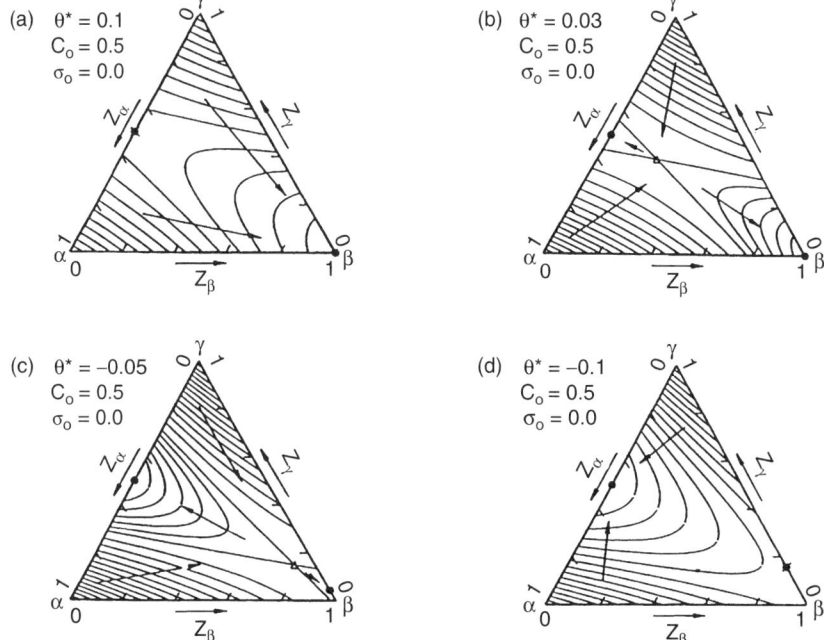

FIG. 36. Isocontours of the free energy are plotted as a function of volume fractions for four different nondimensional temperatures at the bulk alloy composition $c_o = 0.5$ and in the absence of an applied stress.[130] Arrows within the triangle indicate directions of decreasing free energy. The small triangle indicates the point of unstable three-phase equilibrium and corresponds to a saddle point in the free energy. The solid circles identify stable two-phase equilibrium solutions. The solid circles appearing with an X through them indicate points of two-phase equilibrium that are unstable with respect to the formation of a third phase.

two-phase system. However, if the system is cooled reversibly, it would still reside in the state of single-phase β. (The β phase remains stable with respect to formation of the γ phase until the temperature is decreased below the line BD.) Upon further cooling to $\theta^* = -0.1$, Figure 36(d), the saddle point in the free energy disappears and the $\beta + \gamma$ two-phase state becomes unstable with respect to the formation of the α phase. (The $\beta + \gamma$ two-phase state is unstable with respect to the formation of α for temperatures below the line BC.)

Now consider the case in which the same system is heated reversibly from the $\alpha + \gamma$ two-phase state at temperature $\theta^* = -0.1$ to the β single-phase state at temperature $\theta^* = 0.1$. The $\alpha + \gamma$ two-phase state remains (absolutely) stable until crossing the dotted curve and a relative minimum in the free energy until crossing the boundary AC. This indicates the system could remain in the $\alpha + \gamma$ two-phase state until it loses stability with respect to the formation of the β phase. On heating, the system would not necessarily pass through the $\beta + \gamma$ two-phase state as it did during the cooling process. It is important to note that this hysteresis is not a kinetic phenomenon due to a nucleation barrier but an equilibrium characteristic unique to a coherent system.

c. Equilibrium Shape of a Coherent Particle

a. Introduction. In this section, the influence of both an applied stress and a misfit (or transformation) strain on the morphology of one or more coherent particles, determined by minimizing the sum of the interfacial and elastic energies, is examined. In general, both the elastic and interfacial energies of a precipitate embedded in a crystalline matrix depend on the precipitate shape. The morphology that minimizes the particle's elastic energy can be different from the one that minimizes its interfacial energy. Because the ratio of elastic to interfacial energy is proportional to the particle size, the particle possessing minimum energy should tend toward the shape corresponding to the minimum interfacial energy shape when the particle is small, and to the minimum elastic energy shape when the particle is large. Thus, for a particle of given size, there are three possibilities for the particle morphology:

1) The particle shape is the one that minimizes interface energy alone.
2) The particle shape is the one that minimizes elastic energy alone.
3) The particle shape is a compromise shape that reflects a balance between the interface and elastic energies.

The particle size range over which each of these particular cases might hold is examined. For example, it might be expected that case 3 holds for the entire range of precipitate sizes, approaching the other two cases only in a limit of very large or very small particles. We demonstrate, however, that under some conditions, case 1 holds identically for all particle sizes below a critical size. At the critical size, the energy minimum shifts discontinuously to another case; i.e., there is a discontinuous shift change in the particle shape that minimizes the system energy with increasing size of the particle. Such a singularity in the behavior of the energy minimum is called a bifurcation, and the shape transition is referred to as an elastically induced shape bifurcation.

Because the change in precipitate shape reflects a change in symmetry, we first discuss the underlying symmetry principles governing the permissible shape transitions. We then show the difference in scaling between the interfacial and elastic energies with increasing particle volume for a general, isolated particle with application to the special case of a misfitting spherical particle in an isotropic matrix. Finally, we combine these concepts with some basic principles of bifurcation theory to explore elastically induced precipitate shape bifurcations with increasing precipitate size, first for a restricted class of geometric shapes and then for precipitates of arbitrary shape. The stress-induced precipitate shape transitions are shown to be of two types, symmetry conserving and symmetry breaking.

b. Crystal Symmetry and Neumann's Principle. Solid-state precipitates display a number of different equilibrium shapes that often change with the size of the precipitate. The actual shape depends on the material parameters of both the precipitate and matrix phases, including the anisotropy of the interfacial energy, the precipitate misfit strain, and the presence of an external stress field. Some general predictions about precipitate shape evolution can be made by considering the symmetry of the precipitate and matrix phases using Neumann's principle[28] and the relative contributions to the total energy made by the elastic and interfacial energies.

One observation of nature is that when a crystal is grown from its isotropic melt, it will assume a form compatible with its crystal symmetry.[132] This observation, along with many others relating to the symmetry of various crystal properties, is contained within Neumann's principle, which states: *The symmetry elements of any physical property of a crystal must include the symmetry elements of the point group of the crystal.*[28] The point group is the basis for dividing crystals into 32 different classes. A point group consists of all the symmetry elements of the crystal that do not involve translation vectors. The macroscopic symmetry elements in a crystal include a center of symmetry, mirror planes, rotation axes, and inversion axes.

Neumann's principle does not indicate what the symmetry elements of a given property actually are, just that the property must include the symmetry elements of the crystal's point group. For example, the chemical diffusivity in a cubic matrix, a second-rank tensor, must display at least a threefold axis about the $\langle 111 \rangle$ and a fourfold rotation axis about the $\langle 100 \rangle$ crystallographic axes. In actuality, however, the chemical diffusivity is isotropic and, therefore, possesses more symmetry than the crystal's point group.

Neumann's principle can also be applied to the interfacial energy and, consequently, to the equilibrium shape of an isolated, stress-free precipitate in a crystalline matrix.[132] Because two crystals are present, the symmetry of the interfacial energy must include the symmetry elements of the point groups of both the matrix and precipitate phases. Stated mathematically, the interfacial energy associated with an interface separating two crystalline phases must display at least the symmetry of the intersection of the point groups of the precipitate and matrix phases. As an example, consider first the trivial case of two cubic phases with coincident crystallographic axes. The minimum symmetry of the interfacial energy is given

---

[132] J. W. Cahn and G. Kalonji, in *Int. Conf. Solid-Solid Phase Transformations*, eds. H. I. Aaronson, D. E. Laughlin, R. F. Sekerka, and C. M. Wayman TMS-AIME, Warrendale, PA (1982), 3.

by

$$\frac{4}{m}\bar{3}\frac{2}{m} \cap \frac{4}{m}\bar{3}\frac{2}{m} = \frac{4}{m}\bar{3}\frac{2}{m} \quad (7.86)$$

where m denotes a perpendicular mirror plane and $\cap$ is the intersection operator. Equation (7.86) gives the symmetry of the interfacial energy density for a cubic precipitate in a cubic matrix. Any precipitate whose equilibrium shape is determined solely by minimization of the interfacial energy must possess a shape that displays at least the symmetry of Eq. (7.86). Because one symmetry can be represented by many shapes, there are a number of precipitate shapes that a coherent (unstressed) cubic precipitate in a cubic matrix can possess. For example, a sphere, a cube, an octahedron, and a tetrakaidecahedron possess the symmetry given by Eq. (7.86) and all are possible equilibrium precipitate shapes.

Consider now the slightly more complicated case of a tetragonal precipitate nucleated in a cubic matrix. If the axis of tetragonality is assumed to lie along the x_3 axis, then the minimum symmetry of the interfacial energy density is given by

$$\frac{4}{m}\bar{3}\frac{2}{m} \cap \frac{4}{m}mm = \frac{4}{m}mm. \quad (7.87)$$

At this point a distinction must be made between the symmetry of the shape of an isolated precipitate, which we have been discussing, and the symmetry of the two-phase system on a more macroscopic scale; i.e., on a scale large enough to contain many precipitates. Consider again the case of a coherent tetragonal precipitate in a cubic matrix. The symmetry of the interfacial energy density is given by Eq. (7.87) and the equilibrium shape of a precipitate must reflect this symmetry. However, there are three different but equivalent variants to the orientation of the precipitate; the axis of tetragonality would be expected to lie with equal frequency along the x_1, x_2, and x_3 axes. Thus, if a volume of the matrix sufficiently large to contain many precipitates of each of the three variants is considered, the cubic symmetry of the matrix phase is recovered. One would expect the physical properties of the two-phase crystal to still exhibit the cubic symmetry of the matrix phase, the same as for the matrix in its single-phase state. Of course, the values of the physical properties would change with the presence of the tetragonal precipitates.[133]

From the macroscopic point of view, the matrix phase retains its symmetry when second-phase precipitates nucleate and grow coherently within it. The interfacial energy assumes a symmetry commensurate with the intersection of the point groups of the precipitate and matrix crystals. However, the symmetry of the matrix phase is retained on a macroscopic scale owing to the different orientational variants that

---

[133] For example, the yield stress of the two-phase system would still reflect the symmetry of the cubic matrix phase when the tetragonal precipitates assume all orientational variants equally. This would not be true if the axis of tetragonality of all precipitates were parallel.

can be assumed by the individual precipitates. This distinction is important when discussing the symmetry-breaking precipitate shape transitions.

When an external field is present, the interfacial energy density must possess the minimum symmetry given by the intersection of the point groups of the precipitate and matrix phases and the Curie group[134] of the external field.[135] For example, when a uniaxial tensile stress is applied along a fourfold crystallographic axis, the symmetry of the interfacial energy density separating two cubic phases is at least

$$\frac{4}{m}\bar{3}\frac{2}{m} \cap \frac{4}{m}\bar{3}\frac{2}{m} \cap \frac{\infty}{m}\frac{2}{m}\frac{2}{m} = \frac{4}{m}mm. \tag{7.88}$$

The presence of the external field acts to lower the symmetry of the interfacial energy density and the equilibrium precipitate shape. Likewise, all properties of the crystal will be influenced by the presence of the external field.

In the following section, we use these symmetry arguments to define two different types of precipitate shape changes that can be induced by the presence of an elastic stress. Before doing so, it is instructive to first consider the relative contributions of the elastic and interfacial energies to the total energy of the system.

c. Scaling of Elastic and Interfacial Energies. In the absence of elastic stress, the equilibrium shape of a second-phase particle (β) of fixed volume embedded in a matrix (α) is determined by minimizing the interfacial energy. If the precipitate–matrix interfacial energy density is $\gamma(\mathbf{n})$, where $\mathbf{n}$ is the outward-pointing unit normal to the precipitate, the equilibrium shape is that shape that minimizes the free energy, E_t, given by[136]

$$E_t = \int_{S_\beta} \gamma(\mathbf{n}) dS, \tag{7.89}$$

and subject to the constraint

$$V_\beta = \text{constant} \tag{7.90}$$

where S_β and V_β are the surface and volume of the precipitate, respectively. In general, γ depends on the orientation of the interface. γ also possesses a symmetry determined by the symmetries of the precipitate and matrix phases and their relative orientation.[132] Although the total interfacial energy depends on the particle volume, the particle *shape* that minimizes Eq. (7.89) is independent of the particle volume and is that shape given by the Wulff construction.[136] Because the shape

---

[134] The Curie group contains the symmetry operations of the external field.
[135] Y. I. Sirotin and M. P. Shaskolskaya, *Fundamentals of Crystal Physics*, Mir, Moscow (1982).
[136] G. Wulff, *Zeit. f. Kristallog.* **34**, 1901 (1901).

THE THERMODYNAMICS OF ELASTICALLY STRESSED CRYSTALS 177

that minimizes the interfacial energy is independent of its size, no transitions in the equilibrium shape of the precipitate are expected with increasing precipitate size in the absence of elastic stress.[137] A second result that follows from the Wulff construction is that the equilibrium shape must be convex.[138]

If the isolated precipitate possesses a misfit strain, ϵ_{ij}^T, the equilibrium shape is determined by minimizing the sum of the interfacial and elastic strain energies at constant particle volume, assuming that the lattice parameter is independent of composition.[1,2] In this case, the total free energy is given by

$$E_t = \int_{S_\beta} \gamma(\mathbf{n}) dS + \frac{1}{2} \int_{V_\infty} S_{ijkl} \sigma_{ij} \sigma_{kl} dV, \tag{7.91}$$

where the volume integral is over both the matrix and precipitate phases. For an isolated precipitate in an infinite medium, assuming that there is no mechanical load at infinity, using the constitutive equation $\sigma_{ij} = C_{ijkl}(\epsilon_{kl} - \epsilon_{kl}^T)$, $\epsilon_{ij} = (u_{i,j} + u_{j,i})/2$, and the divergence theorem yields

$$E_t = \int_{S_\beta} \gamma(\mathbf{n}) dS - \frac{1}{2} \int_{V_\beta} \sigma_{ij} \epsilon_{ij}^T dV. \tag{7.92}$$

Thus, the elastic energy of the system can be represented as an integral only over the volume of the precipitate phase, V_β. More importantly, Eq. (7.92) shows that the total elastic energy in the system scales with the volume of the precipitate, V_β.[37]

In order to illustrate more clearly the dependence of the total energy of the system on particle size, we nondimensionalize Eq. (3.57) such that the dimensionless quantities are $O(1)$. Choosing the total interfacial energy as the energy scale, the dimensionless total energy of the system is $\tilde{E}_t = E_t/(\gamma_o \ell^2)$, where γ_o is the interfacial energy density averaged over the particle surface, and $\ell = V_\beta^{1/3}$ for a three-dimensional particle. The misfit strain is typically quite small. Thus, to make the misfit strain $O(1)$, we define a scaled misfit strain as $\tilde{E}_{ij}^T = E_{ij}^T/e$, where e is a characteristic strain, such as the trace of misfit strain tensor for a particle with a dilatational misfit. The stress in the system is on the order of Ce^2, where C is an elastic constant, for example the shear modulus in an isotropic system and, consequently, the dimensionless stress is $\tilde{\sigma}_{ij} = \sigma_{ij}/(Ce^2)$. Scaling the area and volume by ℓ^2 and ℓ^3, respectively, and using the nondimensional variables given above yields

$$\tilde{E}_t = \int_{\tilde{S}_\beta} \tilde{\gamma}(\mathbf{n}) d\tilde{S} - \frac{L}{2} \int_{\tilde{V}_\beta} \tilde{\sigma}_{ij} \tilde{\epsilon}_{ij}^T d\tilde{V}, \tag{7.93}$$

---
[137] W. C. Johnson and P. W. Voorhees, *Solid State Phen.* **23**, 87 (1992).
[138] J. E. Taylor, *Bull. Am. Math. Soc.* **84**, 197 (1978).

where $\tilde{\gamma} = \gamma/\gamma_o$ and $L = Ce^2\ell/\gamma_o$. As was seen with the dimensionless form of the Gibbs-Thompson equation, L is a measure of the magnitude of the elastic energy compared to the interfacial energy. Thus the interfacial energy is the dominant energy in setting the equilibrium shape of a particle for small particles, those for which $L \ll 1$, and elastic energy is the dominant energy for particle sizes for which $L \gg 1$. As a particle grows, L increases, because ℓ appears in L. Consequently, the equilibrium particle shape can be expected to change with particle size under certain conditions. Because L is proportional to e^2, small changes in the misfit strain can have large effects on the equilibrium shape of a particle.

The dependence of the system energy on precipitate shape, Eq. (7.92), usually must be obtained numerically. However, before examining numerical results, it is instructive to explore qualitatively the relative contributions of the interfacial and elastic energies to the total free energy for a precipitate of fixed shape, in order to identify underlying principles of the shape transitions. We do so by considering a spherical precipitate in an isotropic system. In this case, the interfacial energy density is independent of interface orientation and

$$\int_{S_\beta} \gamma(\mathbf{n})dS = \gamma \int_{S_\beta} dS = 4\pi R^2 \gamma \tag{7.94}$$

where R is the radius of the precipitate. Using Eq. (191) of Reference [22] for the strain energy associated with a spherical precipitate in an infinite isotropic system with a dilatational misfit strain, $\epsilon_{ij}^T = \epsilon^T \delta_{ij}$, the free energy of the isolated spherical precipitate is given by

$$E_t = 4\pi R^2 \gamma + \frac{18\mu^\alpha K^\beta (\epsilon^T)^2 V_\beta}{(3K^\beta + 4\mu^\alpha)} \tag{7.95}$$

where K and μ are the bulk modulus and shear modulus, respectively, of the indicated phase. The energy per unit volume of precipitate is

$$E_t/V_\beta = \frac{3\gamma}{R} + \frac{18\mu^\alpha K^\beta (\epsilon^T)^2}{(3K^\beta + 4\mu^\alpha)}. \tag{7.96}$$

For sufficiently small precipitate volumes, the interfacial energy makes the dominant contribution to the system energy and the equilibrium particle shape is expected to be dictated primarily by the interfacial energy while, for large precipitate volumes, the elastic energy is the dominant contribution to the total free energy, and the equilibrium particle shape is expected to be dictated primarily by the elastic energy. The interfacial and elastic energy contributions are equal at the critical size R_c, where

$$R_c = \frac{\gamma(3K^\beta + 4\mu^\alpha)}{6\mu^\alpha K^\beta (\epsilon^T)^2}. \tag{7.97}$$

The critical size at which the elastic energy becomes larger than the interfacial energy depends on the elastic constants of both phases, the interfacial energy density, and the square of the misfit strain.

Expressed in terms of the dimensionless parameters previously defined and using $\ell = R$, $C = K^\beta$, and $e = \epsilon^T$, yields for the dimensionless system energy, $\tilde{E}_t$,

$$\tilde{E}_t = 3 + L\frac{18\mu^\alpha}{(3K^\beta + 4\mu^\alpha)} \tag{7.98}$$

where $L = K^\beta(\epsilon^T)^2 R/\gamma$. If Poisson's ratio is taken as 1/3 and the precipitate and matrix elastic constants are equal, the elastic term appearing on the righthand side of Eq. (7.98) is equal to 3/2. Consequently, when $L \ll 1$, interfacial energy dominates total energy while the elastic energy is the dominant contribution to the system energy when $L \gg 1$. The interfacial and elastic energies are equal at the scaled particle radius L_c^* where

$$L_c^* = \frac{(3K^\beta + 4\mu^\alpha)}{6\mu^\alpha}. \tag{7.99}$$

The critical particle size does not necessarily correspond to the particle size at which a shape transition occurs.

d. Symmetry-conserving Shape Transitions. Two general types of stress-induced precipitate shape transitions have been observed experimentally with increasing particle size in two-phase alloys possessing coherent precipitates. The first type is classified as symmetry conserving. This means that the symmetry of both the initial and final precipitate shapes is equal to or greater that the symmetry resulting from the intersection of the point groups of the precipitate and matrix phases.[137,139,140] An example of a symmetry-conserving shape transition was evident in Figure 1. In this case, both the precipitate and matrix phases possess cubic symmetry and are coherent. If the equilibrium shape of an isolated precipitate is determined solely by the interfacial energy, then, according to Neumann's principle, the shape of the γ' precipitate must have a symmetry given by Eq. (7.86). At early times the precipitate is a sphere while at later times it has a fourfold rotation axis, eventually becoming a cube. The symmetry of each of these shapes satisfies Eq. (7.86). This transition from a sphere to a fourfold symmetric shape is also common in other Ni-Al based alloys.[141,142] Other documented symmetry-conserving shape transitions include the transitions

---

[139] W. C. Johnson and J. W. Cahn, *Acta metall.* **32**, 1925 (1984).
[140] W. C. Johnson, M. B. Berkenpas, and D. E. Laughlin, *Acta metall.* **36**, 3149 (1988).
[141] A. J. Ardell and R. B. Nicholson, *Acta metall.* **14**, 1295 (1966).
[142] M. Doi, T. Miyazaki, and T. Wakatsuki, *Matls. Sci. and Engr.* **74**, 139 (1985).

between cubes, octahedra, and tetrakaidecahedra for Co-rich precipitates in Cu-Co alloys.[143,144]

The second type of elastically induced shape transition is termed symmetry-breaking.[139] In this case, the symmetry of the final precipitate shape is less than the symmetry resulting from the intersection of the point groups of the precipitate and matrix phases. The sphere-to-ellipsoid transition in Al-Zn[145] or the cube-to-cuboid transition visible in Figures 1 and 3 are examples of symmetry-breaking shape transitions. It is important to remember that stress-induced precipitate shape transitions do not violate Neumann's principle, because different variants of precipitate orientation will exist. When a macroscopic region of the two-phase system is considered, the symmetry of the matrix crystal is recovered.

In addition to the anisotropy of the interfacial energy, the equilibrium precipitate shape depends on the elastic properties of the precipitate and matrix phases as well as the misfit strain. The effect of anisotropy in the interfacial energy density and elastic heterogeneity on the (three-dimensional) equilibrium precipitate shape can be demonstrated by restricting the particle morphology to certain classes of geometrical shapes, such as ellipsoids[139] or tetrakaidecahedra.[144] Figure 37 shows a tetrakaidecahedra with facets along the $\langle 100 \rangle$ and $\langle 111 \rangle$ directions. $A = t/a$, where t and a are defined in Figure 37, is a shape parameter that indicates the degree of faceting. The range of A is $0 \leq A \leq 2/3$. $A = 0$ corresponds to an octahedron and $A = 2/3$ to a cube with faces perpendicular to the $\langle 100 \rangle$ directions of the cubic matrix. All other values of A represent a tetrakaidecahedron. This class of shapes is observed in the Cu-Co system.[143,144]

The dependence of the elastic strain energy on the shape parameter A is shown in Figure 38 for the case of a coherent, tetrakaidecahedral-shaped Co particle in a Cu matrix.[144] Although the parameter A cannot be applied to a spherical precipitate, the strain energy of a tetrakaidecahedron can be compared with that of a sphere by normalizing its strain energy by the elastic energy of a spherical precipitate. The energy of the sphere thus corresponds to one (horizontal line) in Figure 38. Calculations of the elastic energy were performed using Eq. (7.92). The elastic constants of the matrix and particle are those of Cu and Co, respectively. The misfit is dilatational of magnitude -0.018. The cube ($A = 2/3$) has a lower elastic energy than the sphere, and the sphere a lower energy than the octahedron ($A = 0$) for both an elastically homogeneous system, in which the particle has elastic constants of Cu, and the heterogeneous system considered. Changing the elastic constants of either the precipitate or matrix phases, however, could change the precipitate shape that gives the lowest elastic energy.

---

[143] V. A. Phillips, *Acta metall.* **14**, 271 (1966).
[144] S. Satoh and W. C. Johnson, *Metal. Trans.* **23A**, 2761 (1992).
[145] G. Kostorz, *Physica* **120B**, 387 (1983).

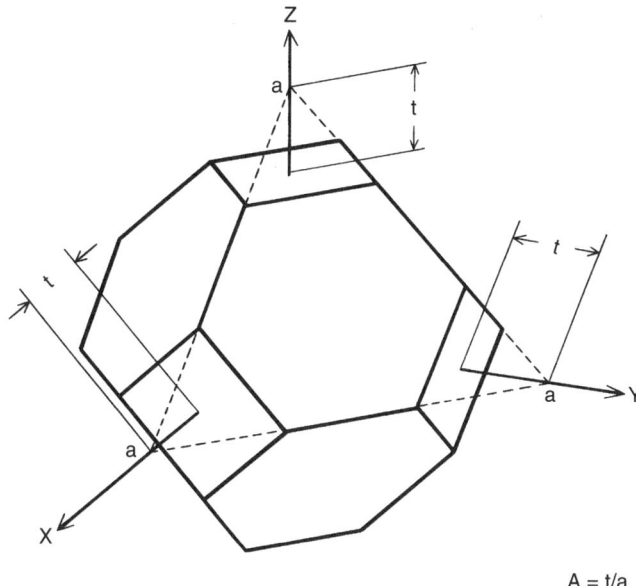

FIG. 37. A tetrakaidecahedron with facets along the ⟨100⟩ and ⟨111⟩ directions of the matrix phase is shown.[144] $A = t/a$ is a shape parameter used to describe the particle. $A = 0$ and $A = 2/3$ correspond to an octahedron and cube, respectively.

Predicted shape transitions for this tetrakaidecahedral system depend strongly on the anisotropy of the interfacial energy density. Figures 39 and 40 show the sum of the elastic and interfacial energies, E_{total}, normalized by the total energy of a spherical precipitate, as a function of particle volume, V_p, for several different shape parameters. When the interfacial energy is assumed to be isotropic with $\gamma_{\text{iso}} = 0.3 \text{Jm}^{-2}$, as in Figure 39, the sphere possesses the lowest energy shape for precipitate volumes less than about 3×10^5 nm^3, while at larger volumes the cube ($A = 2/3$) has the lowest total energy. (The relative energy associated with a spherical precipitate of a given volume is represented by the horizontal line in Figs. 39 and 40.) Figure 39 is the three-dimensional equivalent of the two-dimensional situation depicted in Figure 44 and, on the basis of the two-dimensional calculations, it should be expected that the three-dimensional shape does not necessarily change discontinuously from sphere to cube, but continuously in a manner similar to that depicted in Figure 44. Thus, for the case of an isotropic interfacial energy density and dilatational misfit in a cubic matrix, one would not expect to observe a precipitate shaped as a tetrakaidecahedron, but rather a gradual transition from a sphere to a cube. This sequence of shape transitions is precisely that observed in the Ni-Al alloys of Figures 1–3.

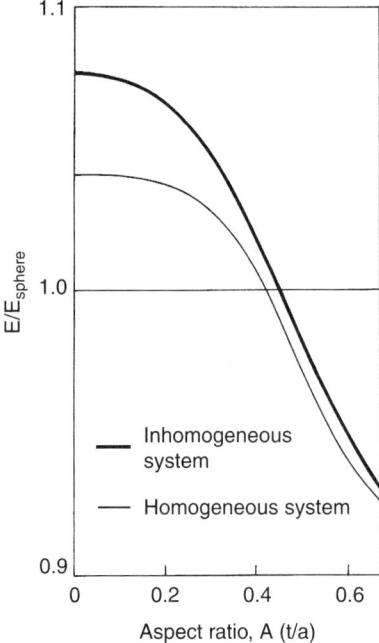

FIG. 38. The elastic strain energy of a tetrakaidecahedron, normalized to the elastic energy of a system with a spherical precipitate having the same elastic constants as the matrix, is shown as a function of the shape parameter, A.[144] The matrix and precipitate elastic constants for the elastically inhomogeneous system are those of Cu and Co, respectively. For the elastic constants employed, the cube ($A = 2/3$) has the lowest elastic energy.

Figure 40 represents the case where all material parameters are the same as those used in Figure 39 but the interfacial energy density is anisotropic. Energy cusps are assumed to exist along the $\langle 100 \rangle$ and $\langle 111 \rangle$ directions with the lowest energy in the $\langle 111 \rangle$ directions. The magnitudes of the energies of various crystallographic orientations are $\gamma_{111}/\gamma_{iso} = 0.7$ and $\gamma_{100}/\gamma_{iso} = 0.8$ where γ_{iso} is the interfacial energy density for all other orientations. At small precipitate sizes, a tetrakaidecahedron of specific shape parameter $A \approx 0.4$ possesses the lowest total energy. The energy extremizing shape changes with increasing particle size; for volumes $V_p > 10^5$ nm^3, the cube becomes the equilibrium precipitate shape. The equilibrium shape at large particle sizes for the examined class of precipitate shapes is a cube in both Figures 39 and 40, even though the interfacial energy densities are different. This is a result of the cube possessing the lowest elastic energy for the permissible particle shapes. Changes in the interfacial anisotropy, misfit strain, and elastic constants of precipitate and matrix change the equilibrium shape.

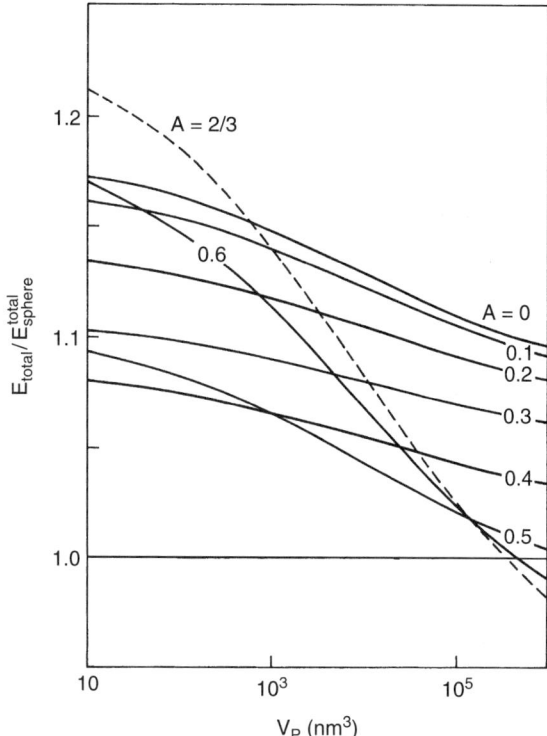

FIG. 39. The sum of the elastic and interfacial energies of a tetrakaidecahedron, normalized to the total energy of an elastically homogeneous sphere, is shown as a function of precipitate size for various shape parameters, A, assuming the interfacial energy is *isotropic*.[144] The shape parameter or amount of facetting changes with increasing particle size. For $V_p < 2 \times 10^5$ nm^3, the sphere (horizontal line) is the lowest energy shape. For large particles, the equilibrium shape is a cube ($A = 2/3$).

e. Symmetry-breaking Particle Shape Transitions. The results just discussed illustrate how the equilibrium particle shape changes with particle size owing to the interplay between the interfacial and elastic energies for a fixed class of symmetry-conserving particle shapes. However, when the precipitate shape that minimizes the elastic energy possesses a symmetry lower than that of the intersection of the point groups of the matrix and precipitate crystal lattices, a symmetry-breaking shape transition can occur during the growth of the precipitate.[139] The nature of the symmetry-breaking transitions (i.e., whether the shape change occurs continuously or discontinuously) can be understood most easily by first restricting the possible precipitate shapes to certain geometrical classes. This approach is certainly not general and the predicted precipitate shapes do not necessarily satisfy all thermodynamic equilibrium conditions. However, this approach does allow

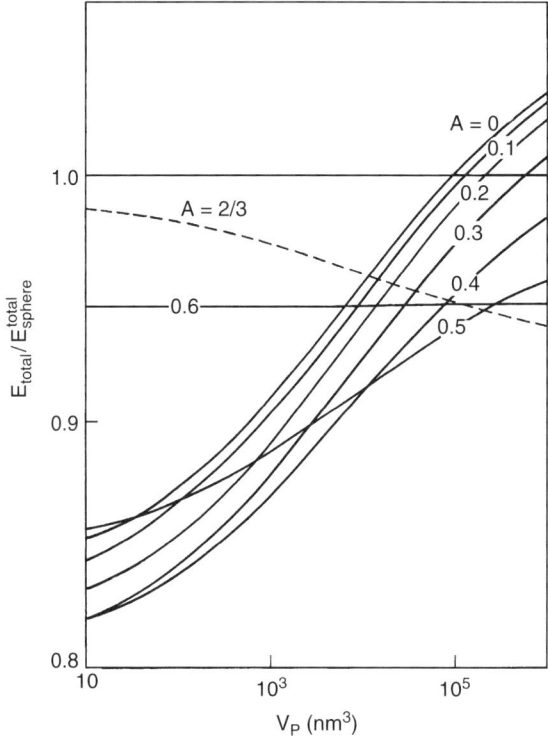

FIG. 40. The sum of the elastic and interfacial energies of a tetrakaidecahedron-shaped Co particle in a Cu matrix, normalized to the total energy of an elastically homogeneous sphere, is shown as a function of precipitate size for various shape parameters.[144] The interfacial energy is *anisotropic* with the interfacial energy in the $\langle 111 \rangle$ and $\langle 100 \rangle$ orientations assumed to be 70% and 80% of that in the other directions ($\gamma_{iso} = 0.3 \text{Jm}^{-2}$), respectively. The shape parameter, or extent of faceting, changes with increasing particle size. For $V_p < 10^5$ nm^3, a tetrakaidecahedron has the lowest energy. For larger particles, the equilibrium shape is a cube ($A = 2/3$).

the qualitative features of the symmetry-breaking transition to become more apparent. It also has the advantage of providing an estimate of the particle size at which the shape transition occurs and the ability to find analytic relationships between various materials parameters that predict how the shape transition can occur.

In order to illustrate the advantages of the bifurcation approach, we consider an isolated precipitate in a cubic matrix whose shape is restricted to be an ellipsoid of revolution.[139] The misfit strain is dilatational, the interfacial energy density is isotropic, and the system can be elastically heterogeneous. The particle's shape and size are completely defined by its major axes or, equivalently, by its volume and a shape parameter, $S(-1 < S < 2)$, given by

$$S = 2(a_3 - a_1)/(a_3 + 2a_1) \tag{7.100}$$

where the a_i are the axes of the ellipsoid (with $a_1 = a_2$). The particle is an oblate spheroid for $S < 0$ and a prolate spheroid for $S > 0$. When $S = 0$, the shape is a sphere.

The total energy, E_t, defined as the sum of the elastic strain (E_{strain}) and interfacial (E_{surf}) energies, is still determined from Eq. (3.57) and, for a given set of material parameters, can be expressed as a function of the particle shape parameters S and V.

$$E_t(S, V) = E_{\text{strain}}(S, V) + E_{\text{surf}}(S, V). \tag{7.101}$$

The energy-extremizing particle shapes are those shapes that render the total energy a minimum (or maximum) for a given precipitate volume. These shapes are given by solutions to

$$\left(\frac{\partial E_t}{\partial S}\right)_V = 0. \tag{7.102}$$

Formal analysis of Eq. (7.102) first requires solving for the elastic field of an ellipsoid of revolution and then calculating the elastic and interfacial energies associated with a precipitate in terms of S and V. The resulting expression is differentiated with respect to S at constant volume and set equal to zero. However, additional insight into the nature of particle shape changes can be obtained by expanding Eq. (7.101) in a Taylor series about $S = 0$ for fixed material parameters and particle volume before differentiation. Assuming a dilatational misfit strain and an isotropic (or cubic) system, symmetry requires[139]

$$E_t = E_0 + \frac{1}{2}E_2(V)S^2 + \frac{1}{6}E_3(V)S^3 + \frac{1}{24}E_4(V)S^4 + \cdots \tag{7.103}$$

where the E_i are the Taylor coefficients. The term linear in S does not appear in the absence of an external stress. Differentiating Eq. (7.103) with respect to S and setting the result to zero yields

$$\left(\frac{\partial E_t}{\partial S}\right)_V = 0 = S\left\{E_2(V)S + \frac{1}{2}E_3(V)S^2 + \frac{1}{6}E_4(V)S^3 + \cdots\right\}. \tag{7.104}$$

One solution to Eq. (7.104) for all particle volumes is a sphere, $S = 0$. Other possible solutions for the extremizing particle shape exist when the term in brackets vanishes. The stability of the solution is determined by the sign of $E_2(V)$.

Application to Isotropic Ellipsoids of Revolution. The analysis just described has been applied to an elastically isotropic and heterogeneous system with a dilatational misfit strain ϵ^T.[139] Expanding the interfacial and elastic energies[146] of an isolated ellipsoid of revolution in an infinite matrix in a Taylor series about $S = 0$, yields the following expression for E_2:

$$E_2(V_\beta) = \frac{8}{5}(1 - \Lambda)4\pi\gamma\left(\frac{3V_\beta}{4\pi}\right)^{2/3}. \tag{7.105}$$

Λ is a dimensionless particle size defined by

$$\Lambda = (V_\beta/V_c)^{1/3} \tag{7.106}$$

where V_c is a critical particle volume defined by

$$V_c^{1/3} = \left(\frac{4\pi}{3}\right)^{1/3} \frac{\gamma[\delta(1 + \nu^\beta) + 2(1 - 2\nu^\beta)]^2[7 - 5\nu^\alpha + 2\delta(4 - 5\nu^\alpha)]}{27\delta(1 - \delta)(1 - \nu^\alpha)(1 + \nu^\beta)^2\mu^\alpha(\epsilon^T)^2}. \tag{7.107}$$

The superscript β denotes the particle phase and $\delta = \mu^\beta/\mu^\alpha$ is the ratio of the particle and matrix shear moduli. The sign of the Taylor coefficient E_2 is determined by the sign of $1 - \Lambda$. If the particle is elastically harder than the matrix, $\delta > 1$, then $\Lambda < 0$ and $E_2(V_\beta) > 0$ for all particle volumes. This result indicates that the sphere is a stable equilibrium shape with respect to ellipsoids of revolution for all particle volumes. For elastically soft particles, $\delta < 1$, $\Lambda > 0$. $E_2(V_\beta)$ is positive when $\Lambda < 1$ and negative when $\Lambda > 1$. Because Λ depends on V_β, $1 - \Lambda > 0$ for small particle sizes and $1 - \Lambda < 0$ for sufficiently large particles. This means that the sphere is a stable shape for smaller particles but will lose its stability with respect to an ellipsoid of revolution when $\Lambda = 1$ or $V = V_c$.

An analysis of the equilibrium shapes shows that, for various combinations of materials parameters, there exists either one or three energy-extremizing particle shapes that are ellipsoids of revolution for a given particle volume.[135,145] This result is depicted in Figure 41 where the energy-extremizing particle shape (S) is plotted as a function of the scaled particle volume, assuming $\delta < 1$. The solid lines in Figure 41 represent energy minima and the dashed lines energy maxima. The absolute minimum is given by the heavy solid line and the relative minima by the thinner solid line. The sphere ($S = 0$) is an energy-extremizing shape for all precipitate sizes: It is an energy minimum for $\Lambda < 1$ and an energy maximum for $\Lambda > 1$. The sphere loses stability with respect to an ellipsoid of revolution at a critical size given by $\Lambda = 1$, when two extremizing solutions to Eq. (7.104) intersect at the bifurcation point.[139] (In this case, one extremizing solution is given

---

[146] D. Barnett, J. Lee, H. Aaronson, and K. Russel, *Scripta metall.* **8**, 1447 (1974).

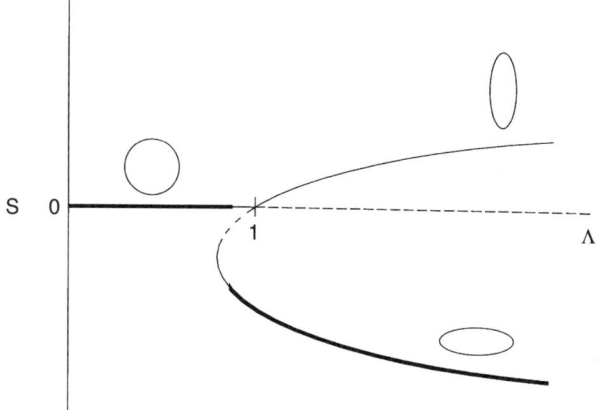

FIG. 41. A bifurcation diagram is shown depicting the three-dimensional shape transition from sphere to ellipsoid of revolution as a function of dimensionless precipitate size, Λ. Heavy solid lines are absolute minima, thin solid lines are relative minima, and broken lines are energy maxima. The sphere is stable at small precipitate sizes and loses stability with respect to a prolate spheroid at the bifurcation point $\Lambda = 1$. However, the oblate spheroid ($S < 0$) has the lowest energy.

by the sphere, i.e., the line $S = 0$, and the other by the parabola-shaped curve corresponding to an ellipsoid of revolution.)

For the three-dimensional problem, the elastically induced shape bifurcation is termed transcritical. The solution is not symmetric about $S = 0$ and the absolute minimum in energy (heavy line) jumps discontinuously from a sphere to an oblate spheroid before the bifurcation point at $\Lambda = 1$ (or $V_\beta = V_c$) is reached. The degree of transcriticality, the distance between the nose to the curve and the $S = 0$ axis is a function of the material parameters. This indicates that the nature of the particle-shape bifurcation is a strong function of values of the elastic constants of the system.

f. External Stress Field: Breaking the Bifurcation. The presence of an externally applied elastic load not only changes the equilibrium shape of the precipitate[147] but also acts to "break" the transcritical[140] and supercritical[148] bifurcations of Figures 41 and 45, respectively. The manner in which the bifurcation is broken can be examined using bifurcation theory in conjunction with Curie's principle. Figure 42 illustrates how the three-dimensional equilibrium precipitate shape for a set of ellipsoids of revolution with axes aligned along the principal directions of the cubic matrix is influenced by a uniaxial stress field applied along the x_3 axis. The shape parameter, S, that extremizes the system energy for a given precipitate

---

[147] J. K. Tien and S. M. Copley, *Metall. Trans.* **2**, 215 (1971).
[148] M. B. Berkenpas, W. C. Johnson, and D. E. Laughlin, *J. Mater. Res.* **1**, 635 (1986).

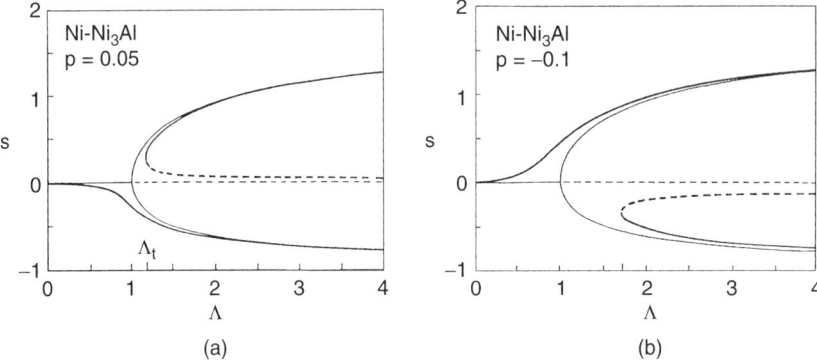

FIG. 42. Two bifurcation diagrams depicting the three-dimensional equilibrium shape parameter S as a function of dimensionless precipitate size (Λ) in the presence of an externally applied uniaxial stress are shown.[140] Solid lines are minima and dashed lines energy maxima. The heavy lines pertain to a system with a uniaxial stress applied along the axis of revolution of the ellipsoidal-shaped precipitates, while the thinner lines, which show a bifurcation point at $\Lambda = 1$, depict the energy extrema in the absence of external stress. The perturbation parameter, p, is positive in (a) and negative in (b). Calculations are based on ellipsoidal-shaped precipitates in a cubic matrix with material parameters corresponding to the $\gamma - \gamma'$ Ni-Al system.

volume is once again plotted as a function of the nondimensional particle size Λ. The analysis used to obtain Figure 42 was performed assuming the cubic elastic constants of the matrix and precipitate phases to be those of nickel (γ) and Ni$_3$Al (γ'), repsectively. The heavy curves indicate the precipitate shapes that extremize the system energy in the presence of the uniaxial applied stress and, for comparison, the fine lines denote equilibrium shapes in the absence of an applied stress. The solid and dashed lines correspond to energy minima and maxima, respectively. The behavior depicted by the fine lines, the shape changes in the absence of applied stress for the elastically anisotropic system, are qualitatively similar to those of Figure 41 for the isotropic system.

As recalled from our earlier discussion of Curie's principle, application of a uniaxial stress reduces the symmetry of the interfacial energy density and, consequently, the symmetry of the precipitate shapes that would be expected to form during precipitate growth. This reduction in symmetry is reflected in the bifurcation plots of Figure 42. The two energy-extremizing solutions no longer intersect to form a bifurcation point, but are broken into two nonintersecting solutions. Hence, the applied stress is said to "break the bifurcation."

The extent to which the applied stress breaks the bifurcation, the direction in which it does so, and the material parameters that determine the break can be obtained by considering a perturbation parameter within bifurcation theory.[149]

---

[149] G. Iooss and D. D. Joseph, *Elementary Stability and Bifurcation Theory*, Springer-Verlag, New York (1980).

The analysis requires the system energy given by Eq. (7.101), including the energy contributed by the interaction of the particle with the applied field, to be expanded in a Taylor series about $S = 0$. This analysis results in the appearance of a linear term in S that is not present in the absence of an applied stress. The linear term is found to be $E_1(V_\beta)S = p\Lambda_S$ where the perturbation parameter, p, is given by[140]

$$p = \frac{\tau(\Delta C_{11} - \Delta C_{12})H}{\epsilon^T} \qquad (7.108)$$

where τ is the applied stress, H is a numerically determined function of the precipitate (β) and matrix (α) elastic constants, and $\Delta C_{ij} = C_{ij}^\beta - C_{ij}^\alpha$. The sign of p gives the direction on the bifurcation diagram in which the bifurcation is broken, while the magnitude of p gives the extent to which it is broken. Changing the sign of the applied stress, precipitate misfit, or difference in elastic constants changes the sign of p and, therefore, the direction in which the bifurcation is broken. For example, if the material parameters are such that $p > 0$, as in Figure 42(a), an oblate spheroid ($S < 0$) is the lowest energy shape for all precipitate sizes. For $\Lambda > \Lambda_t$, another energy-extremizing solution emerges and a prolate spheroid ($S > 0$) is also an energy-minimizing precipitate shape. These solutions do not intersect when an external field is present and, consequently, there is no bifurcation point. The prolate spheroids are metastable in that they give only a local minimum in the energy. Different behavior is predicted for negative values of the perturbation parameter, p, as shown in Figure 42(b). For small precipitate sizes, a prolate spheroid has the lowest energy. A new energy-extremizing solution appears for $\Lambda > \Lambda_t$, corresponding to an oblate spheroid. Depending on the material parameters, the prolate spheroid can remain the lowest energy shape for all precipitate sizes or the lowest energy shape can jump discontinuously from a prolate spheroid to an oblate spheroid when the precipitate is sufficiently large.

As indicated by Eq. (7.108), the sign of the perturbation parameter is determined by the signs of the misfit strain (ϵ^T), the applied stress (τ), and the difference in the precipitate and matrix elastic constants. For example, if $\epsilon^T > 0$ and $\Delta C_{11} - \Delta C_{12} > 0$, then an applied tensile stress renders $p > 0$; the precipitate would be expected to grow as an oblate spheroid with its axis of revolution aligned along the direction of the applied stress according to Figure 42(a). If a compressive stress were applied to the system, $p < 0$, and, according to Figure 42(b), the growth of a prolate spheroid would be favored. Such precipitate shape changes in the presence of a uniaxial stress field have been observed experimentally.[147]

The bifurcation plots predict the equilibrium precipitate shape as a function of precipitate size. The change in stability of a solution (precipitate shape) with increasing precipitate size becomes clearer when the system energy in the presence of an external field is examined. Figure 43 illustrates how the sum of the interfacial

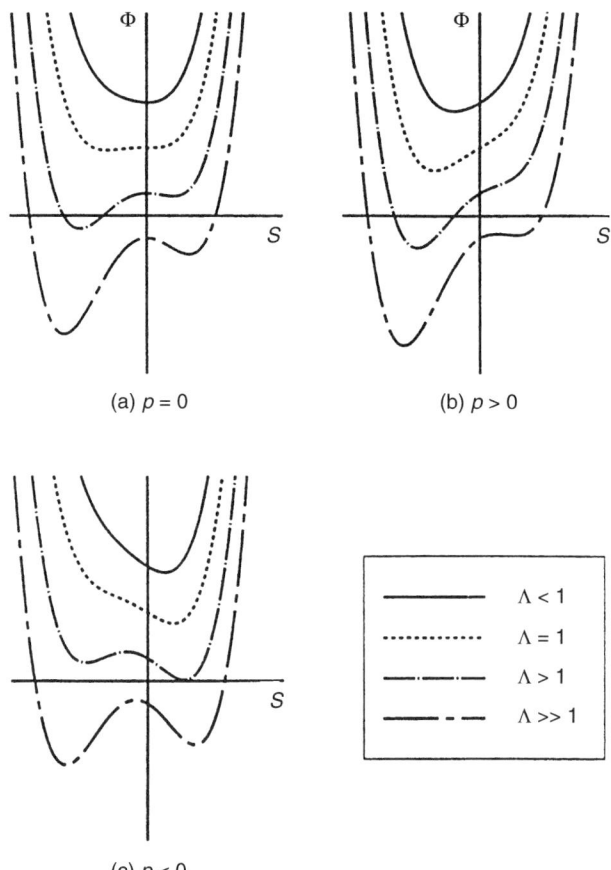

FIG. 43. The sum of the interfacial and elastic energies for a class of misfitting, coherent ellipsoidal precipitates, Φ, is plotted as a function of the three-dimensional shape parameter S for four different precipitate sizes, Λ.[140] In (a) there is no applied field and the equilibrium shape changes discontinuously from a sphere ($S = 0$) to an oblate spheroid ($S < 0$). In (b) the perturbation parameter p is positive and the equilibrium shape is an oblate spheroid for all precipitate sizes. In (c) the perturbation parameter p is negative, and the absolute minimum in the energy jumps discontinuously from a prolate spheroid to an oblate spheroid. A sphere ($S = 0$) is never an energy-extremizing shape when an external field is applied to the system, as seen in (b) and (c).

and elastic energies of the precipitate, Φ, depends on the shape parameter S for four different nondimensional precipitate sizes (Λ). In the absence of applied stress ($p = 0$), Figure 43(a), the sphere is the lowest energy shape until just before the bifurcation point ($\Lambda = 1$), at which point the energy-minimizing shape jumps discontinuously to an oblate spheroid (compare the bifurcation diagram of Figure 41). When the perturbation parameter is positive ($p > 0$), as shown in

Figure 43(b), the oblate spheroid has the lowest total energy for all precipitate sizes. This situation corresponds to Figure 42(a). When $p < 0$, as shown in Figure 43(c), the prolate spheroid has the lowest energy for small precipitate sizes, while at larger sizes the oblate spheroid has the lowest energy. This change in stability would be reflected by a discontinuous jump of the heavy line in Figure 42(b). The presence of an energy barrier separating the two equilibrium shapes suggests that different microstructures might be obtainable in the presence of an external stress. If a precipitate grows in the absence of stress and an external field is applied after the precipitate has exceeded a critical size so that $p > 0$, the precipitate shape would be that of an oblate spheroid. If the external stress field is present from the initiation of precipitate growth, the precipitate shape could be a (metastable) prolate spheroid.

g. Equilibrium Shapes with No Restriction on Particle Morphology. As already illustrated, constraining the shape of a particle to a certain class of geometric shapes allows the elastic energy of the system and the interfacial energy to be determined analytically. Certain geometric shapes, such as ellipsoids, are particularly convenient choices as the stress inside the particle is a constant. A drawback of this approach is that these morphologies are not equilibrium particle shapes; i.e., the particle shapes do not allow all conditions for thermodynamic equilibrium to be satisfied. Consequently, a form of constrained equilibrium, or equilibrium within a constraint of a certain class of shapes, is being determined. As the effects of this constraint cannot be assessed *a priori*, it is necessary to determine equilibrium shapes and particle shape bifurcations for particles satisfying the conditions of thermodynamic equilibrium.

A particularly advantageous approach to calculating numerically the equilibrium shape of a particle is to employ an equilibrium condition that pertains only to the shape of the particle. This approach obviates the necessity of placing computational meshpoints within the bulk phases in order to determine the equilibrium particle shape. It also avoids having to extremize the energy of the system as given in Eq. (3.57), because this expression for the energy involves an integral over the precipitate volume. Satisfying the equilibrium conditions guarantees that the energy is extremized, but does not allow a minimum to be distinguished from a maximum, nor can the energy difference between energy minima be determined.

Governing Equation for the Precipitate Shape. In this section we derive a governing equation that can be used to determine the equilibrium shape of a coherent particle (β phase) in an infinite α matrix. The system is a binary alloy with vacancies. The equilibrium conditions require that, for local equilibrium, the diffusion potentials M_{AV} and M_{BV} must be equal point-to-point across the interface and, for global equilibrium, constant throughout the system. Taking the lattice parameters of the α and β phases to be independent of concentration and using Eqs. (7.42) and

(7.43), the two equilibrium conditions for the equality of the diffusion potentials at the interface can be written in terms of the chemical potentials as

$$\mu_A^{v\alpha} - \mu_V^{v\alpha} = \mu_A^{v\beta} - \mu_V^{v\beta}$$
$$\mu_B^{v\alpha} - \mu_V^{v\alpha} = \mu_B^{v\beta} - \mu_V^{v\beta}. \tag{7.109}$$

Rearranging Eq. (7.109) yields

$$\mu_V^{v\beta} - \mu_V^{v\alpha} = \mu_A^{v\beta} - \mu_A^{v\alpha}$$
$$\mu_V^{v\beta} - \mu_V^{v\alpha} = \mu_B^{v\beta} - \mu_B^{v\alpha}. \tag{7.110}$$

Comparison of Eqs. (7.110) shows that, for local equilibrium to obtain at the interface, the jump in each of the three chemical potentials across the interface must be equal:

$$[\![\mu_A^v]\!] = [\![\mu_B^v]\!] = [\![\mu_V^v]\!]. \tag{7.111}$$

Thus the two interfacial equilibrium conditions on the diffusion potentials can be expressed as one equation, which is a function of the jump in the chemical potentials of the vacancies at the interface. The jump in the vacancy chemical potential at a point on the interface can be expressed in terms of the local strain field and interfacial curvature using the interfacial condition on the jump in the grand canonical free energies, Eq. (7.38), in the limit that the reference pressure vanishes, $P^o = 0$. Thus the following condition must hold at equilibrium:

$$[\![\mu_V^v]\!] = \frac{1}{2}[\![S_{ijkl}\sigma_{ij}\sigma_{kl}]\!] - \sigma_{ij}^\beta[\![\epsilon_{ij}]\!] + 2\gamma\kappa. \tag{7.112}$$

Equation (7.112) is a local equilibrium condition. Global equilibrium is obtained when the righthand side of Eq. (7.112) is a constant everywhere along the particle–matrix interface. If Eq. (7.112) is not constant along the interface, net exchange of atoms or vacancies would occur to lower the free energy. For particles with interfaces that are noncircular, the terms involving the elastic stress must vary with position in such a fashion as to balance the nonuniform $2\gamma\kappa$ term. The value of $[\![\mu_V^v]\!]$ at equilibrium is a function of the size of the precipitate and the elastic stress field. For example, in the limit of no elastic stress, Eq. (7.112) becomes $[\![\mu_V^v]\!] = 2\gamma\kappa$. Because γ is assumed to be isotropic and constant, the mean curvature of the equilibrium particle's interface must also be a constant. This result requires that the equilibrium shape of the particle must be a circle of radius R. Thus, $M_{AV}^e = M_{BV}^e = 2\gamma/R$, and the value of the equilibrium diffusion potential is set by the size of the particle.

When the particle possesses a misfit strain, it is necessary to calculate the corresponding stress and strain fields appearing in the interfacial condition, Eq. (7.112), in order to determine the equilibrium shape of the particle. One approach is to use the elastic Green's functions to calculate the elastic strain field.[150–152] This appraoch allows the stresses and strains to be expressed as integrals over the interface only.[3] For example, in a system where the elastic constants of the precipitate and matrix are identical, the displacement gradient, $u_{j,k}$, is given by

$$u_{j,k} = C_{ilmn}\epsilon^T_{mn} \int_{S_\beta} G_{ij,k}(\mathbf{x} - \mathbf{x}')n'_l dA', \quad (7.113)$$

where the integral is over the particle–matrix interface S_β and, in this equation, the primes denote the source point for the Green's function. Using Eq. (7.113), the relationship between the strain and displacement gradient, and Hooke's law allows the stress and strain fields at the interface appearing in Eq. (7.112) to be replaced by integrals over the interface of the particle.[150] The curvature of the interface involves derivatives of the interfacial shape (see, for example, the section on the stability of a solid–liquid interface under stress). Consequently, Eq. (7.112) can be recast as an integro-differential equation for the particle shape. Qualitatively, this equation reflects both the local effects of the interfacial energy, because it is a function only of the local interfacial curvature, and the nonlocal effects of the elastic stress, because the value of the strain at a point on the interface depends on the entire shape of a particle through the elastic integral over the precipitate–matrix interface. Equation (7.112) is nonlinear owing to the presence of both the curvature and elastic integrals and, consequently, numerical solutions are necessary to calculate the equilibrium particle shape. This approach can be generalized to systems where the elastic constants of the particle and matrix are different.[152,153] For an elastically heterogeneous system, boundary integral equations for the stress and strain fields are obtained, rather than an integral as in Eq. (7.113), which must be solved numerically.

Application to Two-dimensional, Elastically Anisotropic System. Here, Eq. (7.112) is solved for an elastically anisotropic cubic system in two dimensions assuming the interfacial energy is isotropic. To do so, the equation is first nondimensionalized using the interfacial energy density γ as the characteristic energy scale, as before, to give

$$[\![\mu^v_V]\!] = L \left\{ \frac{1}{2} [\![\tilde{S}_{ijkl}\tilde{\sigma}_{ij}\tilde{\sigma}_{kl}]\!] - \tilde{\sigma}^\beta_{ij}[\![\tilde{\epsilon}_{ij}]\!] \right\} + 2\tilde{\kappa} \quad (7.114)$$

---

[150] M. E. Thompson, C. S. Su, and P. W. Voorhees, *Acta metall. mater.* **42**, 2107 (1994).
[151] P. H. Leo, J. S. Lowengrub, and H. J. Jou, *Acta mater.* **46**, 2113 (1998).
[152] I. Schmidt and D. Gross, *J. Mech. Phys. Sold.* **45**, 1521 (1997).
[153] H. J. Jou, P. H. Leo, and J. S. Lowengrub, *J. Comp. Phys.* **131**, 109 (1997).

where $L = (\epsilon^T)^2 C_{44}\ell/\gamma$, ϵ^T is the magnitude of the dilatational component of the misfit strain, which is also chosen to scale the strain, $\tilde{\epsilon}_{ij}$, $\ell = A^{1/2}/\pi$, and A is the area of the particle. We track families of solutions to Eq. (7.114) using different values of L, because L is a function of the particle size and is easily related to experiment. Substituting for the dimensionless form of the stress and scaled strains, and using the dimensionless form of Eq. (7.113) yields the integro-differential equation that must be solved for the equilibrium particle shape.

In order to characterize the shape of the particle, we employ a Fourier expansion of the curvature as a function of interfacial arclength, s,

$$\kappa(s/s_T) = \sum_{n=-\infty}^{\infty} a_n \exp(-2\pi n s/s_T), \qquad (7.115)$$

where s_T is the total arclength. Thus, for particles with a fourfold symmetric shape, the real part of the first nonzero Fourier coefficient, other than a_o, is a_4^R. For particles that are twofold symmetric, the real part of the first nonzero Fourier coefficient, other than a_o, is a_2^R. These parameters thus can be used to track families of solutions to Eq. (7.114) that depend on these symmetries of particle shape.

Because this approach employs the equilibrium conditions to determine the particle shape, it is not possible to determine if the shapes yield either relative or absolute energy minima or maxima. Because the type of the extremum changes only at bifurcation points, this requires that the total energy of the system be determined for at least some particle areas, in particular near bifurcation points. The total elastic energy appearing in Eq. (3.57) can be written as an integral over the interface using the divergence theorem, and thus the total energy is given by

$$E_t = \gamma s_T - \frac{1}{2} \int_{S_\beta} \sigma_{ij} n_j u_i^T dA, \qquad (7.116)$$

where u_i^T is the displacement that yields the transformation strain.

Figure 44. gives numerical solutions to Eq. (7.114) that are symmetry-conserving precipitate shapes. We have taken the system to be a two-dimensional coherent precipitate embedded in a matrix with an axis of fourfold symmetry oriented perpendicular to the plane of the paper.[150] The elastic constants of Ni were employed in the calculation. As the particle size (L) increases, regions of low interfacial curvature appear along the two elastically soft $\langle 10 \rangle$ directions (parallel to the sides of the box), even though the interfacial energy is isotropic. This is because of the elastic energy: A square oriented with sides perpendicular to the $\langle 10 \rangle$ directions has a lower elastic energy than a circle for given material parameters. However, the interfacial energy term diverges at a sharp corner, so an intermediate shape that is square-like but with rounded corners and sides results. As seen here, observation

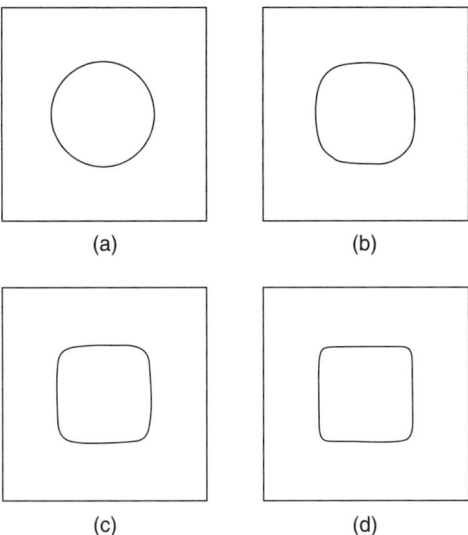

FIG. 44. Symmetry-conserving shape transitions are shown in two dimensions for increasing precipitate size (L).[150] Both precipitate and matrix phases possess cubic symmetry with a fourfold axis perpendicular to the plane of the paper. The sides of the box are parallel to the elastically soft ⟨10⟩ directions. (a) $L = 0$, (b) $L = 4$, (c) $L = 10$, (d) $L = 26$. All shapes are equilibrium shapes, but not necessarily energy minima.

of anisotropic particle shapes does not imply anisotropy of the interfacial energy density. Secondly, the change in equilibrium shape with particle size also shows that the anisotropic shape is not a result of an anisotropic interfacial energy density. Although the equilibrium particle shape changes smoothly with particle size, not all these shapes are energy minima.

If L is sufficiently large, symmetry-breaking precipitate shape transitions become possible. We choose a_2^R to characterize the classes of solutions to the equilibrium shape equation Eq. (7.112). As in the three-dimensional case, either one or three precipitate shapes exist that give constant diffusion potentials or, equivalently, extremize the system energy. One solution corresponds to fourfold symmetric shapes, $a_2^R = 0$, and the other to twofold symmetric shapes, $a_2^R \neq 0$. Owing to the fourfold symmetry of the matrix, the horizontally and vertically oriented shapes are identical and the second solution ($a_2^R \neq 0$) is symmetric about $a_2^R = 0$. For small sizes, the particle retains the fourfold symmetric shape. (These fourfold symmetric shapes are shown in Fig. 45.) In the vicinity of $L = 0$, the equilibrium shape is approximately a circle. However, the particle becomes progressively more square-like as the particle size increases. The fourfold shape loses stability with respect to a twofold symmetric shape at the point where the two energy-extremizing solutions intersect; i.e., at the bifurcation point corresponding to $L = 5.6$ for this

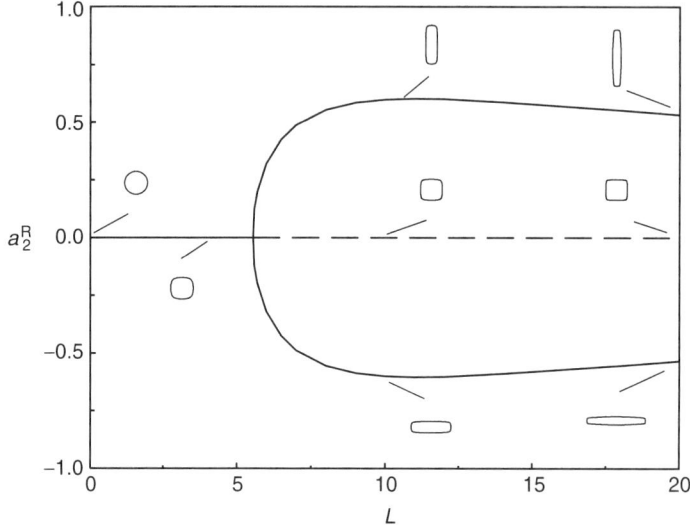

FIG. 45. A bifurcation diagram is shown depicting symmetry-conserving and symmetry-breaking two-dimensional shape transitions from circle to fourfold to twofold symmetric shapes as a function of dimensionless precipitate size, L.[150] Solid lines are minima and broken lines are energy maxima. a_2^R is a shape parameter that is zero for fourfold symmetric shapes and nonzero for twofold symmetric shapes. The shape bifurcation is supercritical in two dimensions.

choice of elastic constants. For $L > 5.6$, the fourfold shape is an energy maximum, depicted with the dashed line, and the twofold shape is an energy minimum. The importance of the elastic energy in setting the equilibrium shape is illustrated by comparing the twofold and fourfold shapes for $L = 10$. The fourfold shape clearly has less interfacial length than the twofold shape. However, the twofold shape is an energy minimum, despite the larger interfacial length, because the elastic energy of the twofold shape is less than that of the fourfold shape. This reflects the fact that the shape that minimizes the elastic energy is an infinitely long thin plate along the $\langle 10 \rangle$. The interfacial energy does not allow such particle shapes and, thus, a compromise is reached as shown with the twofold shape. In the two-dimensional case, the bifurcation is termed supercritical, because the solutions are symmetric about $a_2^R = 0$.

The symmetry of the bifurcation diagram shown in Figure 45 can be destroyed by small changes in the materials parameters that affect the crystallographic symmetry of the system. To illustrate this effect for the two-dimensional system, we fix the elastic constants to be those of Ni, but allow the misfit along the [01] direction to be 5% larger than along the [10] direction. Figure 46 shows that this small degree of tetragonality acts as an imperfection to the supercritical bifurcation; it is broken

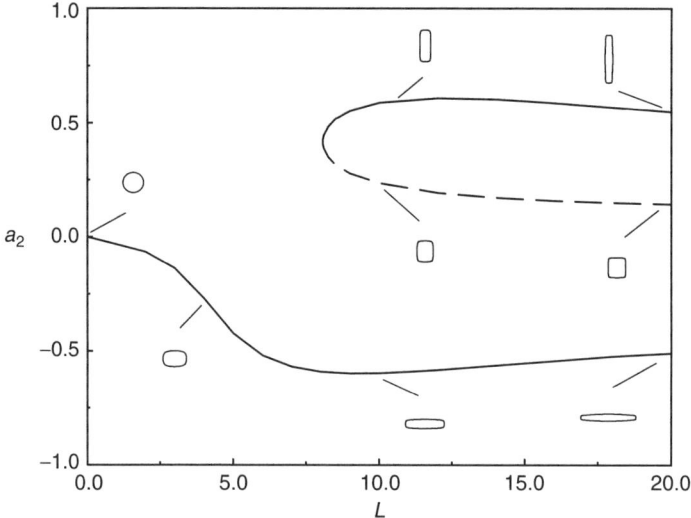

FIG. 46. A bifurcation diagram is shown depicting the evolution in particle shape for a tetragonal particle as a function of dimensionless precipitate size, L.[150] Solid lines are minima and broken lines are energy maxima. The tetragonality acts as a bifurcation to the supercritical bifurcation present for a dilatational misfit.

just as illustrated previously for the three-dimensional ellipsoidal particle under an applied stress. In this case the lower branch of solutions are absolute energy minimia and the upper branch are metastable or relatively energy minima. Thus, if a particle on the upper branch were shrinking, as may occur during Ostwald ripening, the morphology would jump discontinuously when the particle reaches the nose of the upper curve, or turning point, to the lower branch. More interesting is when the misfit along the [01] is 10% larger than along the [10] for $L = 20$. The equilibrium shape is shown in Figure 47. This particle shape is metastable, but is an energy minimum. It is clearly nonconvex, indicating that the equilibrium shape of a particle in the presence of elastic stress does not have to be convex, as is the case when only interfacial energy is present.

While these calculations illustrate the novel phenomena that are possible in systems with misfitting precipitates, they are for a two-dimensional, elastically homogeneous system. Schmidt and Gross have relaxed this assumption. They find that the aspect ratio of twofold-shaped particles increases as the shear modulus of the particle is decreased with respect to that of the matrix.[152] In addition, they find that the equilibrium shape of a particle with a dilatational misfit can be nonconvex, if the particle is sufficiently soft compared to the matrix. There have been limited calculations of the equilibrium shape of precipitates in three dimensions, however,

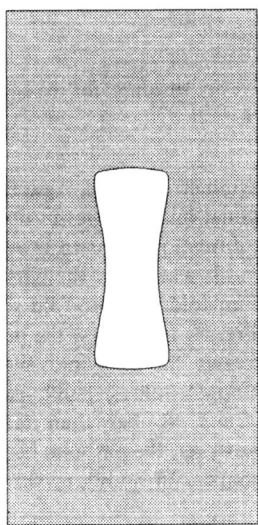

FIG. 47. The calculated particle shape for a two-dimensional particle with a tetragonal misfit strain showing that the equilibrium shape of a particle can be nonconvex.[150]

and they are usually limited to low values of L.[93,154,155] More recently Mueller and Gross have shown that if a three-dimensional particle is softer than the matrix, the equilibrium shape can be nonconvex.[154] Equilibrium particle shapes in three dimensions have also been calculated using a phase field approach.[156] The phase field approach is not based on the thermodynamics previously developed. It has the advantage that the interface between the particle and matrix, and thus the equations, are easier to solve numerically, but the disadvantage that the diffuse interface between the particle and matrix must be resolved.

VIII. Acknowledgements

We gratefully acknowledge the following grants, which have provided the long-term support necessary to undertake this work: The United States Department of Energy through Grant DE-FG02-99ER45771 (WCJ); the MRSEC Center for Nanoscopic Materials Design of the National Science Foundation under Award Number DMR-0080016 (WCJ); the National Science Foundation under Award DMR-9707073 and a NSF-NIRT grant DMR-0102794.

---

[154] R. Mueller and D. Gross, *Z. Angew. Math. Mech.* **80**, S397 (2000).

[155] R. Mueller and D. Gross, *Comp. Mater. Sci.* **16**, 53 (1999).

[156] J. Z. Zhu, Z. K. Liu, V. Vaithyanathan, and L.-Q. Chen, *Scripta Mater.* **46**, 401 (2002).

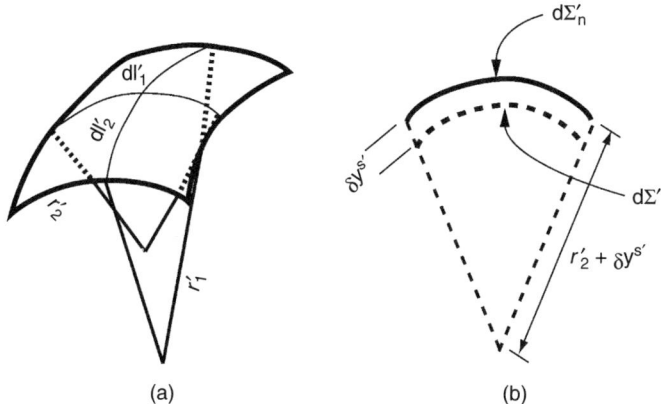

FIG. 48. (a) An element of surface, the elemental area with principal radii of curvature, r'_1 and r'_2, (b) The change in length of the interface with a perturbation along the direction of the principle radius.

IX. Appendix A: Surface Area Change Owing to Accretion

Accretion or dissolution at a crystal interface changes the interfacial area as measured in the referential state. In this appendix, we calculate the change in area per unit referential area owing to the addition of material of thickness $\delta y^{s'}$ to the interface. Consider an element of the interface, $d\Sigma'$, of area $\mathcal{A}'$ with principal radii of curvature R'_1 and R'_2 as shown in Figure 48. If $d\theta_1$ and $d\theta_1$ are elements of arclength, then

$$\mathcal{A}' = d\theta_1 d\theta_2 R'_1 R'_2. \tag{9.1}$$

The area of the surface element after accretion by an amount $\delta y^{s'}$ perpendicular to the surface, $\mathcal{A}'_p$ is

$$\mathcal{A}'_p = d\theta_1 d\theta_2 (R'_1 + y^{s'})(R'_2 + \delta y^{s'}). \tag{9.2}$$

Using Eqs. (9.1) and (9.2), the change in area per unit area to first order in $\delta y^{s'}$ is

$$\frac{\mathcal{A}'_p - \mathcal{A}'}{\mathcal{A}'} = \frac{(R'_1 + R'_2)\delta y^{s'}}{R'_1 R'_2} = 2\kappa' \delta y^{s'} \tag{9.3}$$

where $\kappa' = (\kappa'_1 + \kappa'_2)/2$ is the mean curvature and $\kappa'_1 = 1/R'_1$.

X. Appendix B: Continuity Condition at Two-Phase Crystalline Interface

Consider a point on the planar, coherent interface separating the α and β phases. Let $\delta\mathbf{y}$ be a perturbation in the position of the interface as measured in the actual system. Physically, this perturbation can accrue from two different sources: an elastic deformation of each crystal without undergoing a phase transformation or a phase transformation of one phase into the other (the addition or removal of lattice sites from one phase to another) without elastic deformation.[79,157] (Of course, a combination of the two is also possible.) Because the interface is constrained to be coherent, the α and β phases must remain in contact at the interface after the perturbation. This requires that, for every point on the interface,

$$\delta\mathbf{x}^\alpha + \delta\mathbf{y}^\alpha = \delta\mathbf{x}^\beta + \delta\mathbf{y}^\beta \tag{10.1}$$

or, in indicial notation,

$$\delta x_i^\alpha + \delta y_i^\alpha = \delta x_i^\beta + \delta y_i^\beta \tag{10.2}$$

where $\delta\mathbf{x}$ is the change in interfacial position of the indicated phase owing to a variation in the elastic field and $\delta\mathbf{y}$ is the change in interfacial position owing to a virtual phase transformation, measured in the direction of the outward-pointing normal of the indicated phase. $\delta\mathbf{y}$ is the displacement of the interface due to accretion.

The accretion term can be referred back to the reference state configuration as shown in Figure 48. It is simplest to assume that matter is accreted along the outward-pointing normal direction in the reference state; $\mathbf{n}^{\alpha'}$ for the α phase and $\mathbf{n}^{\beta'}$ for the β phase.

The accretion vectors in the actual and reference states can then be related using the deformation gradient tensor as in Eq. (3.5).

$$\delta y_i^\alpha = F_{ij}^\alpha \delta y_j^{\alpha'} = F_{ij}^\alpha n_j^{\alpha'} \delta y^{\alpha'} \tag{10.3}$$

and

$$\delta y_i^\beta = F_{ij}^\beta \delta y_j^{\beta'} = F_{ij}^\beta n_j^{\beta'} \delta y^{\beta'}. \tag{10.4}$$

Because the interface must remain coherent

$$\delta y^{\alpha'} \mathbf{n}^{\alpha'} = \delta y^{\beta'} \mathbf{n}^{\beta'} \quad \text{or} \quad \delta y^{\alpha'} n_j^{\alpha'} = \delta y^{\beta'} n_j^{\beta'}. \tag{10.5}$$

---

[157] W. Johnson, *J. Amer. Cer. Soc.* **77**, 1581 (1994).

As $\mathbf{n}^{\alpha'} = -\mathbf{n}^{\beta'}$, then

$$\delta y^{\alpha'} = -\delta y^{\beta'}. \tag{10.6}$$

Substituting Eq. (10.5) into Eq. (10.4) gives

$$\delta y_i^{\beta} = F_{ij}^{\beta} n_j^{\alpha'} \delta y^{\alpha'}. \tag{10.7}$$

Substituting Eqs. (10.7) and (10.3) into Eq. (10.2) and recognizing that the variation in the position coordinate of a given material point results from a variation in the displacement field,

$$\delta x_i = \delta(x_i' + u_i) = \delta u_i, \tag{10.8}$$

gives for the continuity condition at the two-phase coherent interface

$$\delta u_i^{\alpha} = \delta u_i^{\beta} + \left(F_{ij}^{\beta} - F_{ij}^{\alpha}\right) n_j^{\alpha'} \delta y^{\alpha'}. \tag{10.9}$$

XI. Appendix C: Continuity Condition at Crystal–Fluid Interfaces

The continuity condition at a crystal–fluid interface is obtained in a similar way to that used for the coherent interface. In performing the variations involving the change of phase at the crystal–fluid interface, the crystal and fluid phases must remain in contact after the virtual variation. This requires, for the varied state,

$$\delta \mathbf{x}^{\alpha} + \delta \mathbf{y}^{\alpha} = \delta \mathbf{y}^{f}. \tag{11.1}$$

Substituting Eqs. (10.3) and (10.8) into Eq. (11.1) gives

$$\delta u_i + F_{ij}^{\alpha} \delta y_j = \delta y_i^{f}. \tag{11.2}$$

Because the phase change is understood to occur normal to the interface in the reference state of the crystal, and normal to the actual state in the fluid,

$$\delta u_i + F_{ij}^{\alpha} n_j^{s'} \delta y = n_i^{f} \delta y^{f}. \tag{11.3}$$

Because $n_i^{f} = -n_i$,

$$-\delta y^{f} = n_i \left(\delta u_i + F_{ij}^{\alpha} n_j^{s'} \delta y\right). \tag{11.4}$$

Solid Solutions of Hydrogen in Complex Materials

REINER KIRCHHEIM

Institut für Materialphysik, Georg August Universität Göttingen Tammannstr. 1, D-37077 Göttingen, Germany

I.	Introduction	203
II.	Fundamental Properties of Hydrogen in Metals	207
	1. Equilibrium Pressure and Solubility	207
	2. Diffusivity	209
	3. Partial Molar Volume and Interaction with Stress and Strain Fields	212
III.	Behavior of Hydrogen in Defective and Disordered Metals	214
	4. Density of Site Energies (DOSE) and Fermi Dirac Statistics (FD-Statistics)	215
	5. Diffusivity	220
	6. H–H Interaction	227
IV.	Interaction of Hydrogen with Defects	228
	7. Interaction with Other Solutes and Vacancies	228
	8. Interaction with Dislocations	229
	9. Interaction with Grain Boundaries	237
	10. Interaction with Metal/Oxide Boundaries	243
	11. Defect Formation Energy	247
	12. Interaction with Crack Tips and Hydrogen Embrittlement	251
V.	Hydrogen in Disordered and Amorphous Alloys	252
	13. Disordered Crystalline Alloys	252
	14. Metal/Nonmetal Glasses	253
	15. Early Transition/Late Transition Metallic Glasses	257
	16. Bulk Metallic Glasses	259
VI.	Other Interstitials in Amorphous Materials	262
	17. Modeling Diffusion	263
	18. Small Molecules in Glassy Polymers	266
	19. Hydrogen in Amorphous Silicon and Germanium	273
	20. Ions in Oxidic Glasses	275
VII.	Hydrogen in Systems with Reduced Dimensions	278
	21. Thin Films	279
	22. Multilayers	283
	23. Clusters	288

I. Introduction

Hydrogen in metals has attracted considerable attention from physicists, chemists, material scientists, and engineers for many decades. Most of the exciting properties relate to the small size of the H atom, which leads to a high mobility in materials. For a metal, its diffusivity is very high at room temperature and may reach values that are the same as for ions in aqueous solutions. High H mobility results from two physical reasons. On the one hand, H atoms are dissolved interstitially and migrate via a direct interstitial mechanism which at dilute concentrations does not require the formation of vacancies. On the other hand a site exchange may occur via quantum mechanical tunneling.

The consequences of high H mobility include

(1) Thermal equilibrium established in rather short times at room temperature between the H dissolved in the metal and either hydrogen gas or protons in aqueous solutions. Thus thermodynamic properties, especially the chemical potential of hydrogen, can be obtained simply by measuring the partial pressure or the electrochemical potential.
(2) Hydrogen storage in metals and its use as an energy carrier becomes possible at room temperature.
(3) Easy redistribution of hydrogen and segregation at defects produced during plastic deformation, i.e. dislocations, crack tips, etc. This interaction gives rise to hydrogen embrittlement.

The small size of the H atom allows a large packing density in those metals that have a high affinity, i.e. large negative heat of solution, for hydrogen. In other metals H concentrations at reasonable H_2 pressures may be very low despite the same number density and size of interstitial sites in these metals. In metal hydrides the atomic density can be even larger than in liquid hydrogen. This property of the hydrides is advantageous when using hydrogen as a fuel, although the energy storage per weight is still too low for many applications. At present conventional storage alloys contain 2 wt.-% H, whereas figures above 3 wt.-% are required for use in vehicles. However, in rechargeable batteries metal hydrides are widely used. It will not be the purpose of this review to discuss issues related to hydrogen economy in general or the special role of hydrides. Research results in this area are published elsewhere.[1–5]

Hydrogen embrittlement is another technological subject of major concern, where the negative role of hydrogen is played at much lower concentrations. In iron base alloys, for example, embrittlement effects are observed at contents as

---

[1] International Symposium on Metal-Hydrogen Systems, Fundamentals and Applications, Stuttgart, Sept. 04–09, 1988, R. Oldenburg Verlag, München (1989).
[2] International Symposium on Metal-Hydrogen Systems: Fundamentals and Applications, Aug. 25–30, 1996, *J. Alloy Comp.* **253** (1997).
[3] International Symposium on Metal-Hydrogen Systems, Fundamentals and Applications, Oct. 04–09, 1998, *J. Alloy Comp.* **295** (1999).
[4] International Symposium on Metal-Hydrogen Systems, Fundamentals and Applications (MH2000), Oct. 01–06, 2000, *J. Alloy Comp.* **330** (2002).
[5] L. Schlapbach and A. Züttel, *Nature* **414**, 353 (2001).

low as a few ppm. Here the interaction of H atoms with lattice defects and their effect on plasticity has to be studied in detail. Again, the high mobility of H atoms is crucial for the phenomena, because hydrogen has to reach or follow moving dislocations or crack tips. It appears as if quantum mechanical diffusion plays a role as well. This effect is especially strongly pronounced for adjacent tetrahedral sites in body-centered cubic metals, which are only a very short distance apart, and leads to the result that ferritic steels (bcc-lattice) are more susceptible to hydrogen embrittlement than austenitic steels (fcc-lattice). Topics of embrittlement phenomena are treated in conferences[6–8] that seldom overlap with the ones on hydrogen storage.

Metal–hydrogen systems are often used as model systems to study physical or chemical properties and how they change with composition. This is often very easy because hydrogen can be doped in a controlled way by either measuring changes of the H_2 pressure in closed systems or by electrochemical deposition on a metal electrode applying Faraday's Law. This advantage of easy alloying is supported by the possibility of obtaining the chemical potential of hydrogen by measuring partial pressures and/or electrode potentials. Cases where metal–hydrogen has served as a model system include

(1) solute–solute interaction measured and interpreted in the framework of a quasi-chemical approach for the first time in the Pd-H system by Lacher,[9]
(2) tunneling as a diffusion mechanism for atoms in solids, discovered and discussed for hydrogen in metals,[10–12]
(3) the behavior of hydrogen in systems with reduced dimensions, studied nicely in metals,[13]
(4) hydrogen interaction with defects in metals as representative for other solute/solvent systems, studied extensively.[14] After knowing the basic features of the interaction, H atoms can be used as probes for the defects.[15,16]

---

[6] "Hydrogen Effects on Materials Behavior", eds. N. R. Moody and A. W. Thompson, TMS, Warrendale, Pennsylvania (1990).
[7] A. W. Thompson, and N. R. Moody, "Hydrogen Effects in Materials," TMS,Warrendale, PA (1994).
[8] International Conference on Hydrogen Effects on Materials Behavior and Corrosion Deformation Interaction, Jackson Lake Lodge, Moran, WY, Sept. 22–26, 2002, in press.
[9] J. R. Lacher, *Proc. Roy. Soc.* (London) A**161**, 525 (1937).
[10] C. P. Flynn and A. M. Stoneham, *Phys. Rev.* B**1**, 3966 (1970).
[11] J. Völkl and G. Alefeld in "Diffusion in Solids," ed. A. S. Nowick, J. J. Burton, Academic, New York (1979).
[12] H. K. Birnbaum and C. P. Flynn, *Phys. Rev. Lett.* **37**, 25 (1976).
[13] P. F. Miceli, H. Zabel, and J. E. Cunningham, *Phys. Rev. Lett.* **54**, 917 (1985).
[14] R. Kirchheim, *Progr. Mat. Sci.* **32**, 262 (1988).
[15] R. Kirchheim, *Encyclopedia of Materials Science*, Suppl. vol. 2, ed. R. W. Cahn, Pergamon Press, Oxford (1990).
[16] T. B. Flanagan, R. Balasubramaniam, and R. Kirchheim, *Platinum Metals Review* **45**, 114 and 166 (2001).

By gradually increasing the H concentration the sites of increasing energy within the defects are saturated successively and, therefore, a kind of spectroscopic method will be available (examples will be discussed throughout this study).

Another peculiarity of the lightest element is the large mass difference of its isotopes, giving rise to pronounced isotope effects in most of its properties. The different scattering length for neutrons, even changing sign from H to D, gave rise to an extensive use of neutron scattering and diffraction techniques in the area of metal–hydrogen systems.[17]

Hydrogen at high concentrations may change the physical properties of a material remarkably. Examples are changes of the magnetic coupling between ferromagnetic layers[18,19] or, even more exciting, a metals–insulator transition in yttrium going from the dihydride to the trihydride.[20] In addition, microstructural changes have been observed during alloying a metal with hydrogen. Examples are the generation of abundant vacancies[21] and dislocations,[22–24] the decomposition of miscible alloys,[25,26] and the improvement of mechanical properties of Ti-alloys by decreasing the grain size[27] or by variation of the α to β volume fraction.[28] Thus hydrogen can be used as a temporary alloying element, in order to set up a desired microstructure. In other materials such as semiconductors hydrogen is used as a permanent alloying addition for the purpose of saturating deep impurity levels or dangling bonds at the Si/SiO_2 interface as well as in amorphous silicon.[29]

The present study will focus on solid solutions of hydrogen in metals, so the properties of hydrides will be omitted and the reader is referred to monographs

---

[17] D. K. Ross, "Hydrogen in Metals III," in *Topics in Applied Physics*, vol. 73, ed. H. Wipf, Springer, Berlin (1997).

[18] D. Nagengast, C. Rehm, F. Klose, and A. Weidinger, *J. Alloy Comp.* **253**, 347 (1997).

[19] V. Leiner, K. Westerholt, A. M. Blixt, H. Zabel, and B. Hjorvarsson, *Phys. Rev. Lett.* **91**, 037202 (2003).

[20] J. N. Huiberts, R. Griessen, J. H. Rector, R. J. Wijnaarden, J. P. Dekker, D. G. deGroot, and N. J. Koeman, *Nature* **380**, 231 (1996).

[21] Y. Fukai and N. Okuma, *Phys. Rev. Lett.* **73**, 1640 (1994).

[22] H. Wenzl and T. Schober in Ref. [31], p. 11.

[23] H. C. Jamieson, G. C. Weatherly, and F. D. Manchester, *J. Less Comm. Met.*, **50**, 85 (1976).

[24] T. Flanagan and J. Lynch, *J. Less-Common Metals* **49**, 25 (1976).

[25] H. Noh, J. D. Clewley, and T. B. Flanagan, *Scripta Mat.* **34**, 665 (1996).

[26] R. Lüke, G. Schmitz, T. B. Flanagan, and R. Kirchheim, *J. Alloy Comp.* **330–332**, 219 (2002).

[27] J. Nakahigashi and H. Yoshimura, *J. Alloy Comp.* **330–332**, 384 (2002).

[28] D. Eliezer, N. Eliaz, O. N. Senkov, F. H. Froes, *Mat. Sci. Eng. A-struc.* **280**, 220 (2000).

[29] S. M. Myers, M. I. Baskes, H. K. Birnbaum, J. W. Corbett, G. G. Deleo, S. K. Estreicher, E. E. Haller, P. Jena, N. M. Johnson, R. Kirchheim, S. J. Pearton, and M. J. Stavola, *Reviews of Modern Physics* **64**, 559 (1992).

including these subjects.[30–33] The major topic will be the behavior of hydrogen in disordered and amorphous systems, including the interaction with defects. The experimental and theoretical findings are relevant for other materials such as polymers, oxidic glasses, or amorphous silicon, which will be treated in a special section of this study.

II. Fundamental Properties of Hydrogen in Metals

1. EQUILIBRIUM PRESSURE AND SOLUBILITY

a. Solubility

Hydrogen molecules interacting with a metal are dissociating on the surface and dissolved as atoms within the metal according to the following reaction:

$$H_2(gas) \to 2H(metal). \tag{2.1}$$

Evidence for this reaction is provided, for instance, by measuring pressure composition isotherms (pc-isotherms) and showing that Sieverts' Law

$$c_H = K\sqrt{p_{H_2}} \tag{2.2}$$

is valid, where c_H is the H concentration within the metal, p_{H_2} is the partial pressure of hydrogen, and K is Sieverts' constant. In rubbery polymers, for instance, the molecules are not dissociated during sorption, and proportionality between concentration and pressure arises. Eq. (2.2) is derived by assuming thermodynamic equilibrium between gaseous and dissolved hydrogen, which requires that the chemical potentials in the two phases are the same. H atoms are occupying interstitial sites in the metal lattice, which are mostly tetrahedral or octahedral sites. This can be proven directly experimentally by ion channeling experiments using single crystals of the metal.[34] Then the configurational entropy s_{cf} is given by

$$s_{cf} = -k_B \ln \frac{N_H}{N_i} = -k_B \ln \frac{N_H}{\beta N_{Me}}, \tag{2.3}$$

where k_B is Boltzmann's constant, N_H, N_i, N_{Me} are the numbers of H atoms, interstices, and metal atoms respectively. Thus β is the ratio of interstitial sites to metal atoms, being 6 for tetrahedral sites in bcc lattices and 1 for octahedral

---

[30] J. Völkl and G. Alefeld, *Hydrogen in metals I*, Springer, Berlin (1978).
[31] J. Völkl and G. Alefeld, *Hydrogen in metals II*, Springer, Berlin (1978).
[32] Y. Fukai, *The metal-hydrogen system*, Springer, Berlin (1993).
[33] H. Wipf, *Hydrogen in Metals III*, Springer, Berlin (1997).
[34] H. D. Carstanjen, *phys. status sol. (a)* **59** (1980).

sites in fcc lattices. Then the chemical potential of ideally dissolved H atoms is expressed as

$$\mu_H = \mu_H^o + k_B T \ln \frac{N_H}{\beta N_{Me}} \equiv \mu_H^o + k_B T \ln \frac{r_H}{\beta}, \qquad (2.4)$$

where μ_H^o is a standard value of the chemical potential. As a measure of hydrogen concentration, the ratio r_H of hydrogen and metal atoms is used most often. The chemical potential of the molecules in the gas phase is

$$\mu_{H_2} = \mu_{H_2}^o + k_B T \ln p_{H_2}. \qquad (2.5)$$

Eq. (2.1) and equilibrium require

$$\mu_H = \frac{1}{2} \mu_{H_2}. \qquad (2.6)$$

Inserting Eqs. (2.4) and (2.5) into Eq. (2.6) yields Sieverts' Law

$$r_H = \beta \sqrt{p_{H_2}} \exp\left(\frac{\mu_{H_2}^o - 2\mu_H^o}{2 k_B T}\right), \qquad (2.7)$$

because the relation between r_H and the H concentration c_H, which is usually defined as the number of moles of H per unit volume, is proportional. Often the H concentration at a given hydrogen pressure and temperature is called solubility. This has to be distinguished from the maximum or terminal solubility of hydrogen in a metal for those cases where a hydride is formed.

The difference of the standard values of the chemical potentials in Eq. (2.7) is equal to the Gibbs free energy of the reaction given in Eq. (2.1) or the Gibbs free energy of absorption (dissolution), respectively, i.e.,

$$\Delta G^o = \Delta H^o - T \Delta S^o = 2\mu_H^o - \mu_{H_2}^o. \qquad (2.8)$$

The major contribution to the entropy change of dissolution stems from the lost degrees of freedom the free H_2 molecules have in the gas phase. Thus ΔS^o is about equal to the standard entropy of gaseous hydrogen at room temperature[35]

$$\Delta S^o \approx S_{295}^o = 131 \ J/K/mol. \qquad (2.9)$$

Because of this rather high and positive entropic contribution to the Gibbs free energy of dissolution, hydrogen can be desorbed at high temperatures.

The enthalpy of dissolution in metals is very negative at the lefthand side of the periodic table and increases to very positive values going to the right via

---

[35] J. A. Dean, *Lange's Handbook of Chemistry*, 15th Ed., McGraw Hills Handbooks, New York (1999), 6.93.

the transition metals.[32,36] An atomistic interpretation is rather involved, as elastic and electronic contribution play an important role. As the metal expands during dissolution of hydrogen (see section II.C), elastic energy has to be paid. Assuming no changes of the electronic structure the additional electron of the hydrogen atom has to be placed in states above the Fermi level, which gives rise to a positive contribution to the enthalpy of dissolution. However, it has been shown[32,37] that by the incorporation of H atoms new energy levels below the Fermi level are generated yielding negative contributions to ΔH^o.

The Gibbs free energies of hydride formation per H atom or molecule are defined as the free energy change during the following reaction:

$$H_2 + xMe \rightarrow Me_x H_2 \quad \Delta G_f^o. \qquad (2.10)$$

Values of ΔG_f^o are very close to the Gibbs free energy of dissolution,[36] which has been defined via the reaction in Eq. (2.1). The similarity of the corresponding entropy changes arises from the fact that in both reactions the major contribution stems from the loss of translational freedom of gaseous hydrogen. The similarity of the enthalpy changes means that contributions of solute–solute interaction in the hydride are small in comparison with the elastic and electronic contributions to the enthalpy of dissolution.

2. DIFFUSIVITY

Diffusion of hydrogen in metals occurs via the direct interstitial mechanism with a diffusion coefficient given by[38]

$$D^* = \frac{l^2 \Gamma}{2d} f \left(1 - \frac{r_H}{\beta}\right) \qquad (2.11)$$

where Γ is the jump frequency, l the jump distance, and d the dimensionality of the lattice (which is mostly 3 but for diffusion along grain boundaries we have $d = 2$). The quantities r_H and β have been defined after Eqs. (2.3) and (2.4) and f is a correlation factor that is unity for $r_H \rightarrow 0$ but will be smaller than unity for $r_H/\beta \rightarrow 1$ where the interstitial lattice becomes filled[39] and the direct interstitial mechanism of diffusion changes gradually to a vacancy mechanism. Then blocking of sites comes into play as well, which is accounted for by the factor in brackets. A tacit assumption for the validity of Eq. (2.11) is that all sites visited during a random walk of the H atom have the same average jump frequency, i.e., the same

---

[36] E. Fromm and E. Gebhardt, *Gase und Kohlenstoff in Metallen*, Springer, Berlin (1976).
[37] A. C. Switendick in Ref. [30], p. 101.
[38] Th. Heumann, *Diffusion in Metallen*, Springer, Berlin, London, New York (1992).
[39] K. Nakazato and K. Kitahara, *Prog. Theor. Phys.* **64**, 2261 (1980).

site and saddle point energy for the case of thermally activated hopping or the same transition probability for tunneling through the potential barrier.

Very often the chemical diffusion coefficient of hydrogen, D_H, is measured via the flux of hydrogen J_H in a concentration gradient according to Fick's First Law.

$$J_H = -D_H \frac{\partial C_H}{\partial x} \qquad (2.12)$$

Using the gradient of the chemical potential of hydrogen as the driving force for diffusion, it can be shown[40] that

$$D = \frac{C_H}{k_B T} \frac{\partial \mu_H}{\partial C_H} D^* = \frac{\partial \ln a_H}{\partial \ln C_H} D^* = \left[1 + \frac{\partial \ln \gamma_H}{\partial \ln C_H} \right] D^*, \qquad (2.13)$$

where a_H is the thermodynamic activity of H and γ_H the activity coefficient ($a_H = \gamma_H C_H$).

The experimental techniques of measuring diffusion coefficients of hydrogen in metals are manifold.[30–33] The ones used to obtain most of the data presented in this study are described as follows.

a. Electrochemical Techniques

The metal sample is immersed in an electrolyte and a current is passed from a counter electrode. Then the amount of hydrogen deposited on the sample surface is simply calculated from Faraday's Law. Whether this hydrogen produced in *statu nascendi*, i.e. in the atomic form, is absorbed by the sample or whether it recombines to H_2 molecules and escapes by dissolution in the electrolyte or as gas bubbles, depends on the H solubility of the metal, the current density, and the permeability of surface oxides.[14] Natural oxides can be removed by sputtering and replaced by a palladium layer as shown for niobium and tantalum.[41] Hydrogen can also be desorbed from a sample by reversing the direction of the current. The H activity on the surface or the chemical potential, respectively, can be obtained by measuring the voltage between sample and a reference electrode. Depending on the boundary conditions with respect to either current or electrochemical potential, transient changes of these parameters can be evaluated in order to obtain a value for the chemical diffusion coefficient.[42] If the sample is mounted in a double cell as suggested in Reference [43], the values of the diffusion coefficient are more reliable.[44] As diffusion occurs in a concentration gradient, values of the chemical diffusion coefficient are obtained. The main advantages of electrochemical techniques are

---

[40] P. G. Shewmon, *Diffusion in Solids*, McGraw Hill, New York (1963).
[41] H. Boes and H. Züchner, *Z. Naturforsch.*, **31a**, 754 and 760 (1976).
[42] R. Kirchheim and R. B. McLellan, *J. Electrochem. Soc.* **127**, 2419 (1980).
[43] M. A. V. Devanathan and Z. Stachurski, *Proc. Roy. Soc.* **A270**, 90 (1962).
[44] H. Züchner, *Z. Naturforsch.*, **25a**, 1490 (1970).

applicability at low H concentrations (down to a few at-ppm), simplicity of equipment, ease of doping, and the possibility of getting values of the (electro-)chemical potential. Drawbacks are a limited temperature range between the freezing and boiling points of the electrolyte, and nonpermeable surface barriers. Because of the latter, most measurements were made with palladium and its alloys.

b. Gorsky Effect [30]

Here the hydrogen-containing sample is bent producing both expanded and compressed regions. This way a gradient of the chemical potential of hydrogen is set up and hydrogen having a positive molar volume migrates from compressed to expanded regions until an equilibrium concentration profile has been established. The corresponding strain remains in the sample after the bending stress is released, i.e., the sample is still bent. Then the concentration gradient is no longer in equilibrium and vanishes by diffusion. The associated strain gradient vanishes too, and can be measured by monitoring the decreasing bending of the sample. In order to get measurable strains the H concentration has to be at least a few tenths of one at.-%. This condition sets a lower limit for the temperature range, because lowering the temperature also decreases terminal solubility, which will finally become higher than the H concentration and allow hydride precipitation to occur. Contrary to the electrochemical technique, surface oxides are advantageous for Gorsky-effect measurements, because they prevent desorption of hydrogen and/or an equilibration via the gas phase. At higher temperatures a limit is set by the increasing permeability of hydrogen through the oxide or the dissolution of oxygen within the metal and the destruction of the barrier.

c. Permeation

Here a pressure difference is set up across a metallic membrane separating two closed compartments and the hydrogen flux through the membrane is measured by monitoring pressure changes. Transient and steady state behavior yield the chemical diffusion coefficient and the permeability (product of diffusion coefficient and difference of H concentrations between entrance and exit surface of the membrane). The technique requires rather high permeabilities (high temperatures, diffusivities, and solubilities) and may be dominated at the lower temperatures by rate-controlling reactions at the interface.

d. Internal Friction (Mechanical Spectroscopy) [45]

Damping of vibrations in a vibrating reed or in a torsional pendulum often results from jumping atoms, if their strain field interacts with the externally applied stress field that excites the vibrations. If the frequency of vibration and the jump frequency are equal, damping reaches its maximum. Measuring damping as a function of sample frequency yields a damping peak called the Snoek peak,[45] which has its maximum at a value equal to the jump frequency. The frequencies of sample vibrations can be changed by changing the sample dimensions or by exciting various modes of vibration. Experimentally, it is easier to change the temperature, thereby changing the jump frequency of the atoms while keeping the frequency of vibration constant. Then the Snoek peak occurs in a damping versus temperature plot. H atoms incorporated in octahedral sites of an fcc lattice (for instance Pd or Ni) cause a strain in the lattice, which has cubic symmetry and, therefore, does not cause any damping. However, H atoms in tetrahedral sites of a bcc lattice (for instance Nb and Ta) give rise to a tetragonal distortion and a Snoek relaxation is expected, though it has not been observed so far—perhaps due to a relaxation strength being smaller than the detection limit of the technique or to the very rapid tunneling of H atoms between the four adjacent tetrahedral sites smearing out the tetragonal distortion. However, for hydrogen in amorphous alloys, an internal friction peak has been discovered[46–49] and it was shown that the jump frequency is in agreement with the one calculated from diffusion coefficients via Eq. (2.11). In addition, hydrogen atoms being trapped in the neighborhood of a foreign solute atom, i.e. at a substitutionally dissolved titanium atom in a niobium lattice,[50] give rise to a Snoek peak.

The diffusion coefficients usually obey an Arrhenius Law when they are measured in a limited temperature range. However, the values obtained from Gorsky effect measurements over a large range down to very low temperatures revealed a pronounced curvature for the Vb metals vanadium, niobium, and tantalum when presented in an Arrhenius plot.[11,30] The curvature corresponds to a decreasing activation energy of diffusion with decreasing temperature, which is in accordance with a quantum mechanical tunneling of the H atom through a potential barrier between two adjacent sites. Thus values as large as $1 \cdot 10^{-5}$ cm^2/s have been measured in α-Fe and V at room temperature, which corresponds to a diffusion length of about 1 cm in one day.

3. Partial Molar Volume and Interaction with Stress and Strain Fields

Partial molar volumes of hydrogen in metals are usually obtained by measuring changes of the lattice parameter as a function of H concentration.[51] Because

---

[45] A. S. Nowick and B. S. Berry, *Anelastic Relaxation in Crystalline Solids*, Academic Press, New York (1972).
[46] B. S. Berry and W. C. Pritchet, *Scripta Metall.*, **15**, 637 (1981).
[47] U. Stolz, *J. Phys. F Met. Phys.* **17**, 1833 (1987).
[48] H. Mitzubayashi, Y. Katoh, and S. Okuda, *phys. stat. sol. (a)* **104**, 469 (1988).
[49] H. Mitzubayashi, S. Murayama, and H. Tanimoto, *J. Alloys Comp.* **330–332**, 389 (2002).
[50] A. Cannelli, R. Cantelli, and F. Cordero, *J. Phys. F Met. Phys.* **16**, 1153 (1986).
[51] H. Peisl in Ref. [30], p. 53.

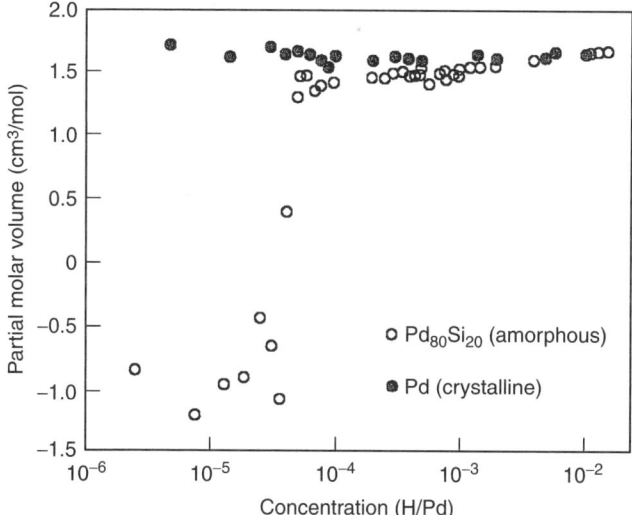

FIG. 1. Partial molar volume of hydrogen in polycrystalline palladium (closed circles) and in a liquid-quenched Pd-Si metallic glass. The negative volume changes for the first 40 at-ppm H are explained by trapping of H atoms in a corresponding fraction of vacancies.[53]

hydrogen is mostly dissolved in interstitial sites with a smaller volume, the lattice parameter increases with increasing H concentration. For many metals and their alloys the partial molar volume is 2.9 Å^3 per H atom or 1.7 cm^3/mol, respectively.[32,51,52] Details of the lattice distortion in the neighborhood of the H atom can be revealed by Huang scattering.[51]

For high H concentrations, measurements of the density of the sample yield the partial molar volume, whereas at low concentrations in-situ dilatometry is more appropriate. An example for the latter case is shown in Fig. 1,[53] where results obtained by electrochemical doping of one sample of polycrystalline palladium and one of an amorphous Pd-Si alloy are presented. Increasing the H concentration step by step and measuring the length change of the ribbon-shaped samples gave the same partial molar volume for the crystalline Pd, whereas large variations were observed for the amorphous alloy. The volume contraction being observed for the first 40 at-ppm of hydrogen was explained by trapping of H atoms in vacancy-like defects.[53] At higher concentrations the expected volume change of about 1.5 cm^3/mol for hydrogen being dissolved in interstitial-like sites was measured. In this concentration range there is a tendency of increasing partial molar volume with increasing H concentration, indicative of filling larger interstices in

---

[52] B. Baranowski, S. Majchrzak, and T. B. Flanagan, *J. Phys. F*, **1**, 258 (1971).
[53] R. Kirchheim, *Acta metall.* **34**, 34 (1986).

the amorphous structure first.[54] In order to provide additional experimental evidence for the unusual negative volume change, the partial molar volume was also measured by measuring changes of the chemical potential under tensile stress as explained in the following.

For a solute atom that leads to an isotropic strain of insertion, the chemical potential changes if an arbitrary stress state σ_{ik} is present according to the relation[55,56]

$$\mu(\sigma) = \mu(\sigma = 0) - \frac{\sigma_{ik} V_s}{3} \quad (2.14)$$

where V_s is the partial molar volume of the solute. For uniaxial tensile stress ($\sigma_{11} = \sigma$ and $\sigma_{ik} = 0$ otherwise), hydrogen as a solute, and measuring changes of the electromotive force ΔE, the last equation becomes

$$\Delta E = \frac{\sigma V_H}{3F} \quad (2.15)$$

where F is Faraday's Constant. By measuring changes of the electrochemical potential of metal–hydrogen systems as a function of the applied stress the partial molar volume V_H was determined via Eq. (2.15) in an independent way when compared with the dilatometric studies.[53] Namely, the negative changes shown in Figure 1 could be reproduced.

Besides allowing a determination of V_H, Eq. (2.14) is much more important in the context of H embrittlement, as discussed in Section IV.12.

III. Behavior of Hydrogen in Defective and Disordered Metals

The behavior of hydrogen in single crystalline metals or polycrystalline metals with a small density of lattice defects has been studied fundamentally and extensively during the 1960s 70s and 80s. In the 1990s the metal hydride batteries replaced the nickel cadmium batteries gradually and the focus of researchers was directed toward metal hydrides and their application. In order to extend the application of hydrogen storage materials to vehicles driven directly with hydrogen or indirectly via a fuel cell, the density of metallic materials or the storage capacity in terms of hydrogen per unit of mass is a matter of serious concern. In addition, the theoretical storage capacity of an alloy is often not reached because of crystalline defects, namely dislocations.[57,58] They are present from the very beginning, or even

---

[54] U. Stolz, U. Nagorny, and R. Kirchheim, *Scripta metall.*, **18**, 347 (1984).
[55] J. C. M. Li, R. A. Oriani, and L. S. Darken, *Z. Phys. Chem.*, **49**, 271 (1966).
[56] R. Kirchheim and J. P. Hirth, *Acta metall.*, **35**, 2899 (1987).
[57] H. Inui, T. Yamamoto, H. Hirota, and M. Yamaguchi, *J. Alloys Comp.* **330–332**, 117 (2002).
[58] E. Wu, E. MacA. Gray, and D. J. Cookson, *J. Alloys Comp.* **330–332**, 229 (2002).

more, they are formed during loading and unloading of the alloy. Grain boundaries play an important role beyond affecting storage capacity, because they increase the kinetics of uptake and release of hydrogen. Thus nanocrystalline alloys are favorable storage alloys.[59]

The interaction of hydrogen with crystalline defects is even more important in the context of hydrogen embrittlement. There are various mechanisms leading to hydrogen embrittlement but they are all related to hydrogen/defect interaction. The most prominent one is the direct action of hydrogen at the crack tip by either a decohesion mechanism[60] or the formation of a brittle hydride.[61] In the first case a single hydrogen atom migrating with the propagating crack tip will be sufficient, whereas in the latter higher H concentrations are required. For ductility being part of the failure mechanism, the interaction of hydrogen with dislocation becomes important. In general, the "storage" of hydrogen in crystal defects and/or the trapping by the defects determines how much "mobile" hydrogen will be available to reach a critical value for embrittlement. As this is a rather complex problem, it has been treated qualitatively only at the beginning of research in this area. The present study will provide a quantitative result by defining a density of site energies (DOSE) or an energy landscape, respectively, and filling the various sites according to Fermi Dirac Statistics (FD-Statistics).

It will be shown in section VI that the concepts of a DOSE and FD-Statistics are useful for other interstitials in other materials besides metallic ones. It also provides the basis for a fundamental study of the interaction of solute atoms with the microstructure of a material. Hydrogen systems are especially suited for these fundamental experiments because of the ease of doping and the ease of measuring their chemical potential, which is of central importance in FD-Statistics. As the system can be studied around room temperature the crystal defects will not be annihilated, i.e. the microstructure remains the same during experimentation.

4. DENSITY OF SITE ENERGIES (DOSE) AND FERMI DIRAC STATISTICS (FD-STATISTICS)

The DOSE is defined as usual in solid state physics as the normalized number of sites $n(E)$ in a given energy window $E, E + dE$ with

$$\int_{-\infty}^{\infty} n(E)\, dE = 1. \tag{3.1}$$

---

[59] S. Orimo, H. Fujii, and K. Ikeda, *Acta mater.* **45**, 116 (1996).
[60] R. A. Oriani *Acta metall.* **18**, 147 (1970).
[61] S. Gahr and H. K. Birnbaum, *Acta metall.*, **26**, 1781 (1978) and D. S. Shih, I. M. Robertson, and H. K. Birnbaum, *Acta metall.*, **36**, 111 (1988).

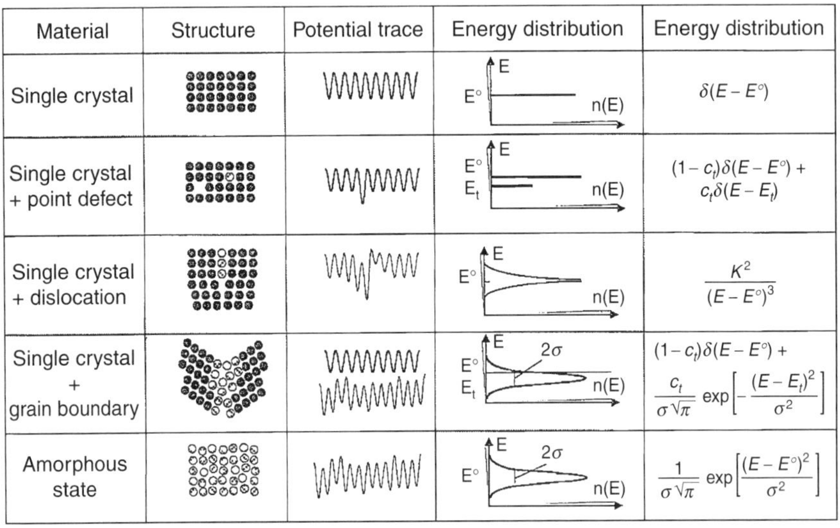

Material	Structure	Potential trace	Energy distribution	Energy distribution
Single crystal		WWWWW	$E°$ — $n(E)$	$\delta(E - E°)$
Single crystal + point defect		WWWMWW	$E°$, E_t — $n(E)$	$(1 - c_t)\delta(E - E°) + c_t\delta(E - E_t)$
Single crystal + dislocation		WWMMWW	$E°$ — $n(E)$	$\dfrac{K^2}{(E - E°)^3}$
Single crystal + grain boundary		WWWWWW / MWMWMWM	$E°$, E_t, 2σ — $n(E)$	$(1 - c_t)\delta(E - E°) + \dfrac{c_t}{\sigma\sqrt{\pi}}\exp\left[-\dfrac{(E - E_t)^2}{\sigma^2}\right]$
Amorphous state		WMMMMMM	$E°$, 2σ — $n(E)$	$\dfrac{1}{\sigma\sqrt{\pi}}\exp\left[\dfrac{(E - E°)^2}{\sigma^2}\right]$

FIG. 2. Schematic presentation of a distribution of site energies, which starts with the most degenerate one, i.e. the single crystal in the top line. In the following lines increasing structural disorder leads to a decreasing degeneracy by increasing the dimensions of lattice defects. For diffusion potential traces or energy landscapes, respectively, including the defect is important. For grain boundaries traces within the grain and within the grain boundary are shown separately, where the latter are present with a volume fraction of c_t. The density of site energies (DOSE), $n(E)$, defined as the number of sites within a given energy window, is presented both graphically and analytically. The various forms of $n(E)$ are derived in sections IV and V.

This definition is equivalent with the density of states (DOS) function used for electrons or other particles in quantum mechanics. For electrons in solids the distribution of energy states in reciprocal space is considered, whereas here we are interested in sites in real space. However, the DOSE does not contain any information about the localization of the sites and spatial correlations between low and/or high energy sites are not accessible. For the energy scale the standard state of gaseous hydrogen at one bar and 298 K is used. In the following a few relevant example of a DOSE of hydrogen in metals will be presented (cf. Fig. 2).

Because of the translational symmetry all the interstitial sites in a single crystal have the same energy and, therefore, the site energy for hydrogen atoms is the same. In other words, the system is totally degenerate. This simple picture has to be modified, because in a crystalline lattice different interstices are available, among which the tetrahedral and octahedral sites are the most prominent ones. The situation becomes even more complex if the single crystal contains different constituents, i.e., a solid solution or an intermetallic compound. Then the degeneracy of the system is partly lifted and a discrete number of site energies have to be considered. Nevertheless, for dilute solutions of hydrogen and single-component crystals the simplification of one interstitial site is usually fulfilled because the site energies of other interstices are too high to be occupied. The mathematical representation of the corresponding DOSE will be a Dirac delta function.

From the one-level system of a single crystal we come to a two-level system by introducing isolated point defects like substitutionally dissolved foreign atoms or vacancies, by restricting the interaction with hydrogen to the nearest interstices. As spatial correlation does not play a role, all the foreign atoms may be combined to a layer between two host metal layers and by neglecting the interface this becomes a two-level system, too.

Going from zero to one- and two-dimensional defects of the lattice, it is more convenient to introduce continuous functions instead of discrete energy levels. This will be discussed in more detail in the following sections dealing with dislocations and grain boundaries. Finally, in a perfect amorphous structure all sites may be considered to be different from each other, representing a system without degeneracy.

The various sites of a DOSE compete for the occupancy with hydrogen. By occupying the sites of lowest energy the system reduces its total energy, whereas distribution among sites being present in large numbers increases the configurational entropy. Both effects reduce the Gibbs free energy. If we allow a site to be occupied by one H atom only, the corresponding minimization can be treated in the framework of the Fermi Dirac Statistics (FD-Statistics).[62] Here the configurational entropy in each energy window of the DOSE is calculated under the assumption of single occupancy. During this procedure it does not matter whether electrons are distributed among energy states in reciprocal space or particles among sites in real space. The minimum of Gibbs free energy for all energy windows yields the following result for the thermal occupancy of energy level E_i or the corresponding site, respectively:

$$o(E_i) = \frac{1}{1 + \exp[(E_i - \mu)/k_B T]} \quad (3.2)$$

where μ is the derivative of Gibbs free energy with respect to particle concentration and, therefore, it is the chemical potential of the particles despite being called Fermi energy in this context.

Integration over the DOSE with the correspondent thermal occupancy yields the total fraction N/N_o of sites occupied with particles, where N is the number of dissolved particles and N_o is the total number of available sites.

$$\frac{N}{N_o} = \int_{-\infty}^{\infty} \frac{n(E)\,dE}{1 + \exp\left(\frac{E-\mu}{k_B T}\right)} \quad (3.3)$$

---

[62] T. L. Hill, *Introduction to Statistical Thermodynamics*, Addison-Wesley, London (1962), 432.

The term particle instead of hydrogen atoms has been used before, in order to stress the fact that the concept of a DOSE and FD-Statistics is rather general and not restricted to hydrogen in metals. It has been applied before in the framework of heterogeneous adsorption[62] and the interaction of solute atoms with dislocations.[63,64]

a. Solubility

The relation between hydrogen concentration in the metal expressed as the ratio $r_H = N_H/N_{Me}$ and its partial pressure in the gas phase is obtained by using Eq. (2.5)

$$\frac{r_H}{\beta} = \frac{N_H}{\beta N_{Me}} = \int_{-\infty}^{\infty} \frac{n(E)\,dE}{1 + \frac{1}{\sqrt{p}}\exp\left(\frac{E-(\mu^o_{H_2}/2)}{k_B T}\right)}. \tag{3.4}$$

For a single crystal with $n(E) = \delta(E - E^o)$ and dilute concentrations ($r_H \ll 1$) the well-known result (cf. Eq. (2.7))

$$r_H = \sqrt{p}\,\beta \exp\left(\frac{\mu^o_{H_2} - 2E^o}{2k_B T}\right) \tag{3.5}$$

is obtained. A closed solution of the integral in Eq. (3.4) will be possible for simple forms of $n(E)$ only. In order to understand the competition between energy decrease and entropy gain the following approximation based on the step or $T = 0$ approximation of the Fermi-Dirac function (i.e. $o(E) \approx 1$ for $E < \mu$ and $o(E) \approx \exp[(\mu - E)/kT]$ for $E > \mu$) will be applied

$$c = \int_{-\infty}^{\mu} n(E)\,dE + \int_{\mu}^{\infty} \frac{n(E)\,dE}{\exp[(E-\mu)/kT]} \equiv c_1 + c_2 \tag{3.6}$$

where c will be used for the concentration instead of r_H, in order to be consistent with previous publications of the author and to refer to the general validity of the equations. Thus c refers to the fraction of interstices occupied by particles. The first term c_1 on the righthand side of the equation is that part of the total concentration c arising from particles below the Fermi level and c_2 corresponds to sites above μ. Occupying sites below the Fermi level the system decreases its energy, whereas for the sites above it increases its configurational entropy. Thus for $c_1 > c_2$ energy plays a more important role than entropy and vice versa.

---

[63] D. N. Beshers, *Acta metall.*, **6**, 521 (1958).
[64] N. Louat, *Proc. Phys. Soc., London*, **B69**, 459 (1956).

For $c_2 \ll c_1 \approx c$, which is always fulfilled at low temperatures because of the step-type behavior of the Fermi-Dirac function, we get from the last equation

$$\frac{\partial c}{\partial \mu} = n(\mu). \tag{3.7}$$

The last equation can be used to get the DOSE from measured values of μ as a function of c (for instance by the electrochemical method).

b. Validity of Henry's Law

For $c \ll 1$ or $\mu \ll 0$, respectively, and $c_1 \ll c_2 \approx c$ we obtain from the last equation

$$c \approx \int_{-\infty}^{\infty} \frac{n(E)\,dE}{\exp[(E-\mu)/kT]} = \exp\left[\frac{\mu}{kT}\right] \int_{-\infty}^{\infty} n(E) \exp\left[-\frac{E}{kT}\right] dE = a\gamma \tag{3.8}$$

Then Henry's Law $a = \gamma c$ is fulfilled with the thermodynamic activity $a = \exp(\mu/kT)$ being proportional to c.

At low concentrations the condition $c_1 \ll c_2$ is not fulfilled for all site energy distributions. For a Gaussian distribution it was shown that the condition is valid for $\mu \ll -\sigma^2/kT$, where σ is the width of the distribution.[14] Then $\gamma \to \exp[\sigma^2/(2kT)^2]$ for $c \to 0$. However, if we consider a continuous exponential distribution, $n(E) = \sigma^{-1}\exp[E/\sigma]$ for $E \leq 0$ and $n(E) = 0$ otherwise, Eq. (3.8) yields

$$c_1 = \exp(\mu/\sigma) \quad \text{and} \quad c_2 = \frac{\exp(\mu/kT) - \exp(\mu/\sigma)}{1 - \sigma/kT} \tag{3.9}$$

where independent of μ (being negative) c_2 is smaller than c_1 for $\sigma \gg kT$. In this limiting case the system decreases its free energy because it lowers its energy by occupying sites of lowest energy only. Thus we have $c_1 \approx c$ and the activity coefficient becomes

$$\gamma = \exp\left[\frac{\mu}{kT} - \frac{\mu}{\sigma}\right] = c^{\sigma/kT-1}. \tag{3.10}$$

This power law has been derived in a different context as well[65] and it shows that γ approaches 0 for $c \to 0$. The activity coefficient depends on c and, therefore, Henry's Law is not fulfilled. However, as all the samples have a finite number of sites the exponential DOSE has to have a cut-off at the low energy side and it can be shown that this leads to the validity of Henry's Law again.

---

[65] T. Wichmann, K. G. Wang, and K. W. Kehr, *J. Phys.* **A27**, L263 (1994).

Very often a sample contains a small fraction f of crystal defects only. Then the majority of sites are the same as in a single crystal and the DOSE can be written as

$$n(E) = (1-f)\delta(E-E^o) + fn_f(E) \text{ with } f \ll 1, \tag{3.11}$$

If the average energy of $n_f(E)$, i.e. its first moment, is above E^o, the defects have a negligible effect and the H atoms are almost all in sites of the single crystalline region. For the reverse case of the average energy being smaller than E^o, integration over the first part of the DOSE in Eq. (3.11) yields the concentration in the single crystalline region

$$c_{fr} = \frac{1-f}{1+\exp\left(\frac{E^o-\mu}{k_B T}\right)} \approx \exp\left(\frac{\mu-E^o}{k_B T}\right) \approx \sqrt{p}\exp\left(\frac{2E^o - \mu^o_{H_2}}{2k_B T}\right). \tag{3.12}$$

The subscript fr refers to these H atoms as being free and not bound to crystalline defects. Then the activity of hydrogen or its partial pressure, respectively, is solely determined by the free hydrogen.

5. Diffusivity

a. Tracer Diffusion

Diffusion of tagged particles in a system with a DOSE and constant concentration can be obtained simply from averaging over the jump frequencies Γ and calculating D^* with the jump distance l via Eq. (2.11) written as

$$D^* = \frac{l^2}{6}\langle\Gamma\rangle = \frac{l^2}{6}\left\langle\Gamma_o \exp\left(-\frac{Q^o + E^o - E_1}{k_B T}\right)[n(E_1)o(E_1)][1-n(E_2)o(E_2)]\right\rangle \tag{3.13}$$

where $\langle\Gamma\rangle$ is the average jump rate. The jump rate for particles hopping from sites of energy E_1 into sites of energy E_2 is written as the product of

1. a constant prefactor Γ_o, which is equal for all sites,
2. a Boltzmann factor $\exp[-(Q^o + E^o - E_1)/k_B T]$ because thermally activated jumps are considered (cf. Fig. 3),
3. the partial concentration in sites of energy $E_1[=n(E_1)o(E_1)]$ and
4. the availability of empty sites of energy $E_2[=n(E_2)\{1-o(E_2)\}]$.

Note that a constant saddle point energy has been assumed, in order to allow for an uncorrelated random walk. The effect a distribution of saddle point energies

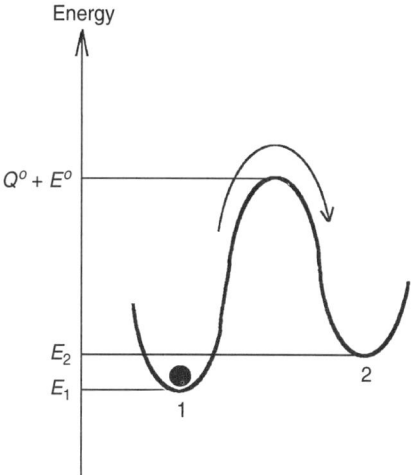

FIG. 3. Potential trace for an interstitial jumping from site 1 with energy E_1 to site 2 with energy E_2. E^o is an average energy of the DOSE and Q^o is the activation energy with respect to this reference. Thus particles in a material containing sites of energy E^o only have a jump frequency of $\Gamma = \Gamma^o \exp[-Q/k_B T]$.

has on the diffusion coefficient will be discussed in section V. By averaging over E_1 and E_2 one obtains

$$\langle \Gamma \rangle = \frac{1}{c} \iint \Gamma_o \exp\left(-\frac{Q^o + E^o - E_1}{k_B T}\right) \frac{n(E_1)}{1 + \exp\left(\frac{E_1 - \mu}{k_B T}\right)}$$
$$\times \left[n(E_2) - \frac{n(E_2)}{1 + \exp\left(\frac{E_2 - \mu}{k_B T}\right)} \right] dE_1 \, dE_2 \qquad (3.14)$$

Thus Q^o is the activation energy of diffusion in the reference material. Integration of the last term in brackets immediately yields one factor $(1 - c)$. By using the identity

$$\frac{n(E) \exp\left(\frac{E}{k_B T}\right)}{1 + \exp\left(\frac{E - \mu}{k_B T}\right)} = \exp\left(\frac{\mu}{k_B T}\right) \left[n(E) - \frac{n(E)}{1 + \exp\left(\frac{E - \mu}{k_B T}\right)} \right] \qquad (3.15)$$

integration of Eq. (3.14) with respect to E_1 gives

$$\langle \Gamma \rangle = \frac{\Gamma_o}{c} \exp\left(-\frac{Q^o}{k_B T}\right) \exp\left(\frac{\mu - E^o}{k_B T}\right) (1 - c)^2. \qquad (3.16)$$

Inserting in Eq. (3.13) yields a simple result

$$D^* = \frac{l^2}{6}\Gamma_o \exp\left(-\frac{Q^o}{k_B T}\right)(1-c)^2 \gamma^o = D^o(1-c)^2 \gamma^o \qquad (3.17)$$

with D^o being a tracer diffusion coefficient in a material that contains sites of energy E^o only and γ^o being an activity coefficient with respect to this material as a reference state. One of the factors $(1-c)$ is due to blocking of sites, whereas the second factor takes care of the fact that γ^o approaches $(1-c)^{-1}$ for $c \to 1$ in a single crystal.[66]

The temperature dependence of the tracer diffusion coefficient is solely determined by the exponential dependence of the average jump frequency according to Eqs. (3.16) and (3.17). Thus at a given concentration the effective activation energy is $Q^o + E^o - \mu$, which is the energy difference between saddle point energy and Fermi energy. The Fermi energy μ will not change very much with temperature for broad DOSE and, therefore, D^* obeys an Arrhenius Law despite a distribution of site energies or jump rates, respectively. This is a consequence of the step behavior of the FD-function, where the occupancy at low temperatures and a broad DOSE is smeared out around the Fermi energy without changing the latter very much.[67]

The concentration dependence of the effective activation energy $Q^o + E^o - \mu$ is very pronounced as μ increases with increasing concentration because of the second derivative of Gibbs free energy being always positive in equilibrium. Thus an increasing concentration always leads to a decreasing activation energy and an increasing diffusivity, if we neglect the effect of c in the denominator of Eq. (3.16). Again the atomistic interpretation is very simple. With increasing concentration, sites of higher energy have to be occupied and, therefore, the Fermi energy rises and comes closer to the saddle point energy. Thus an increasing number of particles experience a smaller activation barrier.

b. Diffusion in a Concentration Gradient

In order to derive an expression for the chemical or intrinsic diffusion coefficient D we consider two adjacent lattice planes of distance l (=jump distance). The x-axis is parallel to the normal of the planes and the average concentrations c_1 and c_2 within the planes are different. Thus we have a concentration gradient and D is defined by Fick's First Law as

$$J = -D \frac{\partial c}{\partial x} \frac{\beta}{\Omega} = -D \frac{c_2 - c_1}{l} \frac{\beta}{\Omega}, \qquad (3.18)$$

---

[66] J. Gegner, G. Hörz, and R. Kirchheim, in Ref. [7], p. 35.
[67] Ch. Kittel, *Introduction to Solid State Physics*, 3rd ed., John Wiley & Sons, New York (1967), 204.

where J is the flux of particles from plane 1 at x to plane 2 at $x+l$, Ω is the atomic volume of the metal, and β is the number of interstices per metal atom. The ratio β/Ω on the righthand side of Eq. (3.18) has to be included because c was defined as the fraction of occupied interstices, whereas in Fick's Law concentration is defined as particles per volume (cf. Eq. 2.12). The flux J is also obtained by averaging over the jumps in between the planes from sites of energy E_1 in plane 1 to sites of energy E_2 in plane 2 and vice versa.

$$J = \frac{\beta l}{6\Omega} \iint \left[\frac{\Gamma_1(x)n(E_1)}{1+\exp\left(\frac{E_1-\mu(x)}{k_B T}\right)} - \frac{\Gamma_2(x+l)n(E_2)}{1+\exp\left(\frac{E_2-\mu(x+l)}{k_B T}\right)} \right] dE_1 dE_2 \quad (3.19)$$

where the jumps occur in all directions and only 1/6 of them to the plane under consideration. The first term in brackets corresponds to the jumps of particles in sites of energy E_1 out of plane 1 at x and the second term accounts for particles in sites of energy E_2 in plane 2 at $x+l$, where the jump frequencies Γ_i have to be weighted with the fraction of occupied sites of a given energy $n(E_i)o(E_i)$. Jump frequencies are calculated as before for thermally activated processes over saddle points of constant energy $E^o + Q^o$ (cf. Fig. 3), including blocking of sites by a factor of $n(E_i)[1 - o(E_i, \mu)]$

$$\Gamma_1 = \Gamma_o n(E_2)[1 - o(E_2, \mu(x+l))] \exp\left(-\frac{Q^o + E^o - E_1}{k_B T}\right) \quad (3.20)$$

$$\Gamma_2 = \Gamma_o n(E_1)[1 - o(E_1, \mu(x))] \exp\left(-\frac{Q^o + E^o - E_2}{k_B T}\right) \quad (3.21)$$

In Ref. [68] similar calculations were presented for the first time and the fact that the occupancy depends on the chemical potential (cf. Eq. (3.2)), which depends on position x, was not taken into account properly, leading to a slightly different expression for D. Inserting Eqs. (3.20) and (3.21) in to Eq. (3.19) and partly integrating yields

$$J = \frac{\beta l \Gamma_o (1-c)}{6\Omega} \exp\left(-\frac{Q^o + E^o}{k_B T}\right)$$
$$\times \left[\int \frac{\exp\left(\frac{E_1}{k_B T}\right) n(E_1) dE_1}{1+\exp\left(\frac{E_1-\mu(x)}{k_B T}\right)} - \int \frac{\exp\left(\frac{E_2}{k_B T}\right) n(E_2) dE_2}{1+\exp\left(\frac{E_2-\mu(x+l)}{k_B T}\right)} \right] \quad (3.22)$$

---

[68] R. Kirchheim, *Acta metall.* **30**, 1069 (1982).

Using Eq. (3.15) and expanding $\mu(x+l)$ in a Taylor series gives

$$J = \frac{\beta l \Gamma_o (1-c)}{6\Omega} \exp\left(-\frac{Q^o}{k_B T}\right) \exp\left(\frac{\mu - E^o}{k_B T}\right)$$
$$\times \left[(1-c) - (1-c)\left(1 + \frac{l}{k_B T} \frac{\partial \mu}{\partial x}\right) \right] \quad (3.23)$$

and

$$J = -\frac{\beta D^o (1-c)^2}{\Omega} \exp\left(\frac{\mu - E^o}{k_B T}\right) \frac{1}{k_B T} \frac{\partial \mu}{\partial x}. \quad (3.24)$$

This way we have derived an expression that is in agreement with irreversible thermodynamics, stating that the gradient of the chemical potential is the driving force for diffusion. Using the expression derived for the tracer diffusion coefficient D^* (Eqs. (3.16) and (3.17)) gives

$$J = -\frac{D^* C}{k_B T} \frac{\partial \mu}{\partial x} = -D^o (1-c)^2 \frac{\partial a}{\partial x}, \quad (3.25)$$

where C is the concentration as usually defined, i.e., moles per unit volume. Thus the factor in front of the gradient term is the mobility of the particles as expected from the Einstein-Stokes relation.[38]

Thus a self-consistent derivation is provided for both diffusion coefficients D^* and D. In order to apply the concept to diffusion and permeation of hydrogen through materials, the material and its defects have to be characterized first and described by a DOSE. Then Eq. (3.3) is used to calculate $\mu(c)$ numerically or as a closed solution. This function allows us to calculate D^* and D via Eqs. (3.17) and (2.13). With appropriate boundary conditions Fick's Second Law has to be solved with a concentration-dependent diffusivity D. Instead of using calculated values of μ measured ones can be inserted in Eqs. (2.13) and (3.17) as well.

c. General Random Walk

It will be shown in the following that the assumption of constant saddle point energies can be replaced by less restrictive conditions. This has been done for thermally activated hopping in a previous study,[14] including a Gaussian distribution of saddle point energies. However, one may argue that thermally activated hopping is not appropriate for H atoms because they may migrate via quantum mechanical tunneling (cf. section II). Therefore we approach the problem of diffusion in an energy landscape from a different perspective.

We consider P particles in N sites of a lattice with a given DOSE. The equilibrium distribution of the particles according to FD-Statistics shall be stationary in space and one tagged particle shall do a random walk on the empty lattice. It might not be really necessary to assume a stationary distribution for the remaining particles. What might be the important effect is that the moving particle will not be able to occupy all lattice sites as they are differently blocked according to Eq. (3.2). One might call this simplification the one-particle approximation resembling the one-electron approximation of solid state physics. Further on the walk is uncorrelated besides the few blocking events and, therefore, the mean distance R after Z jumps is given according to simple random walk theory[38,40]

$$R^2 = Zl^2, \qquad (3.26)$$

where l is the jump distance. Then a diffusion coefficient can be defined by[38,40]

$$D^* \equiv \lim_{t \to \infty} \frac{R^2}{6t} = \frac{l^2}{6} \lim_{t \to \infty} \frac{Z}{t}, \qquad (3.27)$$

where t is the total time required for the walk. The time t is the sum of all the times of residence τ_m the particle stayed in the various sites visited during the walk, if the time interval required for the site exchange is small compared to τ_m. Thus we get

$$t = \sum_{m=1}^{Z} \tau_m. \qquad (3.28)$$

This way the jump rate as the reciprocal of the residence time does not have to be described by a special mechanism such as thermally activated hoping, for instance. Among the various sites, we combine those having the same energy E_i and average over the distribution of empty sites, i.e. sum over all $N - P$ sites. Hereby it is assumed that the Z sites that have to belong to the empty category are as representative for empty sites as are the $N - P$ ones. Thus we obtain

$$t = \sum_{m=1}^{Z} \tau_m = \sum_i \sum_{E_m = E_i} \tau_m \equiv \frac{Z}{N-P} \sum_{i=1}^{N-P} t_i, \qquad (3.29)$$

where t_i is the time the particle spends in sites of energy E_i. It is tacitly assumed that the number of jumps Z and the number of lattice sites N are large enough in order to apply the laws of random walk and statistical mechanics. According to the ergodic hypothesis the fraction of time a particle spends in sites of type i is equal to the fraction of particles in this type of site, yielding

$$\frac{t_i}{t} = \frac{n'_i o(E_i)}{P} = \frac{n_i[1 - o(E_i)]o(E_i)}{P}, \qquad (3.30)$$

where n'_i is the number of free sites of energy E_i, which is obtained from the site energy distribution by multiplying with $[1 - o(E_i)]$. For the dilute solution most of the free sites have a low occupancy [$o(E_i) \ll 1$] and Eq. (3.2) gives

$$o(E_i) \approx \exp\left[\frac{\mu - E_i}{k_B T}\right]. \qquad (3.31)$$

This way FD-Statistics is replaced for the empty sites by Boltzmann statistics. For convenience we choose an arbitrary reference site having energy E^o and a low value of $o(E^o)$. Then the following relation is derived from Eqs. (3.30) and (3.31)

$$\frac{t_i}{t^o} = \frac{n_i[1 - o(E_i)]}{n^o[1 - o(E^o)]} \exp\left[\frac{E^o - E_i}{k_B T}\right], \qquad (3.32)$$

where t^o is the fraction of the total time t particles reside within sites of energy E^o. Inserting Eq. (3.32) in Eq. (3.29) with [$1 - o(E^o)] \approx 1$ and Eq. (3.2) gives

$$t = \frac{Zt^o}{(N-P)n^o} \sum_{i=1}^{N-P} n_i \exp\left[\frac{E^o - E_i}{k_B T}\right]\left[\frac{\exp[(E_i - \mu)/k_B T]}{1 + \exp[(E_i - \mu)/k_B T]}\right] \qquad (3.33)$$

or

$$t = \frac{Zt^o \exp[(E^o - \mu)/k_B T]}{N(1 - c)n^o} \sum_{i=1}^{N-P} \frac{n_i}{1 + \exp[(E_i - \mu)/k_B T]} \qquad (3.34)$$

where the last sum on the righthand side is nothing else than the discrete form of Eq. (3.3) and, therefore, it can be expressed by the concentration c. Then the following simple result is obtained:

$$t = \frac{cZt^o \exp[(E^o - \mu)/k_B T]}{(1 - c)n^o} = \frac{Zt^o}{\gamma^o(1 - c)n^o} = \frac{Z\tau}{(1 - c)\gamma^o} \qquad (3.35)$$

where τ is the mean residence time as defined by

$$\tau = \frac{1}{n^o}\sum_{s=1}^{n^o} \tau_s \quad \text{for } E_s = E^o. \qquad (3.36)$$

Inserting Eq. (3.35) into Eq. (3.27) yields

$$D^* = \frac{l^2(1 - c)\exp[(\mu - E^o)/k_B T]}{6\tau c} = \frac{l^2(1 - c)}{6\tau}\gamma^o \qquad (3.37)$$

If we consider the hypothetical material with sites of energy E^o only, the residence time τ as defined by Eq. (3.36) is enlarged compared to the empty lattice due to blocking of sites. The dilute residence time is then $\tau^o = \tau(1 - c)$ and the diffusion coefficient in the dilute regime of the reference lattice is given by

$$D^o = \frac{l^2}{6\tau^o} = \frac{l^2}{6\tau(1-c)}. \tag{3.38}$$

By combining the last two equations the same result as in Eq. (3.17) is obtained and, therefore, the result is independent of thermally activated hopping. However, the concentration dependence of the diffusion coefficient, despite containing the same exponential term $\exp(\mu/k_B T)$, has to be interpreted differently. Independent of the atomistic mechanism of diffusion the occupancy of sites with higher energy at high concentration is accompanied by a decrease of the time of residence (ergodic hypothesis) and a concomitant increase of mobility.

d. Low Concentration Limit

The previous discussion of solubility at low concentrations is relevant for the diffusion coefficient, too. For those cases where Henry's Law is obeyed the activity coefficient is independent of concentration and so are both the tracer and the chemical diffusion coefficient, if in the dilute regime the $(1-c)$ term is neglected. However, for the academic case of an exponential DOSE discussed in Section III.4.b, the tracer diffusion coefficient D^* will always decrease with decreasing concentration. More realistic cases of a DOSE, such as the Gaussian one, and experimental results show that Henry's Law is fulfilled and, therefore, D^* becomes independent of c at $c \to 0$. The obvious contradiction to the discussion at the end of the last section is overcome by realizing that at low concentrations the step behavior of the FD function no longer holds. The step behavior is a consequence of energy minimization, but at very low concentrations maximization of entropy is more important. Thus most of the particles go in sites above the Fermi level without filling these up to saturation and, therefore, gain configurational entropy. For a Gaussian distribution it has been shown that the average energy of the occupied sites becomes independent of concentration at $c \to 0$ and, therefore, the average time of residence and D^* are independent of c. This delicate balance between energy and entropy is discussed in more detail in Ref. [14, 69].

6. H–H Interaction

Although in the framework of this study only dilute H-systems are considered, H–H interaction has to be taken into account for those cases where segregation at extended defects takes place and, therefore, locally high concentrations occur. Then the chemical potential may be written in a first-order

---

[69] R. Kirchheim and U. Stolz, *J. Non-Cryst. Solids* **70**, 323 (1985).

approximation as[70]

$$\mu = \mu_{id} + W c_{loc}, \qquad (3.39)$$

where W is an interaction parameter. Again it is reasonable to assume that we have a bimodal DOSE like the one in Eq. (3.11) yielding[70]

$$c_{loc} = f \int_{-\infty}^{\infty} \frac{n_f(E)\,dE}{1 + \exp\left(\frac{E-\mu-Wc_{loc}}{k_B T}\right)} \quad \text{and} \quad c_{fr} = \exp\left(\frac{\mu - E^o}{k_B T}\right). \qquad (3.40)$$

With the trivial equation $c = c_{fr} + c_{loc}$ we can solve, analytically or numerically depending on the form of $n_f(E)$, Eq. (3.40) to get the relation $\mu(c)$.

IV. Interaction of Hydrogen with Defects

7. Interaction with Other Solutes and Vacancies

An appropriate DOSE for the interaction of hydrogen with point defects is a two-level system

$$n(E) = n_o(E) + n_t(E) = (1 - c_t)\delta(E - E^o) + c_t \delta(E - E_t) \qquad (4.1)$$

which is a special form of the DOSE in Eq. (3.11). The traps provided by the defects are present with a concentration c_t and have all the same binding energy $E_t - E^o$ with respect to the "normal" sites. Hydrogen residing in normal sites will be called free and integration of $n_o(E)$ according to Eq. (3.3) yields the concentration of free hydrogen

$$c_f = \frac{(1 - c_t)}{1 + \exp\left(\frac{E^o - \mu}{k_B T}\right)} \quad \text{or} \quad \mu = E^o + k_B T \ln \frac{c_f}{1 - c_t - c_f}. \qquad (4.2)$$

For the dilute case $c_t \ll 1$ and $c_f \ll 1$ Eq. (3.25) gives the same result as derived by Oriani[60]

$$J = -\frac{\beta D^o}{\Omega} \frac{\partial c_f}{\partial x} \qquad (4.3)$$

stating that the flux of hydrogen is proportional to the concentration gradient of free particles. This assumption is valid beyond the two-level case as shown in the following. The derivation of the last two equations is independent of the functional form of $n_t(E)$ as can be seen by comparing Eqs. (3.12) and (4.2).

---

[70] R. Griessen, *Phys. Rev.* **B 27**, 7575 (1983).

The interaction of hydrogen with foreign atoms both substitutional and interstitial is rather weak with a binding energy of about 10 kJ/Mol.[29,71] This has been determined by a large variety of experimental techniques measuring permeation, internal friction, resistivity, and neutron scattering. Because of the small value of the interaction energy it is difficult to decide how much of it is elastic or electronic interaction. However, interaction with vacancies is very strong with measured binding energies between 30 and 100 kJ/Mol. These are in good agreement with calculated values using the effective medium theory.[29]

The hydrogen atom is considered to be dissolved within the vacancy although slightly off center. The metal hydrogen distances are so large that the interaction potential is in the attractive region, giving rise to a volume contraction. This is in agreement with experimental findings in heavily deformed Pd, where negative volume changes occurring at low concentrations have been attributed to vacancies formed during plastic deformations.[53]

8. Interaction with Dislocations

Theoretical models describing the interaction of solute atoms with dislocations[72] are based on an elastic interaction between the stress field around dislocations and the strain caused by a solute atom (cf. Section II.3). The interaction is strongest for edge dislocations because of the hydrostatic stress field. Thus solute atoms changing the volume during dissolution, i.e. containing a nonvanishing trace in their strain tensor, experience a strong elastic interaction with edge dislocations. Hydrogen in metals causes volume expansion and, therefore, interacts strongly with the stress field of edge-type dislocations. As stresses of dislocations are calculated by applying continuum theory they are less reliable within the dislocation core and, therefore, trapping of hydrogen in dislocation cores has to be treated separately. In the following the DOSE is calculated based on these considerations for edge dislocations.

The hydrostatic part of the stress field of an edge dislocation is given by[72]

$$p = \frac{\sigma_{ii}}{3} = \frac{Gb(1+v)}{3\pi(1-v)}\frac{\sin\theta}{r}, \qquad (4.4)$$

where G is the shear modulus, v Poisson's ratio, b the magnitude of the Burger's vector, θ and r are cylindrical coordinates as defined in Fig. 4 with the z-axis along the dislocation line.

---

[71] P. Vargas and H. Kronmüller, *Z. Phys, Chem. N.F.*, **143**, 229 (1985).
[72] J. P. Hirth and J. Lothe, *Dislocations in Solids*, McGraw Hill, New York (1968).

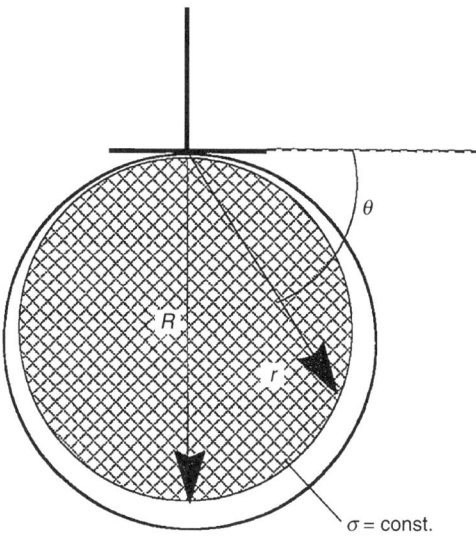

FIG. 4. Circles of constant hydrostatic stress at an edge dislocation as calculated in continuum mechanics (cf. Eq. (4.4)). The elastic interaction energy with hydrogen is constant at these lines and the corresponding DOSE is the number of sites within the two circles.

Then the interaction energy with H atoms on a circle of constant pressure is obtained from

$$pV_H = \frac{Gb(1+v)}{6\pi(1-v)R} V_H \equiv \frac{AV_H}{R}, \quad (4.5)$$

where R is the radius of the cylinder and A is defined by (4.5).

For H in Pd the number of octahedral sites chosen by hydrogen is the same as the number of Pd atoms. We assume that as in the β-phase of Pd only the fraction α (being ca. 0.6 at room temperature) is occupied. Then in a material containing ρ dislocations per unit area the number of sites, n, in a cylinder of radius R and unit length is $\rho\pi R^2$ and the DOSE becomes

$$n(E) = \alpha\rho \frac{dn}{dE} = \alpha\rho \frac{dn}{dR}\frac{dR}{dE} = \alpha\rho\pi R \frac{AV_H}{E^2} = \frac{\alpha\rho\pi AV_H}{E^3}. \quad (4.6)$$

Inserting this in Eq. (3.3), using the step approximation, i.e. the first part on the righthand side of Eq. (3.6), and solving for μ gives

$$\mu = \sqrt{\frac{\alpha\rho\pi AV_H}{c}}. \quad (4.7)$$

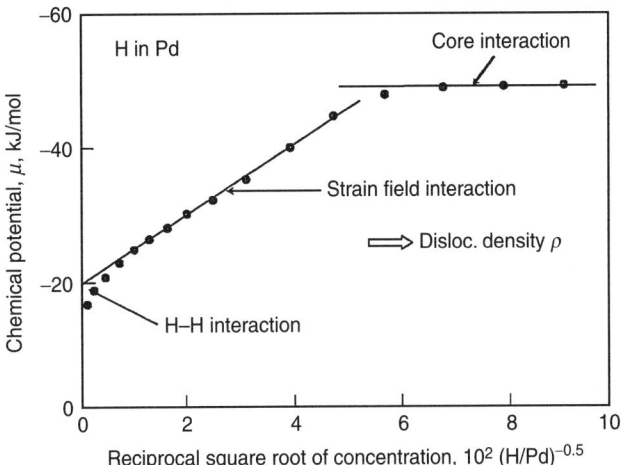

FIG. 5. Electrochemically measured chemical potentials of hydrogen in heavily deformed palladium plotted according to Eq. (4.7) versus $1/\sqrt{c}$. The various regions of the data points represent regions of a predominant interaction mechanism.

Equation (4.7) is checked by plotting in Figure 5 measured values of μ for H in cold-rolled Pd[73] versus the reciprocal square root of concentration.

Unlike the theoretical prediction, the interaction energy (difference of chemical potentials between deformed and annealed samples) becomes constant at very low concentrations, which is explained in terms of a direct interaction with the dislocation core not included in Eq. (4.5). The value of about -50 kJ/Mol-H has also been determined for H in Fe.[74] The linear dependence on $1/\sqrt{c}$ in Figure 5 yields a slope that corresponds to reasonable dislocation densities for heavily deformed metals ($2 \cdot 10^{11}$ cm^{-2}). Contrary to Eq. (4.7), the straight line corresponding to the interaction with the long-range stress field of the dislocation does not intercept the ordinate at 0 but at a value of about -20 kJ/Mol-H. This is attributed to a direct H–H interaction. It can be calculated from the values obtained for this interaction in well-annealed Pd at high H concentrations where the parameter, W, defined in Section III.6 was measured to be -30 kJ/mol-H.[75] With a maximum local concentration of $\alpha = 0.6$ the contribution to μ from H–H interaction becomes -18 kJ/mol-H in good agreement with the intercept in Figure 5.

During the electrochemical measurements of μ the standard value was defined such that

$$\mu = k_B T \ln c_f, \qquad (4.8)$$

---

[73] R. Kirchheim, *Acta metall.* **29**, 835 and 845 (1981).
[74] A. J. Kummnick and H. H. and Johnson, *Acta metall.* **28**, 33 (1980).
[75] E. Wicke and J. Blaurock, *Ber. Bunsenges. Phys. Chem.* **85**, 1091 (1980).

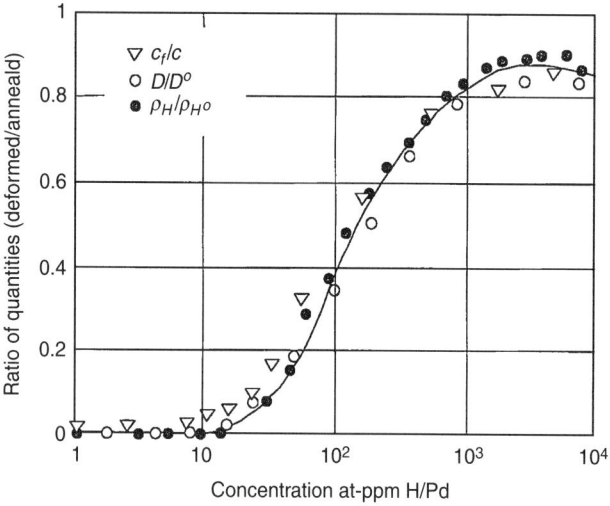

FIG. 6. Hydrogen resistivity increment, hydrogen diffusivity, and the concentration of free hydrogen in heavily deformed palladium divided by the corresponding values in well-annealed palladium. These ratios are plotted versus H concentration and in agreement with Eq. (4.9) they are about the same over the whole range of concentrations within the α-phase of Pd.

where according to Eq. (3.12) $E^o = 0$. This is also the chemical potential for the single crystalline metal, where the concentration of free hydrogen c_f is equal to the total concentration. Besides measuring μ for a metal with a high dislocation density in comparison with a single crystal, the concentration of free hydrogen can be determined also from measurement of the electrical resistance. It has been shown for Pd[76] that hydrogen trapped as a hydride at the dislocation lines contributes to the resistivity to a negligible part only when compared to the same number of H atoms distributed homogeneously. Thus the resistivity increment ρ_H caused per unit of H concentration divided by the same increment in a single crystal ρ_H^o yields the fraction of the H atom that is free. The same is true for a dislocated metal and its tracer diffusion coefficient $D^* = D^o \gamma$ (cf. Eq. (3.17) with $c \ll 1$) where D^o has a well-defined operational meaning, being the tracer diffusion coefficient in the single crystal. Therefore, we have

$$\gamma = \exp\left(\frac{\mu}{k_B T}\right) = \frac{c_f}{c} = \frac{\rho_H}{\rho_H^o} = \frac{D^*}{D^o} \qquad (4.9)$$

It is shown in Figure 6 that this simple relation holds for hydrogen in strongly deformed palladium. At very low H concentrations the fraction of free hydrogen is negligible and, therefore, all the hydrogen is trapped at dislocations.

---

[76] J. A. Rodriges and R. Kirchheim, *Scripta metall.* **17**, 159 (1983).

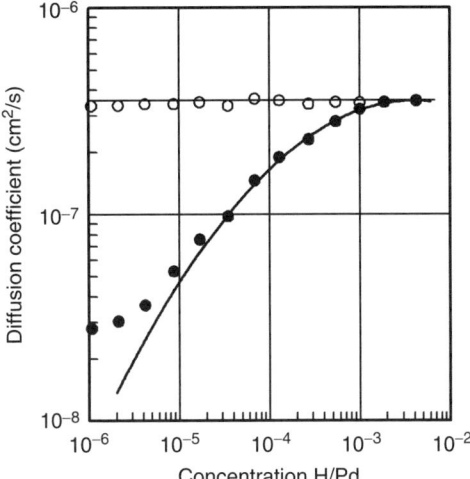

FIG. 7. Diffusion coefficient of hydrogen in heavily deformed Pd (closed circles) and well-annealed Pd (open circles). The line following the values of deformed Pd was calculated from measured chemical potentials without a fitting parameter (cf. text).

At intermediate concentrations hydrogen atoms are partitioned between sites far away from dislocations and those close to them. But sites at dislocations never become saturated because of both the long-range elastic interaction and an attractive H–H interaction. Thus the corresponding cylinder being enriched in hydrogen is steadily growing.

The concentration dependence of the chemical diffusion coefficient of H in deformed crystalline Pd is shown in more detail for very low concentrations in Figure 7. Using measured values of the chemical potential for the same deformed Pd and a diffusion coefficient obtained for a single crystalline Pd the lines in Fig. 7 were calculated by using Eqs. (2.13) and (3.17). Thus, without a fitting parameter, the concentration dependence of D is obtained in very good agreement with experimental results. Deviations occurring at very low concentrations are either due to an enhancement of diffusion along the dislocation core or arise because of difficulties measuring chemical potentials at very low H concentrations.

The pronounced changes of H activity, diffusivity, and resistivity in crystalline Pd, which are caused by plastic deformation and arise from the presence of dislocations, do not occur after cold rolling of amorphous PdSi alloys.[77] The absence of a detectable H trapping in deformed amorphous alloys is considered to reveal the absence of edge-dislocation-like defects.

---

[77] R. Kirchheim, A. Szökefalvi-Nagy, U. Stolz, and A. Speitling, *Scripta metall.* **19**, 843 (1985).

By measuring volume changes caused by dissolved hydrogen in severely deformed crystalline palladium (99% reduction in cross section by cold rolling) it was shown[53] that samples contracted for the first 50 to 100 at-ppm of H. This was attributed to trapping in vacancies (cf. Section IV.7), which supposedly form during cold rolling. After saturating the vacancies hydrogen was trapped in dislocation. In this concentration range the molar volume of hydrogen was slightly smaller than in single crystalline Pd. This behavior is in accordance with the assumption of hydrogen being dissolved in the expanded region below the glide plane of edge dislocations. With increasing concentration the dislocations became saturated and additional hydrogen was predominantly dissolved in normal octahedral sites far away from dislocations. As a consequence the hydrogen partial molar volume approached that of the single crystal.

So far the segregation of hydrogen at edge dislocation and the concomitant formation of hydride cylinders below the glide plane have been proven in an indirect way only. Direct evidence can be provided by small-angle neutron scattering (SANS). Scattering by randomly oriented cylinders has to be described by the following macroscopic cross section:[78]

$$\frac{d\Sigma}{d\Omega} = \frac{2\pi^3 \rho R_0^4 \Delta g^2}{Q} \exp\left[-\frac{1}{4} Q^2 R_0^2\right], \qquad (4.10)$$

where ρ is the dislocation density, R_o the radius of the cylinders, Δg the difference of scattering length densities, and Q the magnitude of the scattering vector. By plotting the logarithm of the product of Q and measured values of the macroscopic cross section (after appropriate subtraction of background and incoherent scattering) versus Q^2, straight lines are expected according to Eq. (4.10). This is in agreement with experimental findings as shown in Figure 8. The slope of the straight line yields the radius of the cylinders and the intercept with the ordinate yields the dislocation density. Similar plots have been evaluated for different H concentrations and the results are presented in Figure 9.

However, the natural choice for SANS is deuterium instead of hydrogen because the former has a larger cross section for coherent scattering. In addition, hydrogen gives rise to pronounced incoherent scattering, i.e. raises the background. The larger coherent scattering of the isotope deuterium leads to a larger macroscopic cross section as shown in Figure 10. However, they are larger by a factor of 1.5 only, whereas Eq. (4.10) predicts a factor of 3.2 corresponding to the squared ratio of scattering length densities. This discrepancy is overcome by taking into account that there are two contributions to the scattering contrast. The first is due to the H or D atoms having a higher concentration at the dislocation

---

[78] M. Maxelon, A. Pundt, W. Pyckhout-Hintzen, J. Barker, and R. Kirchheim, *Acta mater.*, **49** 2625 (2001).

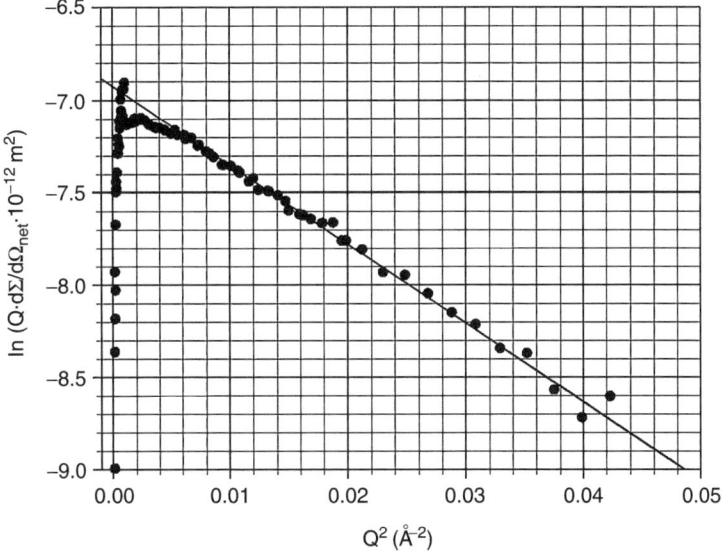

FIG. 8. A modified Guinier Plot of the macroscopic scattering cross section $d\Sigma/d\Omega$ (cf. Eq. (4.10)) measured by small-angle neutron scattering for deformed Pd with 1 at.-% H. The slope of the linear part yields the radius of hydrogen enriched cylinders formed by segregation at the dislocation line.

lines. The second is a consequence of this segregation, because both isotopes expand the Pd lattice and, therefore, reduce the scattering contrast of Pd with respect to the matrix far away from dislocations. H atoms have a negative scattering length for neutrons and the corresponding negative contrast (difference of scattering

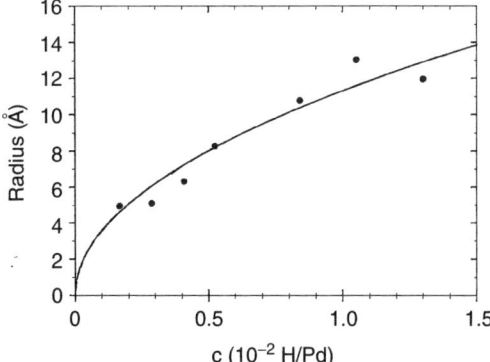

FIG. 9. Radii obtained from small-angle neutron scattering for a deformed Pd sample at various H concentrations. The line is calculated by assuming the formation of a hydride of cylindrical shape in the strain field of dislocations. The dislocation density ($\rho = 2 \cdot 10^{11}$ cm^{-2}) was used as a fitting parameter to obtain agreement with experimental data.

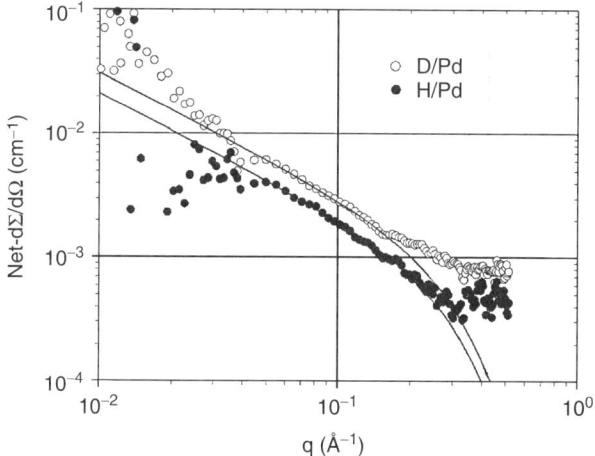

FIG. 10. Macroscopic cross section for a deformed Pd sample containing 0.8 at.-% hydrogen (closed circles) or deuterium (open circles). The lines are calculated by using Eq. (4.10). The difference of $d\Sigma/d\Omega$ between the two isotopes is less than expected from their scattering length, which is evidence for a volume expansion during the "hydride" formation at the dislocation lines (cf. text).

length density) is exaggerated by the lattice expansion. For deuterium with a positive scattering length the opposite is true. Thus the peculiar scattering behavior of the hydrogen isotopes yields additional insight into the segregation at dislocations.

The high local concentrations of hydrogen at the dislocations may be treated as suggested in Section III.6. However, it is simpler to apply a thermodynamic model treating the pronounced segregation as a hydride formation in a hydrostatic stress field. Without stresses ($p = 0$) and for equilibrium between the hydride and saturated solid solution, the chemical potentials in the two phases have to be the same, i.e.,

$$\mu_{(hydride, p=0)} = \mu^0 + kT \ln c_{ts} \qquad (4.11)$$

where c_{ts} is the terminal solubility of hydrogen in Pd in equilibrium with the β-phase (of composition PdH$_\alpha$). As the terminal solubility of H is 0.01 H/Pd at room temperature, the ideal solution approach for the configurational entropy (logarithmic term in Eq. (4.11)) is justified. At the border between the cylindrical hydride and the solid solution a constant hydrostatic pressure p (cf. Eq. (4.4)) is present, and the chemical potential is changed to

$$\mu = \mu^0 + kT \ln c_{ts} + pV_H. \qquad (4.12)$$

Far away from the dislocation where hydrogen is free and where it has a local concentration of c_f, the chemical potential is given by Eq. (4.8) (here with the standard value μ^o). Then Eqs. (4.5), (4.8) and (4.12) yield

$$c_f = c_{ts} \exp\left(-\frac{Gb(1+v)V_H}{6\pi(1-v)kTR}\right) = c_{ts}\exp\left(-\frac{C}{R}\right) \quad (4.13)$$

where C is 1.0 nm for edge dislocations of $b = 0.275$ nm in Pd. Besides c_f, the hydrogen trapped as a cylindrical hydride of composition $\alpha = 0.6$, radius R, and length ρ_d contributes to the total concentration c_{tot}. Thus we have in terms of H/Pd

$$c_{tot} = \alpha\rho_d\pi R^2 + c_f = \alpha\rho_d\pi R^2 + c_{ts}\exp\left(-\frac{C}{R}\right). \quad (4.14)$$

If we use this implicit function of $R(c_{tot})$ and compare it in Figure 9 with measured values, a good agreement is obtained by using a dislocation density of $2.2 \cdot 10^{11}$ cm^{-2}. This is considered to be additional evidence for an extended segregation of hydrogen, which requires taking into account both elastic and solute/solute interaction.

The experimental results presented in this section are in qualitative agreement with a variety of studies by other groups.[24,44,79,80]

9. INTERACTION WITH GRAIN BOUNDARIES

Segregation of solute atoms at grain boundaries is often studied by breaking a sample in a UHV chamber, where in the case of intercrystalline fracture the crack runs along the grain boundaries and the solute atoms are exposed to surface analytical techniques such as AES, XPS, and SIMS. For hydrogen this is difficult to achieve, as only the latter method is able to detect H. In addition the high H mobility at room temperature allows surface segregation to be established on the former grain boundary before the measurement of the original grain boundary coverage takes place.

Again the concept of a site energy distribution is useful to study H segregation at grain boundaries. Similar to dislocations, the number of traps provided by the grain boundaries is rather small. But to study them by gradually filling, we need a large density of them, i.e., a small grain size. This will be the case for nanocrystalline

---

[79] B. J. Heuser, J. S. King, G. S. Summerfield, F. Boué, and J. E. Epperson, *Acta metall. mater.*, **39**, 2815 (1991) and B. J. Heuser and J. S. King, *J. Alloys Comp.* **261** 225 (1997).
[80] D. K. Ross and K. L. Stefanopoulus, *Z. Phys. Chem.*, **183**, 29 (1994).

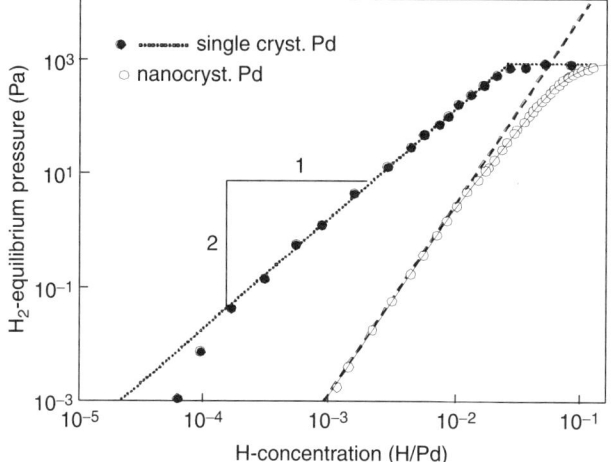

FIG. 11. H_2-equilibrium pressure versus H concentration for single crystalline (●) and nanocrystalline Pd (○)[81] at 295 K. The dotted line has a slope of 2 within the α-phase according to Sieverts' Law for $c < 0.01$ and it becomes a plateau within the $\alpha + \beta$ two-phase region for $c > 0.015$. The solid and dashed lines for the nanocrystalline sample are calculated assuming a distribution of site energies (cf. Fig. 12) and including or excluding H–H interaction as explained in the text.

metals. Electrochemical measurements[81] of the chemical potential μ that were converted into partial pressures are presented in Figure 11 for a nanocrystalline sample and a single crystal of Pd. In the latter case Sieverts' Law is fulfilled in the solid solution range.

The behavior of the nanocrystalline sample is described by a Gaussian distribution of site energies for sites within the boundaries and a single level for sites within the grains (cf. Fig. 12). With respect to the large variety of grain boundaries in a nanocrystalline material and a variety of structural units within a certain grain boundary the concept of a continuous distribution of segregation energies appears to be more reasonable. For the sake of simplicity the sites within the grains are assumed to have the same energy as sites in a single crystal. This is still a good approximation for higher concentrations, where the interfacial stress affects the site energy within the grains.[82] Thus at a given chemical potential or partial pressure, respectively, the concentration in the grains has to be the same as in the single crystal and, therefore, its contribution to the total concentration can be subtracted yielding the amount segregated at the boundaries. Fitting Eq. (3.3) to the experimental results presented in Figure 11 yields values for E_{seg} and σ. A more detailed description of the procedure and the results is given in Refs. [14] and [81]. For H in nanocrystalline Ni experimental values are not available over the same large

---

[81] T. Mütschele and R. Kirchheim, *Scripta metall.* **21**, 135 (1987).
[82] J. Weismüller and C. Lemier, *Phys. Rev. Lett.* **82**, 213 (1999).

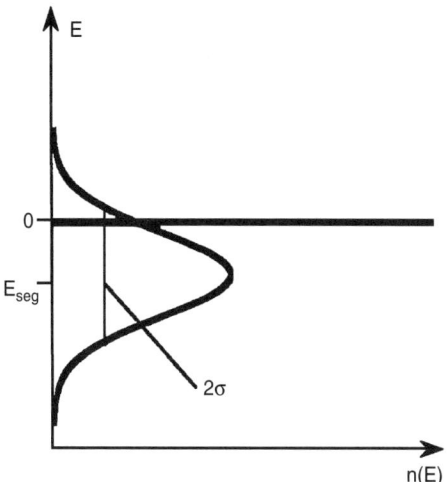

FIG. 12. Site energy distribution for H in nanocrystalline Pd used in order to calculate the line in Figure 11. The bimodal distribution covers sites within the grains with the same energy $E = 0$ by definition and sites with a Gaussian distribution within the grain boundaries. The Gaussian has the average site energy E_{seg} and the width σ.

range of H concentration but they can be described within the same framework of a distribution of segregation energies.[83,84]

Although two parameters (E_{seg} and σ) are available, experimental results at large H concentrations cannot be fitted (dashed curve in Fig. 11). The discrepancy arises from neglecting H–H interaction. This can be included via a quasichemical approach without introducing a new fitting parameter (cf. sections III.6 and IV.8), because the interaction parameter W obtained from pressure composition isotherms of coarse-grained Pd was used. Then the solid curve in Figure 11 is obtained in excellent agreement with experimental data. It is interesting to note that the width of site energies for the grain boundaries $\sigma = 15$ kJ/Mol H is between the ones obtained from fitting data of liquid-quenched and sputtered amorphous Pd-Si alloys, where the width is 11.5 or 17.5 kJ/Mol H, respectively (cf. Section V.14).

H diffusion in nanocrystalline Pd was measured via a time-lag method,[14,81] where the pellet-shaped samples were electrochemically charged with hydrogen from one side and the delayed response of the electrochemical potential at the adjacent side was monitored (cf. Section II.2). The concentration was raised in small steps, in order to minimize errors arising from the concentration dependence of the diffusion coefficient. As the transport of H through the sample is a mixture of grain boundary and bulk diffusion, the numbers evaluated from the time lag

---

[83] D. R. Arantes, X. Y. Huang, C. Marte, and R. Kirchheim, *Acta metall. et mater.* **41**, 3215 (1993).

[84] R. Kirchheim, I. Kownacka, and S. M. Filipek, *Scripta metall. et mater.* **28**, 1229 (1993).

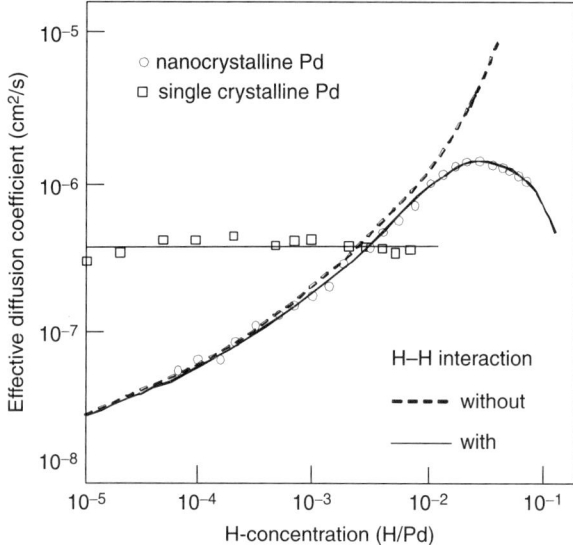

FIG. 13. Effective diffusion coefficient of hydrogen at 295 K in nanocrystalline Pd (○) and single crystalline Pd (□) as a function of the total H concentration. The horizontal line through the single crystalline data corresponds to H diffusion of non-interacting H atoms. The curves through the nanocrystalline data are calculated using Eqs. (2.13) and (3.17) and assuming that the effective diffusion coefficient corresponds to the grain boundary diffusion coefficient (cf. text). H–H interaction is included by adding a term Wc_{gb} to the chemical potential, where W equals -30 kJ/Mol as in polycrystalline Pd[75] and c_{gb} is the local concentration in the grain boundaries. Note that grain boundary diffusion of interstitials at low concentrations is slower than in single crystals.

were called effective diffusion coefficients (see discussion following). The results are presented in Figure 13.

For atoms diffusing along grain boundaries being perpendicular to the surface three limiting cases are discussed,[85] as shown in Fig. 14. Case (A) corresponds to the condition $D_g t \gg d^2$, where D_g is the diffusion coefficient within the grains, d is the distance of grain boundaries, and t is the time. For the results presented in Figure 13 $D_g = 3 \cdot 10^{-7}$ cm^2/s, $d \approx 10$ nm, and t is between a few seconds and a few minutes and, therefore, the condition for type A diffusion is always fulfilled. Then the effective diffusion coefficient is given by[85]:

$$D_{eff} = fD_{gb} + (1-f)D_g \qquad (4.15)$$

where f is the volume fraction of grain boundaries, which (for the case shown in Fig. 14 with additional boundaries running parallel to the drawing plane) is

---

[85] I. Kaur and W. Gust, *Fundamentals of Grain and Interphase Boundary diffusion*, 2nd ed., Ziegler Press, Stuttgart (1989), 77 ff.

SOLID SOLUTIONS OF HYDROGEN IN COMPLEX MATERIALS

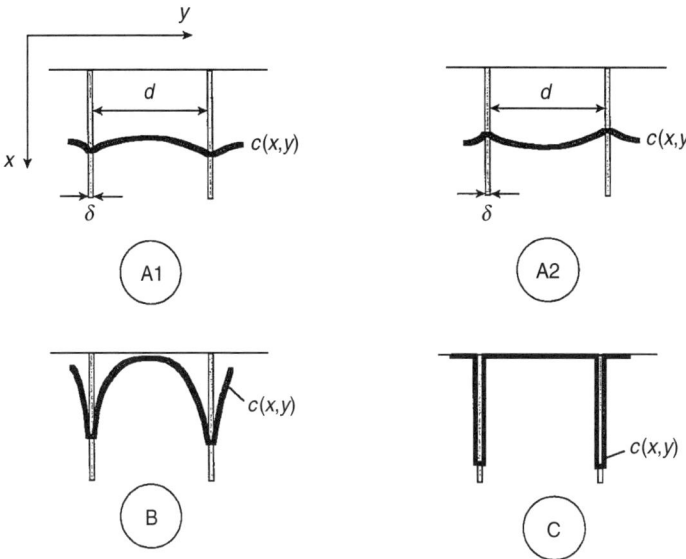

FIG. 14. Schematic presentation of three limiting cases of grain boundary diffusion with boundaries perpendicular to the surface, where d is the distance of grain boundaries or the grain size, respectively. δ is the thickness of grain boundaries and $c(x, y)$ is the concentration profile. With diffusion coefficients D_g and D_{gb} for grains and grain boundaries and t as the diffusion time the conditions for the three limiting cases are:[85] A1 and A2: $D_g t \gg d^2$, B: $100\delta < \sqrt{D_g t} < d/20$ and C: $\sqrt{D_g t} > \delta/20$.

given by

$$f = \frac{2\delta}{d + 2\delta}. \qquad (4.16)$$

The case of solute segregation described by a factor $S = c_{gb}/c_g$ can be formally included by replacing δ by $S\delta$.[85] Then the ratio of the two contributions to the effective diffusion coefficient becomes

$$\frac{fD_{gb}}{(1-f)D_g} = \frac{2S\delta D_{gb}}{dD_g} \qquad (4.17)$$

The contribution of the grain boundaries can be dominant or determine the transport, respectively, for two reasons:

A1: Grain boundary diffusion is much faster than bulk diffusion ($D_{gb} \gg D_g$) but the grains are fed from the grain boundaries because of the short diffusion length $d \ll \sqrt{D_g t}$ and

A2: Grain boundary diffusion is slower than bulk diffusion ($D_{gb} < D_g$) but the segregation factor S is much larger than unity. The concentration front is moving ahead within the grains but the boundaries act as sinks retarding the transport through the grains.

The latter case A2 applies to the low concentration results presented in Figure 13, because from the data of Figure 11 the segregation factor S can be obtained from the trivial relation

$$c_{tot} = fc_{gb} + (1-f)c_g = fSc_g + (1-f)c_g \qquad (4.18)$$

and a grain boundary thickness δ of about 0.5 nm. The concentration within the grains c_g is equal to the values of the single crystalline sample at a given chemical potential, because they are assumed to have the same site energies. Thus the ratio S in Eq. (4.18) is much larger than unity and, therefore, the measured H-transport through the nanocrystalline membrane is determined by grain boundary diffusion and $D_{eff} = D_{gb}$.

Case A2 is not discussed in textbooks on diffusion because grain boundary diffusion is always considered to be faster than diffusion through the grains. This judgment is based on the decreased density of atoms in the boundaries which gives rise to a lower formation energy of vacancies or vacancy like defects as prerequisites for substitutional diffusion. However, interstitial diffusion does not require vacancies and it is slowed down in the presence of a distribution of site energies as the interstitials are trapped in low energy sites at low concentrations. At higher concentrations the traps are gradually saturated and interstitial diffusion takes advantage of the more open structure of the boundary as well.

The concentration dependence of D_{gb} can be calculated via Eq. (2.13) and (3.17) neglecting the contribution from grains, i.e. considering the Gaussian distribution in Fig. 12 only. H-concentration has to be replaced by the grain boundary part c_{gb} and γ^o is obtained from the measured chemical potential. Thus only one fitting parameter, the reference diffusivity D^o, can be changed to obtain agreement with experimental data. With the logarithmic scale used in Fig. 13 the calculated curves are moved up and down in the direction of the ordinate without changing their slope and curvature. The steady increase of D with increasing concentration is equivalent to amorphous materials (cf. chapter VI) because a Gaussian distribution has been used there as well. However, different to the glassy materials concentrations in the boundaries become so large that blocking of sites and H–H interaction has to be taken into account (without additional fitting parameters). The concentration in the boundaries is about 0.25 H/Pd at $c_{tot} = 0.01$ for $\delta = 0.5$ nm. As the H–H interaction is attractive ($W = -30$ kJ/mol-H), the diffusivity is finally decreased at high H concentrations. In agreement with the generally accepted wisdom, the average or reference diffusion coefficient D^o of the grain boundaries is larger than the bulk value due to a lower activation energy for H atoms arising from the lower metal density in grain boundaries. The concept of interstitial diffusion in grain boundaries used in this study and the interpretation of the concentration dependence have been embedded in a general context.[86]

The increased H solubility in the α-phase of nanocrystalline Pd has been confirmed by other experimental methods, i.e. doping samples from the gas phase and measuring pressure drops[87] and/or lattice parameters.[82] As in Ref. [87], pressure–composition isotherms or lattice parameters were also measured at high hydrogen concentrations using an electrochemical technique.[88] In all studies[82,87,88] it was shown that the miscibility gap between α- and β-phase is remarkably reduced in nanocrystalline Pd. The corresponding increase of the terminal solubility in the α-phase is generally accepted to be due to segregation at the grain boundaries, whereas the reduction of the lower limit of H solubility in the β-phase is interpreted differently. In Ref. [88] it is argued that the composition of the grain boundaries does not change within the miscibility gap, because the chemical potential of hydrogen has to be constant in this region. Then all of the grains are transformed into β-phase whereas the H occupancy of the grain boundaries does not change, i.e., remains at a value which is larger than in the α-phase (segregation) but smaller than in the β-phase. Thus an overall decrease of the total H concentration occurs at the $(\alpha + \beta)/\beta$ boundary of the miscibility gap. Following this reasoning an average grain boundary thickness of 0.7 to 1 nm can be calculated without data fitting.[88] However, one has to take into account[82] that the inhomogeneous distribution of hydrogen between grains and grain boundaries gives rise to a corresponding inhomogeneous distribution of mechanical stresses, which affect the chemical potential. Thus segregation in grain boundaries leads to compressive stresses in these regions whereas the grains go into tension. Hydrogen atoms are redistributed and corresponding changes of the lattice parameter occur. At the $(\alpha + \beta)/\beta$ boundary of the miscibility gap the stress distribution is reversed because of the higher H concentrations in grains. These effects do not change the data evaluation presented in this study because at the dilute region the corresponding mechanical stresses are too low to change the chemical potential to a measurable extent.

10. Interaction with Metal/Oxide Boundaries

Hydrogen trapping at metal/ceramic interfaces has been studied extensively[89–91] because of its relevance in the area of hydrogen embrittlement of high-strength

---

[86] A. P. Sutton and R. W. Balluffi, *Interfaces in Crystalline Materials*, Clavendan Press, Oxford (1995), 510.
[87] J. A. Eastman, L. J. Thompson, and B. J. Kestel, *Phys. Rev.* **B48**, 84 (1993).
[88] T. Mütschele and R. Kirchheim, *Scripta metall.* **21**, 1101 (1987).
[89] J. Gegner, G. Hörz, and R. Kirchheim, *Interface Science*, **5**, 231 (1997).
[90] X. Y. Huang, W. Mader, and R. Kirchheim, *Acta metall. mater.* **39**, 894 (1991).
[91] X. Y. Huang, W. Mader, J. A. Eastman, and R. Kirchheim, *Scripta metall.* **22**, 1109 (1988).

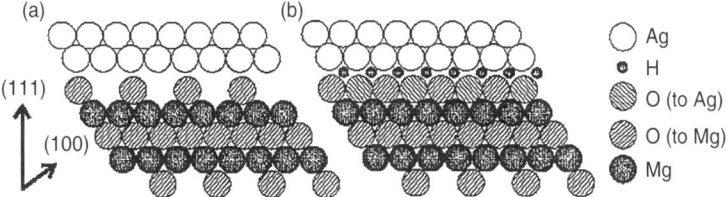

FIG. 15. Model of the (111) Ag/MgO interface at low (a) and high (b) partial pressure of oxygen. In (a) structural vacancies are formed in the terminating oxygen layer, in order to maintain stoichiometry of MgO (cf. text). At high oxygen pressures the vacancies are filled with excess oxygen, which gets its 2 electrons from Ag forming a kind of silver oxide at the interface. In the presence of excess oxygen strong trapping of hydrogen occurs at the metal/oxide interface.

steels, where hydrogen is trapped at the metal/carbide interface.[92] Due to the large H solubility and the ease of measuring chemical potentials, palladium alloys are used as model alloys again. For some metal/oxide interfaces high-resolution electron microscopy and analytical field ion microscopy[93–96] revealed that most of the terminating layers are dense-packed oxygen planes (cf. Fig. 15). As a consequence small oxide precipitates have an excess of oxygen when compared with the stoichiometry resulting from charge neutrality. Thus the negative charge of the excess oxygen at the interface has to be delivered by the surrounding metal as was shown recently by electron energy loss spectroscopy.[97,98] In addition, it has been shown[89] that the excess oxygen at the Ag/MgO interface is bound there with an energy that corresponds to the formation energy of Ag_2O in agreement with the structural model of Figure 15 and electron energy loss spectroscopy for the same[98] and for a similar boundary Cu/MgO.[97]

The concept of varying stoichiometry is in agreement with experimental findings, where "irreversible" and reversible trapping of hydrogen was observed (dependent on the oxygen activity during sample annealing treatments).[90,91] The term irreversible trapping was used because the corresponding part of trapped hydrogen could not be removed by prolonged anodic polarization of the sample, i.e. the binding energy to the traps was so high that the corresponding reduction of H mobility did not allow a removal of the trapped hydrogen. Raising the temperature above 300°C finally leads to a depletion of these traps.

---

[92] B. G. Pound in Ref. [7], p. 115.
[93] W. Mader and G. Necker, in *Metal-Ceramic-Interfaces*, eds. M. Rühle, A. G. Evans, M. F. Ashby, and J. P. Hirth, Acta-Scripta Metallurgica Proc. Series, vol. 4, Pergamon Press, Oxford (1990), 222.
[94] H. Jang and D. N. Seidman, *Interface Science* **1**, 61 (1993).
[95] G. Necker and W. Mader, *Phil. Mag. Letters* **58**, 205 (1988).
[96] H. Jang, D. N. Seidman, and K. L. Merkle, *Scripta metall. mater.* **26**, 1493 (1992).
[97] M. Backhaus-Ricoult and S. Laurent, *Mat. Sci. For.* **294–296**, 325 (1999).
[98] E. Pippel, J. Woltersdorf, J. Gegner, and R. Kirchheim, *Acta mater.* **48**, 2571 (2000).

The amount of irreversibly trapped H corresponds to about one monolayer at the phase boundary and can be ascribed to the formation of O-H bonds at the interface according to the following relation:

$$2H \text{ (in Pd)} + MgO + PdO \rightarrow Mg(OH)_2 + Pd, \qquad (4.19)$$

where the oxygen of the PdO corresponds to the excess oxygen within the structural vacancies of the terminating O^2 layer of the precipitate (cf. Fig. 15 for the analogous case of Ag/MgO). Calculating Gibbs free energy of the reaction and the volume change yields data for the trapping energy and partial molar volume of H, which are in agreement with experimental findings.[90,91] Thus the formation of a $Mg(OH)_2$ layer at the interface gives rise to a remarkable volume change, about two times as much as for H atoms dissolved in octahedral sites of Pd. Although a higher elastic energy for lattice distortion has to be paid by segregation at the oxide/metal interface, it takes place, because the gain of chemical energy by forming the OH bonds is much larger than the corresponding elastic energy. It is interesting to note that the crude approximation of the chemistry at the interface by bulk behavior gives reasonable results.

Because of the higher oxygen mobility in silver it was possible to internally oxidize an Ag-1 at.-% Mg at rather low temperatures leading to very small precipitates of MgO (1.6 to 5 nm in diameter). These were analyzed using a tomographic atom probe.[99] The results confirmed an excess of oxygen at the interface with the metal. However, an analysis of hydrogen turned out to be difficult because of the residual hydrogen in the vacuum chamber. Therefore, small-angle neutron scattering (SANS) was applied, in order to get additional information about segregation of both excess oxygen and hydrogen or deuterium, respectively. Again, the Ag/MgO samples were advantageous because of the smallness of their precipitates.

As the solubility of hydrogen in silver is very low, a temperature of 400°C was chosen, in order to have a flux of hydrogen sufficient to fill all the traps at the oxide/metal interfaces. This makes it possible to do SANS experiments with 3 types of samples: (i) after internal oxidation, (ii) internally oxidized plus doped with hydrogen and (iii) internally oxidized plus doped with deuterium.[100] The macroscopic cross section $d\Sigma/d\Omega$ obtained for these samples is presented in Figure 16. After the background and the incoherent scattering are subtracted, three different regimes can be distinguished for all samples. At very low values of the scattering vector Q large MgO precipitates at the grain boundaries give rise to a steep decrease of $d\Sigma/d\Omega$. Then a plateau region follows, which describes the scattering of the small precipitates within the grains at low Q values. The plateau is followed for $Q > R^{-1}$ by a Guinier regime that allows the evaluation of the average radius R of the oxide particles assuming a spherical shape.[100]

---

[99] C. Kluthe, T. Al-Kassab, and R. Kirchheim, *Mater. Sci. Eng.* **A327**, 70–75 (2002).
[100] R. Kirchheim, A. Pundt, T. Al-Kassab, F. Wang, and C. Kluthe, *Z. Metallkd.*, **94**, 266 (2003).

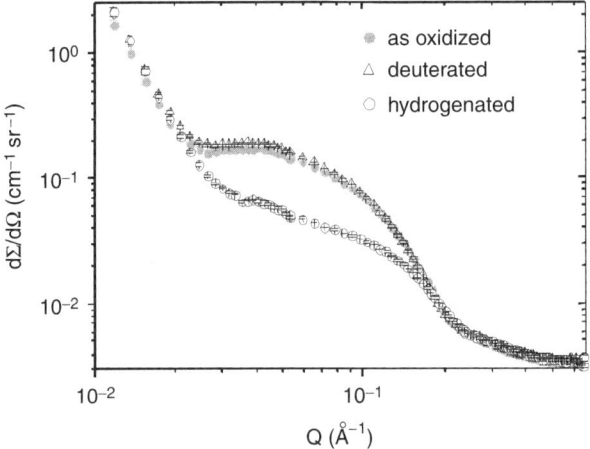

FIG. 16. Macroscopic scattering cross section $d\Sigma/d\Omega$ versus scattering vector, Q, for an internally oxidized Ag-1 at.-% Mg alloy (closed circles) and its changes after exposing to hydrogen (open circles) and to deuterium (triangles). The steep decrease at the lowest Q values is due to large oxide precipitates at grain boundaries. The following plateau and the decrease at large Q values stems from small oxide particles (average radius $\approx$ 1.6 nm) within the grains. The pronounced changes caused by hydrogen are caused by both segregation of hydrogen at the metal/oxide interface and displacement of Ag atoms from the interface (cf. text).

Both H and D doping change $d\Sigma/d\Omega$ considerably. MgO has a higher scattering length density than Ag and, therefore, segregated hydrogen (deuterium) having a negative (positive) scattering length decreases (increases) the macroscopic cross section. The observation that hydrogen has a much more pronounced effect than deuterium cannot be explained by the difference of the scattering lengths, but is due to Ag atoms being repelled from the interface. The corresponding decrease of scattering length density is exaggerated by H, with its negative scattering length, whereas the positive contribution of deuterium is compensated this way. This is in accordance with the observation for dislocations, where the contrast variations for H and D were smaller than expected because of a decrease of the packing density of Pd atoms.

At first sight the changes presented in Figure 16 are astonishing, because they are assumed to be caused by the segregation of one monolayer of H atoms at the interface only. However, one has to take into account that the number of H atoms on the surface of an oxide particle as small as 3 nm corresponds to about 50 percent of all the ions in the oxide. In addition, the decreasing macroscopic cross section in the Guinier regime is determined by the radius of gyration with respect to scattering contrast and, therefore, any changes of contrast in the periphery of an oxide particle are exaggerated. In addition to the changes observed for

H and D segregation, desorption of the excess oxygen can also be detected with SANS.[101]

A quantitative analysis of the SANS data was performed for a set of samples with different average radii and different coverages at the interface.[100,101] The results yield particle radii in agreement with measurements made with the tomographic atom probe.[99] The coverage of the interface with excess oxygen is half of the coverage with hydrogen or deuterium, respectively. This experimental finding supports the model proposed in Figure 15 or the corresponding reaction described by Eq. (4.19). However, the total values of coverage are about half of what the model predicts for a (111) surface of the MgO. These crystallographic planes of the oxide are adjacent to the (111) planes of Ag[93,95] and, therefore, the oxide precipitates should have an octahedral shape. However, as a consequence of minimization of interfacial area the actual shapes are truncated octahedra with (100) planes. For these planes the rock salt structure of MgO predicts 50% Mg and 50% O ions, i.e., no structural vacancies on the oxygen sublattice. Thus the fraction of (111) planes is reduced and so is the average coverage with excess oxygen or hydrogen, respectively. For more results and a detailed analysis see Ref. [101].

11. Defect Formation Energy

It is well known for surfaces and grain boundaries that their energy γ can be reduced by solute segregation leading to an excess concentration Γ_A at the interface. The corresponding change of energy is expressed by the Gibbs Adsorption Equation[102] as

$$\left.\frac{\partial \gamma}{\partial \mu_A}\right|_{T,\mu_B} = -\Gamma_A, \qquad (4.20)$$

where μ_A is the chemical potential of solute A and μ_B the one for solvent B. In a related study Carl Wagner[103] defined the excess Γ_A by the amount of solute dn_A one has to add or subtract from a system where the interfacial area is changed by da with a constant chemical potential of A and a constant number of moles of the solvent B. Thus we have

$$\Gamma_A = \left.\frac{\partial n_A}{\partial a}\right|_{T,P,V,n_B}. \qquad (4.21)$$

---

[101] C. Kluthe, T. Al-Kassab, J. Barker, W. Pyckhout-Hintzen, and R. Kirchheim, *Acta mater.*, in print.
[102] J. W. Gibbs, *Collected works*, vol. I, Yale University Press, New Haven (1948), 219 ff.
[103] C. Wagner, *Nachrichten der Akademie der Wissenschaften in Göttingen, II. Mathematisch-Physikalische Klasse* **Nr. 3**, 1 (1973).

According to the system being partly open (for A) and partly closed (for B) a new characteristic function was introduced[103]

$$F - n_A\mu_A \tag{4.22}$$

using the free energy F. The differential of the new function is

$$d(F - n_A\mu_A) = -SdT - PdV + \gamma da + \mu_B dn_B - n_A d\mu_A. \tag{4.23}$$

Upon differentiating $F - n_A\mu_A$ once with respect to a, once with respect to μ_A, one obtains

$$\frac{\partial^2(F - n_A\mu_A)}{\partial a \partial \mu_A} = -\frac{\partial n_A}{\partial a}\bigg|_{T,V,\mu_A,n_B} = \frac{\partial \gamma}{\partial \mu_A}\bigg|_{T,V,n_B,\mu_A}, \tag{4.24}$$

which yields the classic Gibbs Adsorption Equation via Eq. (4.21).

In the present study this procedure is generalized, in order to include other defects as well. For the sake of the same formalism we introduce an appropriate defect density ρ (grain boundary area, dislocation length, or number of vacancies per volume) with the specific energy γ now being the energy of formation of the defect per area, length, or number, respectively. Then Eq. (4.23) becomes

$$d(F - n_A\mu_A) = -SdT - PdV + \gamma d(\rho V) + \mu_B dn_B - n_A d\mu_A \tag{4.25}$$

The equivalent definition of solute excess at the defect is defined in analogy to Eq. (4.21) as

$$\Gamma_A^{(\rho)} = \frac{1}{V}\frac{\partial n_A}{\partial \rho}\bigg|_{T,P,V,n_B}. \tag{4.26}$$

Because n_A is proportional to ρ, it follows from the last equation that

$$\Gamma_A^{(\rho)} = \frac{1}{V}\frac{\Delta n_A}{\rho}, \tag{4.27}$$

where Δn_A is the total excess due to all the defects in volume V. A generalized adsorption equation is derived by following the same lines of derivation as before

$$\frac{\partial \gamma}{\partial \mu_A}\bigg|_{T,\mu_B} = -\Gamma_A^{(\rho)}. \tag{4.28}$$

With measurements of the chemical potential of hydrogen and measurements of the excess amount of hydrogen at dislocation and grain boundaries as presented in Sections IV.8 and IV.9, the change of the formation energy of these two types of defects can be calculated. Figure 17 shows the changes of concentration as a

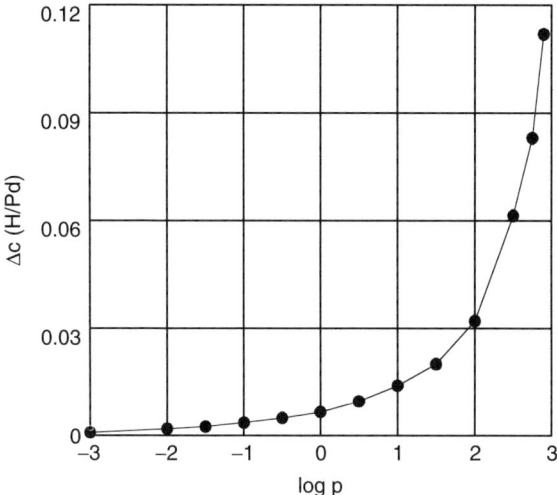

FIG. 17. Difference of H concentration between nanocrystalline and single crystalline Pd, Δc, at a given partial pressure of hydrogen, p. The values are obtained from the data set in Figure 11.

function of the hydrogen pressure for nanocrystalline Pd at room temperature. The following relations have been used to obtain the appropriate quantities:

$$\Delta c = \left(\frac{n_H}{n_{Pd}}\right)_{nano} - \left(\frac{n_H}{n_{Pd}}\right)_{single} = \frac{a\Gamma_H}{n_{Pd}} = \frac{a\Gamma_H \Omega_{Pd}}{V} = \frac{3\Gamma_H \Omega_{Pd}}{g} \quad (4.29)$$

with the last equation being valid for spherically shaped grains and g being the grain size and Ω_{Pd} the atomic volume of Pd. Numerical integration of the Gibbs Adsorption Equation for the data in Figure 17 yields

$$\Delta\gamma = \int_{-3}^{3} \Gamma_H d\left(\frac{2.3 k_B T \log p}{2}\right) = -0.89 \, J/m^2. \quad (4.30)$$

This value is of the order of the energy of the grain boundaries, meaning that segregation may finally lead to zero formation energy. Then no driving force for grain growth would be present. This phenomenon of zero grain boundary energies is discussed in more detail in Ref. [104].

In order to estimate the effect hydrogen has on the reduction of the line energy of dislocations, the simple relation of Eq. (4.14) derived in Section IV.8 for the hydride cylinder formed in the expanded region below the glide plane of an edge

---

[104] R. Kirchheim, Acta mater., 50, 413 (2002).

dislocation will be used.

$$\Delta c = \alpha \rho 4\pi r^2 = \frac{\rho \Gamma_H^{(\rho)} \Omega_{Pd}}{V} \tag{4.31}$$

Then integration of Eq. (4.28) yields

$$\Delta \gamma = -\int_{\mu_1}^{\mu_2} \frac{\alpha 4\pi r^2}{\Omega_{Pd}} d\mu_H = -\int_{c_1}^{c_2} \frac{\alpha 4\pi r^2}{\Omega_{Pd}} k_B T d \ln c_f = \int_{r_1}^{r_2} \frac{\alpha 4\pi r^2}{\Omega_{Pd}} k_B T d \frac{1.1 nm}{r}$$

$$= -\int_0^r \frac{\alpha 4\pi}{\Omega_{Pd}} k_B T (1.1 nm) \, dr = -\frac{\alpha 4\pi k_B T r}{\Omega_{Pd}} (1.1 nm). \tag{4.32}$$

For a radius of $r = 1$ nm as determined experimentally (cf. Section IV.8 and Fig. 9) the last equation gives $\Delta \gamma = 2 \cdot 10^{-9}$ J/m. This is about three times the line energy of a dislocation, if we calculate this value from the empirical relation $\gamma b \approx 1$ eV ($b =$ burger's vector $= 0.275$ nm for Pd).[105] The reduction of the line energy by hydrogen segregation may be overestimated because of the various approximations made during the derivation of Eq. (4.28). Nevertheless, a reduction of the energy of dislocation formation is expected and it may be the reason why higher dislocation densities can be produced by cycling palladium between the α and β phases than by severe cold rolling.[106] Crossing the α/β phase boundary, dislocation rings are punched out, in order to accommodate the misfit between the β and α phases.[22,23] This occurs at a high chemical potential, that is, a high excess concentration and, therefore, a reduced energy of dislocation formation.

For a hydride the situation is different, because the local concentration in the expanded region around an edge dislocation is saturated and, therefore, it is the same as far away from a dislocation. The formation of an edge dislocation leads to a negative H excess because hydrogen is now repelled from the compressed region above the glide plane. Therefore, the energy of dislocation formation is increased, leading to a loss of ductility. This may explain the extreme brittleness of metal hydrides. The commonly accepted theory that solute drag decreases dislocation mobility may not apply to hydrogen at room temperature, because it is highly mobile even in hydride phases.

Finally we discuss the effect hydrogen has on the formation energy of vacancies. In this case the defect density is the number of vacancies per unit of volume. Then the excess defined by Eq. (4.26) has the simple meaning of number of H atoms trapped around one vacancy N_{HV}. Assuming that this number is constant over a

---

[105] P. Haasen, *Physical Metallurgy*, Springer, Berlin (1994).
[106] T. Kuji, T. Flanagan, Y. Sakamoto, and M. Hasaki, *Scripta metall.*, **19**, 1369 (1985).

certain pressure range Eq. (4.28) yields

$$\Delta \gamma = -N_{HV} \Delta \mu_H = -N_{HV} \frac{k_B T}{2} \Delta \ln p_{H_2}. \quad (4.33)$$

As the interaction energies of hydrogen with vacancies in metals are rather high (ca. 1 eV[14]), hydrogen atoms may all be trapped at low concentrations (i.e. 10 at-ppm). Three orders of magnitude in concentration are covered by conditions upto terminal solubility in Pd at room temperature (ca. 10^{-2}), which corresponds to 6 orders with respect to pressure (cf. Eq. (2.12)). With an excess of 1 H atom per vacancy ($N_{HV} = 1$) Eq. (4.33) leads to $\Delta \gamma = -0.18$ eV. By increasing temperature and pressure, and a possible higher excess, N_{HV}, it is easily concluded that the formation energy becomes zero. This is in agreement with high-pressure experiments by Fukai et al.,[21] where abundant vacancies have been detected in Pd.

Discussing the effect solute atoms have on the formation energy of a defect again demonstrates how useful the knowledge of the chemical potential is. Again, hydrogen metal systems are model systems because it is rather easy to measure μ for them.

12. Interaction with Crack Tips and Hydrogen Embrittlement

A crack tip in a sample under external stress according to mode I, that is, tensile load perpendicular to the crack surface, attracts H atoms. The hydrostatic stress in front of the crack tip is enhanced and, therefore, H atoms can lower their chemical potential according to Eq. (2.14). A quantitative treatment of this elastic interaction in terms of a DOSE is presented in Ref. [107]. However, the strength of the interaction is rather weak and the corresponding concentration enhancement is small. Nevertheless, the presence of hydrogen at the crack tip gives rise to the following effects:

(1) For atoms right at the crack tip a small fraction of H atoms is sufficient to occupy and weaken the stretched metal–metal bonds at the tip. After rupture of these bonds hydrogen migrates to adjacent bonds and continues to weaken these bonds, too. Under these circumstances the fracture will be a brittle one. This scenario of hydrogen embrittlement is called decohesion[60] and may apply to those metals with a low H solubility, such as iron.
(2) Due to H–H interaction the segregation of hydrogen will be enhanced near the crack tip because of the concentration enhancement stemming from the elastic interaction. When the terminal solubility is reached a hydride forms at the crack tip and the crack can advance through the brittle hydride. Again, H

---

[107] R. Kirchheim and J. P. Hirth, Scripta metall. **16**, 475 (1982).

atoms easily redistribute and follow the propagating tip. This mechanism of hydrogen embrittlement is expected to be relevant for metals with a high H solubility, such as group Vb and IVb transition metals. Hydride formation in front of a crack has been observed in-situ in an electron microscope.[61]

(3) Opening a crack gives rise to the formation of fresh surfaces and, therefore, the work required for crack growth includes a surface energy term. In the presence of hydrogen the surface energy can be reduced by surface segregation of H atoms,[108] similar to the reduction of grain boundary energy discussed in Section IV.11. This way the energy for propagating a crack tip is reduced.

(4) For those cases where ductility plays an important role during fracture the interaction of hydrogen with dislocation has to be taken into account as well. During TEM observations of dislocations it was observed that the motion of these defects was accelerated in the presence of hydrogen.[109,110] This enhanced local plasticity increases the growth rate of a crack. Therefore, on a macroscopic scale it appears to be an embrittlement phenomenon. The reason for the increased dislocation velocity, as suggested by Birnbaum et al.,[109,110] is an elastic interaction between dislocations and the strain field of solute atoms such as carbon. In light of the decreased line energy of a dislocation by segregated H atoms, as discussed in Section IV.11, it may also be that the rate of generation of dislocations is increased in the presence of hydrogen because their formation energy is decreased.

There are other hydrogen effects, besides the direct interaction of hydrogen with the crack tip that severely alter the mechanical behavior. A recent review is given in Ref. [111].

V. Hydrogen in Disordered and Amorphous Alloys

13. Disordered Crystalline Alloys

For a crystalline alloy A_{1-x}, B_x, where hydrogen occupies tetrahedral sites and A and B atoms are distributed randomly, an appropriate DOSE is

$$n(E) = f \sum_{i=1}^{4} \binom{4}{i} x^i (1-x)^{4-i} \delta(E - E_i), \tag{5.1}$$

where the factor f is equal to the number of tetrahedral sites per metal atom that can be occupied by hydrogen. However, due to the repulsive interaction between the nearest hydrogen atoms, some of the sites remain empty and, therefore, actual

---

[108] J. P. Hirth and J. R. Rice, *Metall. Trans.* **A 11**, 1502 (1980).
[109] H. K. Birnbaum, *MRS Bull.* **28**, 479 (2003).
[110] H. K. Birnbaum and P. Sofronis, *Mat. Sci. Eng.* A **176**, 191 (1994).
[111] In Ref. [33], p. 215.

values of f are less than the total number of tetrahedral sites per metal atom. Five different types of tetrahedral sites (A_4, A_3B, A_2B_2, AB_3, and B_4) have to be distinguished according to the occupancy of their corners with either A or B atoms. For a random distribution of A and B, these five types are present in concentrations of $\binom{4}{i} x^i (1-x)^{4-i}$. Their site energies for hydrogen are labeled E_i where $i = 0, 1, \ldots, 4$ is the number of B atoms distributed among the corners of the tetrahedron. For an alloy dilute in B the DOSE of Eq. (5.1) reduces to the one given by Eq. (4.1), because the number of tetrahedra having more than one B atom becomes negligible.

The DOSE proposed for a concentrated alloy has been used by C. Wagner[112] in order to model the thermodynamic activity of interstitial oxygen in liquid iron alloys. Measurements of hydrogen solubility were performed by Feenstra et al.[113] in niobium-vanadium alloys over a wide range of alloy compositions and hydrogen concentrations. Because of the large data set they were able to show unambiguously that the fraction of the various tetrahedra follows the binominal distribution of a random alloy. Hydrogen prefers tetrahedra having a higher number of V atoms at their corners in agreement with the vanadium hydride being a stronger hydride former when compared with niobium (i.e., $E_4 < E_3 < \cdots < E_0$ where the subscript refers to the number of V atoms. Besides the expected behavior of the site energies with respect to i, they also depend on alloy composition, x. This is explained[113] by an overall change of the size of tetrahedral sites following the changes of the lattice parameter. According to the positive partial molar volume of hydrogen, the site energy is lowered when the site volume increases. Therefore, a V_4 site has a lower site energy in a niobium-rich alloy in comparison with a vanadium-rich alloy. All of these results are compiled in Figure 18 as the DOSE of the Nb-V alloy.

14. METAL/NONMETAL GLASSES

A large number of metal/nonmetal glasses have a concentration of about 20 at.-% nonmetal. Among these amorphous alloys the palladium-silicon alloys were mostly used to measure hydrogen solubility and diffusivity.[46,114–116] In some studies it was considered to be appropriate to use a two-level system[114,116] for the DOSE of the amorphous alloys, whereas others preferred a Gaussian distribution

$$n(E) = \frac{1}{\sigma\sqrt{\pi}} \exp\left[-\left(\frac{E - E^0}{\sigma}\right)^2\right] \quad (5.2)$$

---

[112] C. Wagner, *Acta Metall.*, **21**, 1297 (1973).
[113] R. Feenstra, R. Brower, and R. Griessen, *Europhys. Lett.* **7**, 425 (1988).
[114] R. S. Finocchiaro, C. L. Tsai, and B. C. Giesen, *J. Non-Cryst. Solids*, **61–62**, 661 (1984).
[115] A. Szökefalvi-Nagy, S. Filipek, and R. Kirchheim, *J. Phys. Chem. of Solids*, **48** 613 (1987).
[116] D. Richter, G. Driesen, R. Hempelmann, and I. S. Anderson, *Phys. Rev. Lett.*, **57**, 731 (1986).

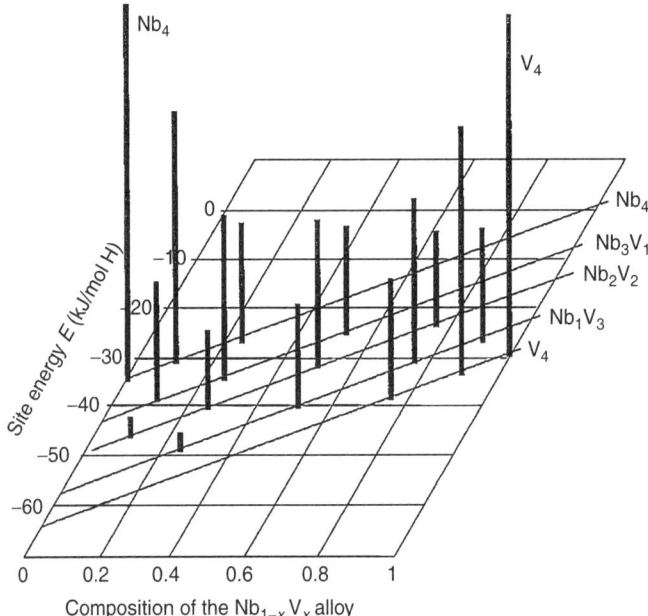

FIG. 18. Density of site energies for the various $Nb_{1-x}V_x$ tetrahedra in crystalline Nb-V alloys as a function of composition.[113]

where σ is the width and E^o the average value of this function. Inserting this DOSE in Eq. (3.3) and applying the step approximation yields

$$c = \frac{1}{2}\text{erfc}\left(\frac{E^o - \mu}{\sigma}\right). \qquad (5.3)$$

Solving the last equation for μ gives

$$\mu = E^0 - \sigma \, \text{erf}^{-1}(1 - 2c). \qquad (5.4)$$

where the inverse error function erf^{-1} was used. By measuring the chemical potential at low concentrations by an electrochemical technique and at high concentrations with high-pressure equipment, it was possible to cover a range of hydrogen pressures extending over 18 orders of magnitude,[115] as shown in Figure 19. In this case the concentration was defined as the ratio H/Pd, which in crystalline Pd is equivalent to the ratio of the number of H atoms to the number of sites (octahedral ones in fcc-Pd).

Over the range of pressures shown in Figure 19 it will be impossible to fit a two-level system to the data points. Besides the good agreement between experimental

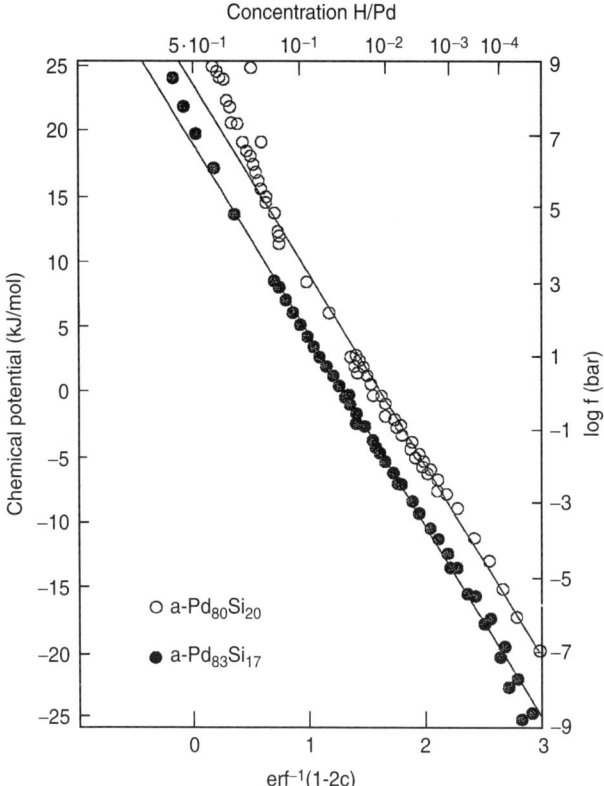

FIG. 19. Pressure-concentration isotherms for hydrogen in two amorphous Pd-Si alloys obtained by various experimental techniques at 295 K.[115] The results are plotted in accordance with Eq. (5.4) and the fugacity of hydrogen, f, is used at high chemical potentials instead of the partial pressure. The slope of the straight lines yields a value for the width σ of the Gaussian DOSE.

results and a Gaussian DOSE there is also a solid theoretical foundation for this DOSE.

In a first-order approximation[14] the DOSE of amorphous PdSi alloys was derived from the distribution of atomic distances, which is given by the first peak in the radial distribution function. This distribution of atomic distances may be considered as a distribution of strain. Multiplying the latter with a factor of 3 yields a distribution of the volume of interstices. Its product with the bulk modulus and the partial molar volume of hydrogen results in a distribution of site energies. As the first peak of the metal/nonmetal glasses has a Gaussian shape, the foregoing simple consideration will lead to a Gaussian DOSE. In a more rigorous treatment,

P. Richards[117] showed that the assumption of an interaction potential between hydrogen and metal of radial symmetry is sufficient to come to the same conclusion. Then the width σ of the DOSE can be calculated from measurable and known quantities, yielding a value only about 30 percent larger than the experimental one.

The pressure composition isotherms shown in Figure 19 do not show a pressure plateau, which is present in crystalline alloys stemming from the constant chemical potential within a miscibility gap or a two-phase region of a solid solution and a hydride. The absence of a miscibility gap or the missing hydride formation in amorphous alloys is treated by R. Griessen,[70] showing that there is an interesting analogy with ferromagnetism and the Stoner criterion. Thus, if the H–H interaction energy is smaller than the width of the Gaussian DOSE, no hydride formation occurs; that is, the H atoms prefer to be distributed among the low-energy sites instead of being forced to occupy high-energy sites in a concentrated phase, in order to profit from a small H–H interaction energy.

Examples of chemical diffusion coefficients in amorphous $Pd_{80}Si_{20}$ measured by an electrochemical technique are presented in Figure 20. Depending on the methods of preparation of the Pd-Si alloy, the values differ by up to two orders of magnitude, although the radial distribution function as measured with X-rays revealed no differences. The chemical diffusion coefficient D of hydrogen in amorphous matrices can be calculated by using Eqs. (3.17) and (2.13) in combination with Eq. (5.2). Experimental values are always in good agreement with the calculated ones,[14,46,118] as shown in Figure 20. The different results in Figure 20 can be explained partly by different distributions of site energies, but they are also affected by the distribution of saddle-point energies as discussed in Section VI.17. The latter effect changes the reference diffusivity D^o. However, the concentration dependence, that is, the slope and curvature of the calculated curves in Figure 20, is solely determined by the measured chemical potentials. The only free-fitting parameter D^o will move the curves up and down, because a logarithmic ordinate has been used in Figure 20 and because D^o is a factor in Eq. (3.17).

In Figure 20 the curve with the steepest slope corresponds to $Pd_{80}Si_{20}$, which was prepared by sputtering and had the broadest DOSE as obtained from fitting measured values of the chemical potential. For a broad distribution an incremental change of the concentration dc gives rise to larger changes of the Fermi energy $d\mu$ when compared with a narrower DOSE (cf. Eq. (3.7) with $n(\mu)$ being smaller for a broad DOSE). For the extreme case of zero width or a crystalline alloy, respectively, D is independent of c as shown in Figure 20 for a crystallized $Pd_{80}Si_{20}$-alloy.[14,119]

---

[117] P. M. Richards, *Phys. Rev.* **B27**, 2095 (1983).
[118] M. Hirscher and H. Kronmüller *J. Less-Common Met.* **17**, 658 (1991).
[119] Y. S. Lee and D. A. Stevenson, *J. Non-Cryst. Solids*, **72**, 249 (1985).

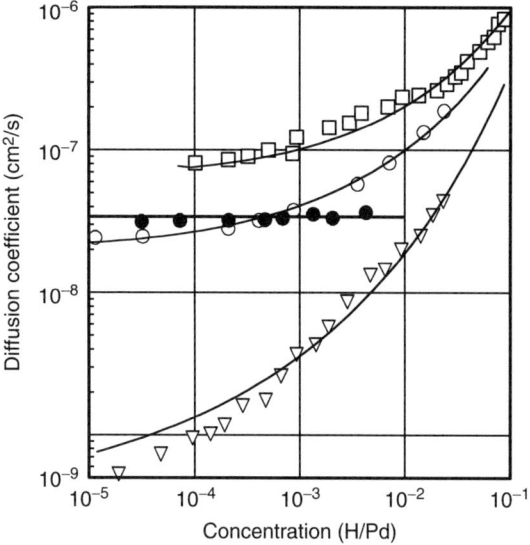

FIG. 20. Hydrogen diffusion coefficient as a function of H concentration in amorphous $Pd_{80}Si_{20}$ alloys prepared by the double piston technique (squares), melt spinning (open circles), and sputtering (triangles). The concentration dependence vanishes after crystallization of the melt spun alloy (closed circles).[14,119]

15. EARLY TRANSITION/LATE TRANSITION METALLIC GLASSES

For an amorphous alloy A_{1-x},B_x the same considerations as in Section V.13 have to be applied in order to obtain the fractions of the various $A_i B_{4-i}$ tetrahedra. However, unlike the crystalline case, each type of tetrahedral site has a broad distribution of site energies. Assuming a Gaussian one for each of them with an average energy E_i and a width σ_i, Eq. (5.1) becomes

$$n(E) = \frac{f}{\sigma_i \sqrt{\pi}} \sum_{i=1}^{4} \binom{4}{i} x^i (1-x)^{4-i} \exp\left(-\frac{(E-E_i)^2}{\sigma_i^2}\right). \quad (5.5)$$

Often binary metallic glasses are alloys of an early- and a late-transition metal. The affinity of these metals may be obtained from their hydride formation energies. A compilation of these values[36] shows that in almost all cases they increase within the transition series from left to right (Pd being an exception). Thus, early-transition metals have a high affinity to H whereas the late ones have a much lower affinity. The average site energy E_i for the $A_i B_{4-i}$ tetrahedron increases with increasing i for A being the late transition component. This expectation is in agreement with

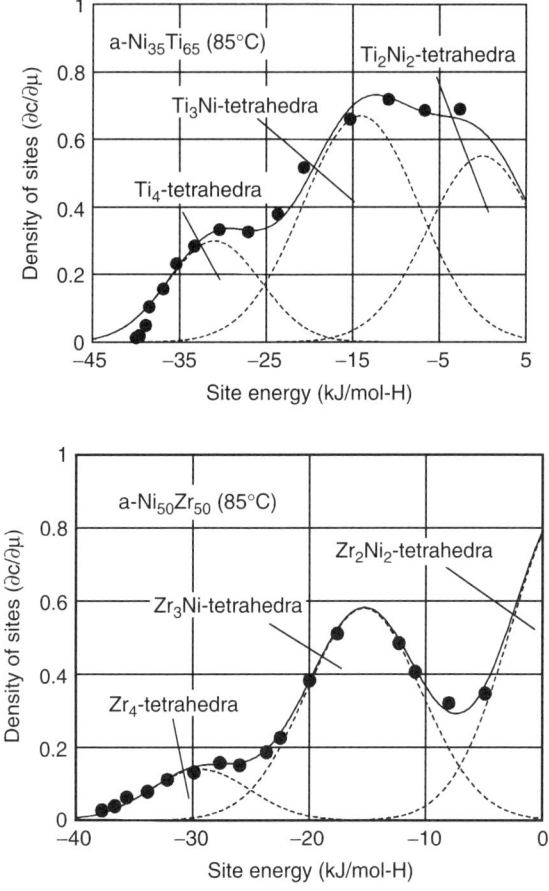

FIG. 21. Measured density of site energies $n(E) \approx \partial c/\partial \mu$ (cf. Eq. (3.7)) for an amorphous Ni-Ti alloy (filled circles) and the contributions from different tetrahedral sites (dashed Gaussian curves and their sum as a solid line). The Gaussian curves have about the same width and the ratio of their areas corresponds to the binomial distribution (cf. Eq. (5.5)). (b) same as (a) but for a Ni-Zr alloy.

experimental findings.[120,121] By using the step approximation of the FD-Statistics (Eq. (3.7)), the DOSE is obtained directly from measurements of $\mu(c)$ as shown in Figure 21.

The DOSE for the amorphous Ti-Ni and Zr-Ni alloys shows a structure that arises from the varying composition of the next neighbor atoms around the H atom. For the sake of simplicity the same width σ was assumed for each type

---

[120] F. Jaggy, W. Kieninger, and R. Kirchheim, *Z. Phys. Chem. N.F.* **163**, 431 (1989).
[121] J. H. Harris, W. A. Curtin, and M. A. Tenhover, *Phys. Rev.* **B36**, 5784 (1987).

of tetrahedron. The position and the area of the various Gaussian functions yield values for f and E_i. In a first-order approximation the average energies are linear combinations of the site energies for A_4 and B_4, and the site energies for the latter are about the same as the hydride formation energies of the crystalline metals A and B. The ratio of the areas of the various Gaussian distributions is independent of f and is determined in agreement with Eq. (5.5) by the alloy composition only. This agreement is considered to be a piece of evidence for a random distribution of Ni and Ti or Zr atoms, respectively. The situation is totally different for a $Mg_{50}Ni_{50}$ alloy, where the pressure composition isotherms resemble those of a crystalline metal (i.e., have rather flat portions like in the ones in the miscibility gap or the two-phase region.[122] Thus the majority of sites have to be very similar and according to the composition they are most probably Mg_2Ni_2 tetrahedra.

The chemical diffusion coefficient of hydrogen in early-transition/late-transition metallic glasses is increasing with increasing concentration.[120] However, in the ascending curves of D vs. c plots there is a dip at those concentrations, where a subset of tetrahedra of the type A_iB_{4-i} is filled and the energetically less favorable type $A_{i+1}B_{3-i}$ has to be occupied. Again this feature provides some information about the structure of a metallic glass where H atoms act as probes. In addition, it is an indication of a correlated random walk through the amorphous structure and, therefore, the concentration dependence of D calculated from measured chemical potentials deviates remarkably from measured data.[14]

16. BULK METALLIC GLASSES

Bulk metallic glasses can be quenched into the glassy state with rather low cooling rates.[123,124] This can be achieved by increasing the number of components. In addition, a larger variety of atomic radii appears to be favorable for glass formation. There are only a few studies on the behavior of hydrogen in bulk metallic glasses.[125-127] This may be due to a low H diffusivity (see following) and the extreme brittleness of hydrogenated samples. Thus electrochemical techniques, where two electrochemical cells are separated by the sample acting as a membrane, fail because cracks form and hydrogen is permeating quickly along these cracks.

Recently a new technique of measuring H diffusion has been invented[128] to circumvent these problems. Thin-film preparation methods were applied in order to

---

[122] S. Orimo, K. Ikeda, H. Fujii, S. Saruki, and T. Fukunaga, *J. Jap. I. Met.*, **63** 959 (1999).
[123] W. L. Johnson, *JOM-J. Min. Met. Mat.*, **54** 40 (2002).
[124] A. Inoue, T. Zhang, and T. Masumot, *Mater. Trans.*, **JIM 31**, 425 (1990).
[125] D. Zander, U. Köster, N. Eliaz, and D. Eliezer, *Mat. Sci. Eng.* **294–296**, 112 (2000).
[126] D. Suh and R. H. Dauskardt, *Met. Trans.*, **42**, 638 (2001).
[127] N. Ismail, M. Uhlemann, A. Gebert, and J. Eckert, *J. alloys comp.*, **198**, 146 (2000).
[128] J. Bankmann, PhD thesis, University of Göttingen (2003).

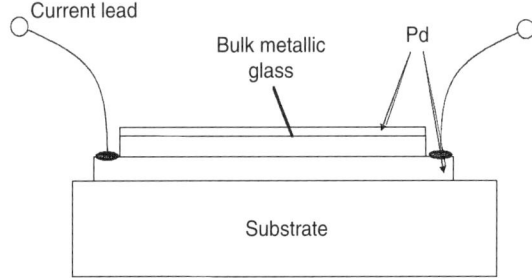

FIG. 22. Bulk metallic glass sandwiched between two Pd layers. The top layer allows permeation of gaseous hydrogen into the glass, and resistivity changes in the bottom layer are a measure of hydrogen transport through the glass. The transient behavior of this transport yields H diffusivity.[128]

produce a multilayer of Pd/bulk-glass/Pd on a substrate (cf. Fig. 22). The covering Pd layer is very thin and serves the purpose of facilitating hydrogen entry. After passing through the bulk metallic glass, hydrogen is dissolved in the Pd layer below. Because of the higher electrical resistance of bulk metallic glass when compared to Pd, and because of the thickness of the various layers, the bottom Pd layer determines the overall resistivity of the package. Then changes of the resistivity as a function of time can be used to calculate the diffusion coefficient of hydrogen, because the rate determining step is diffusion through the bulk metallic glass.

Experimental results of D in a $Zr_{66.8}Al_{17.4}Ni_{7.2}Cu_{8.6}$ glass and the equilibrium pressures are presented in Figures 23 and 24 as a function of H concentration. The equilibrium pressures were obtained by electrochemical measurements as well but

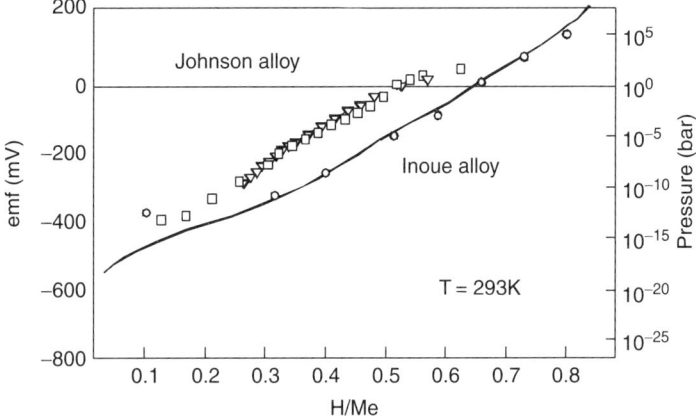

FIG. 23. Pressure composition isotherms for H in two bulk metallic glasses (Johnson glass: $Zr_{46.8}Ti_{8.2}Cu_{7.5}Ni_{10}Be_{27.5}$, squares for a first run and triangles for a second run; and Inoue glass: $Zr_{66.8}Al_{17.4}Ni_{7.2}Cu_{8.6}$, circles). The line was calculated for the Inoue glass as described in the text.

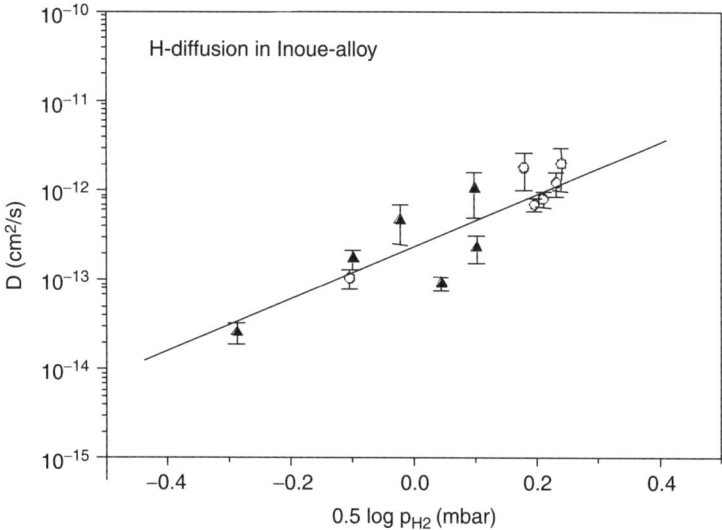

FIG. 24. Diffusion coefficients for hydrogen in the Inoue glass (two samples) as a function of the hydrogen pressure. The straight line is an approximate behavior as expected from the measured chemical potentials.[128]

without the bottom Pd layer shown in Figure 22. In agreement with the behavior in other metallic glasses the diffusion coefficient increases with increasing hydrogen content and the increase can be calculated from the measured values of the chemical potential.[128]

However, modeling the behavior of the chemical potential or the equilibrium pressures, respectively, turned out to be rather complicated. Because of the increased number of components the number of different tetrahedra increases from 5 for a binary alloy to 15 for a ternary and 35 for a quarternary alloy. Assigning different widths and average energies to each of the tetrahedra would result in a corresponding large number of fitting parameters. Thus the measured isotherms were modeled with the smallest reasonable number of parameters by assuming that all Gaussians had the same width $\sigma_i = 8$ kJ/mol-H and an average site energy for a tetrahedron is obtained from the participating corner atoms and their hydride formation energies E_f as

$$E_k = v_{k1} E_{f1} + v_{k2} E_{f2} + \cdots + v_{kn} E_{fn} \quad \text{and} \quad \sum_{j=1}^{n} v_{kj} = 4, \quad (5.6)$$

where v_{kj} is the number of j atoms sitting at the corners of a tetrahedron labeled k and n is the total number of components. The DOSE is given by the sum of Gaussians with these parameters multiplied with the frequency of their occurrence, which depends on concentration. Then the only fitting parameter left is the total number of tetrahedra, f (cf. Eq. (5.5)), which moves the calculated pressure composition isotherms in a double logarithmic plot parallel in the direction of the log c axis but does not affect slope and curvature. Calculating isotherms this way resulted in good agreement with experimental results for the $Zr_{66.8}Al_{17.4}Ni_{7.2}Cu_{8.6}$ alloy as shown in Figure 23. However, the agreement was not achieved by this procedure for the $Zr_{46.8}Ti_{8.2}Cu_{7.5}Ni_{10}Be_{27.5}$ (Vitralloy 1) alloy. This failure of the simple concept may arise because of two reasons: (1) the atoms of the various components are not distributed randomly but a pronounced short-range order exists, and (2) the distances between a H atom and the atom of one of the components is very different from the distances in the pure component (H atoms may actually occupy nontetrahedral sites). The latter effect then has to be treated in a similar way as in the crystalline Nb-V alloys (cf. Section V.13).

VI. Other Interstitials in Amorphous Materials

It has been stated several times in Section III that the concepts of a DOSE and FD-Statistics are not applicable to hydrogen alone but to all small particles dissolved interstitially in a defective crystalline or an amorphous material. The term interstitially requires a new definition for an amorphous matrix, and an attempt is made in Section VI.18. As a consequence of the generalized treatment Eq. (5.4) can be used successfully for other amorphous materials, too, as shown in Figure 25. This way the solubility of interstitials in amorphous materials can be treated in a universal way. However, in some cases the solute particles change the DOSE like H in amorphous silicon (see Section VI.19) or alkali ions in oxidic glasses

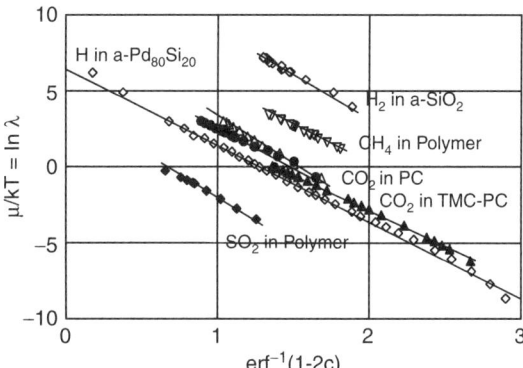

FIG. 25. Chemical potential or logarithm of the thermodynamic activity, λ, of various small solute molecules or atoms in various amorphous matrices plotted in accordance with Eq. (5.4) versus inverse error function of $(1 - 2c)$.

(see Section VI.20) and Eq. (5.4) becomes inapplicable. Before these special systems are considered in more detail, a general treatment of interstitial diffusion in amorphous systems is presented first.

17. MODELING DIFFUSION

During the derivation of Eq. (3.17) it had to be assumed that diffusion occurs via an uncorrelated random walk. As Eq. (3.17) has been successfully used in many cases to describe the concentration dependence of measured chemical diffusion coefficients, correlation effects are apparently not changing the concentration dependence in these cases. However, they can change the magnitude of the D values. Correlation effects can arise in extended defects where solvent particles migrate predominantly within the defect. This can be circumvented as demonstrated for grain boundaries in Section IV.9 by neglecting the material surrounding the defect and considering the DOSE of the defect alone. If this diffusion within the defect is rate determining, the overall effective diffusion coefficient is obtained. Besides these spatial correlations there may be energetic correlations as well. It is easy to comprehend that the saddle point energy between two sites of low energy is reduced on the average when compared with the average saddle point energy. If, for instance, the site energy is reduced because the site volume is large, then atomic distances of the neighboring atoms are larger than the average value. Thus the corresponding distances for the saddle point configuration are large, too, which gives rise to a lower saddle point energy.

Monte-Carlo (MC) simulations have been conducted[14,129–131] to determine the effect of a distribution of saddle point energies. The advantage of studying diffusion by MC simulations is the freedom of choosing various kinds of DOSE and various kinds of distributions of saddle point energies (DOSPE). The dependence of the simulated tracer diffusion coefficient D^* on concentration, c, and temperature, T, is summarized in a schematic way in Figure 26.

The following four cases are distinguished:

1) *Delta function for both DOSE and DOSPE (first column in Fig. 26)*
   This simulates the behavior of an interstitial in an ideal single crystal. D^* does not depend on c unless the lattice becomes filled. For $c \to 1$ ($\log c = 0$), blocking of sites and vacancy correlation effects (preferred jumps back into the site just left) lead to a decrease of D^*. The temperature dependence follows an Arrhenius Law. These dependencies are shown for comparison as dashed curves in Figure 26 for the following cases.

---

[129] R. Kirchheim and U. Stolz, *Acta metall.* **35**, 281 (1987).
[130] A. F. McDowell and R. M. Cotts, *Z. Phys. Chem. NF*, **183**, 65 (1994).
[131] M. Hirscher, J. Mossinger, and H. Kronmüller, *J. Alloys Comp.*, **231**, 267 (1995).

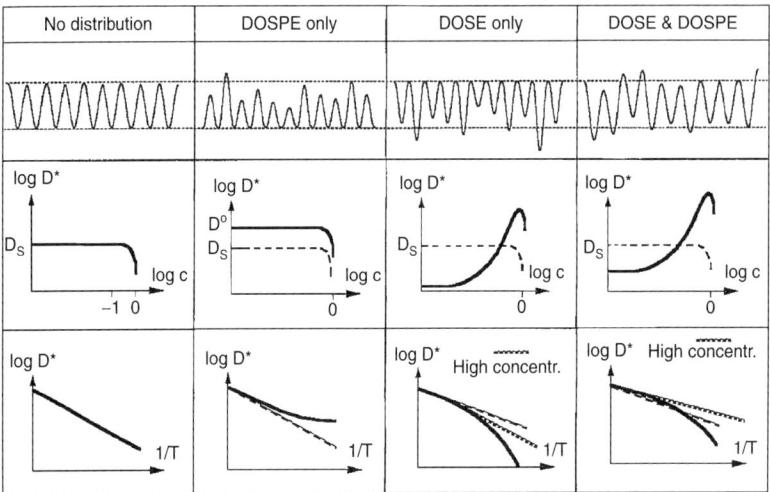

FIG. 26. Schematic presentation of results of Monte-Carlo simulations for interstitial diffusion in a 3-dimensional energy landscape with and without Gaussian distributions of site and saddle point energies. The concentration and temperature dependence of D in a perfect crystal (first column on the left) is shown in the other columns as a dashed line for the sake of comparison. The temperature dependence for high concentrations is shown as a dotted line in the last two columns. The different dependencies are explained in the text.

2) *Delta function for DOSE and Gaussian function for DOSPE (second column in Fig. 26)*

For a Gaussian distribution of saddle point energies having the same average value as the constant value in case (1), the tracer diffusion coefficient is increased, because the interstitials prefer jumps over lower barriers. In two and three dimensions there will always be passes through the lattice having a lower average activation barrier than the mean value. Thus in agreement with MC simulations the effective activation energy is reduced. At very low temperatures and the concomitant short diffusion lengths the lowest barriers allowing a percolated walk through the lattice are preferred even more, giving rise to the convex curvature in an Arrhenius diagram when viewed from the $1/T$-axis. As there is no DOSE D^* does not depend on c unless $c \to 1$ and like in case (1) blocking and correlation effects decrease D^*.

3) *Delta function for DOSPE and Gaussian function for DOSE (third column in Fig. 26)*

As there is no distribution of saddle point energies in this case, there is no correlation between successive jumps and the results derived in section III can be used. At very low concentrations D^* is independent of c because Henry's Law is fulfilled and $\gamma^o = \text{const}(c)$. Here the particle distribution among the sites of the DOSE is controlled by maximizing entropy and, therefore, sites above the Fermi energy are filled predominantly. Then the mean energy of the particles is given by[69]

$$\langle E \rangle = \int_{-\infty}^{\infty} \frac{E \exp[-(E - E^o)^2/\sigma^2]}{1 + \exp[(E - \mu)/k_B T]} \approx E^o - \frac{\sigma^2}{4k_B T} > \mu \quad \text{for } c \to 0. \quad (6.1)$$

Then the activation energy is the difference between the constant saddle point energy and this average energy. Because of the temperature dependence of $\langle E \rangle$ given by Eq. (6.1) the D^* values at low concentrations exhibit a concave curvature in an Arrhenius plot when viewed from the abscissa. By increasing c the Fermi energy μ increases and finally becomes larger than $\langle E \rangle$. Then minimizing of energy controls occupancy of sites and the activation energy is the difference between the constant saddle point energy and μ. This activation barrier decreases with c as μ increases and, therefore, D^* increases with c. This increase of D^* is a consequence of FD-Statistics because low-energy sites will be saturated and high-energy ones have to be occupied. This way, trapping by low-energy sites vanishes and the activation barrier is decreased. The c dependence disappears, if the condition of single occupancy of a site is not taken into account during a MC simulation. Then FD-Statistics is replaced by Boltzmann Statistics. Finally, for $c \to 1$ D^* starts to decrease because of the same reasons discussed in case (1) and (2). Increasing temperature in the intermediate concentration range does not change μ very much, despite an increasing spread between full and empty sites of the order of $k_B T$. Thus the activation energy does not depend on temperature and straight lines are obtained for the MC results in an Arrhenius diagram.

4) *Gaussian functions for both DOSE and DOSPE (fourth column in Fig. 26)*
The case of having a Gaussian distribution in site and saddle point energies appears to be the one that is the most appropriate for real amorphous materials. Evaluating the MC results of this case shows that the effects of both distributions can be superimposed according to the relation

$$D^* = f_{sp}(T) f_{st}(c, T) D^o, \quad (6.2)$$

where the various factors are obtained from the limiting cases (1), (2) and (3), (i.e. $D^o = D^*$ in case (1), $f_{sp} = D^*/D^o$ calculated from the MC data of case (2), and $f_{st} = D^*/D^o$ calculated from the MC data of case (3)). There is also theoretical support for the validity of Eq. (6.2).[65] Thus the c dependence of D^* in case (4) is the same as in case (3) but the values are increased by the factor $f_{sp} > 1$.

Despite the fact that both activity and diffusivity depend strongly on c, the permeation, P, being the product of both, shows a minor c dependence in defective and amorphous materials only. For the stationary flux through a film of thickness d, a permeation coefficient P is defined according to the following equation:

$$J = P\frac{\Delta a}{d}, \qquad (6.3)$$

where d is the membrane thickness and Δa is the activity difference (or partial pressure difference) between entrance and exit sides of the film. According to Eq. (3.25) steady state means

$$J = -D^o(1-c)^2\frac{\Delta a}{d}. \qquad (6.4)$$

Thus the permeation becomes $P = D^o(1-c)^2$, which is independent of c at dilute solution. It is astonishing to note that this result, although it is derived for a distribution of site energies, is equivalent with permeation through a matrix that contains sites of energy E^o only. This result is caused by a compensation effect, that is, any increase or decrease of D is accompanied by a equivalent decrease or increase of the solubility and, therefore, both changes have no effect on the concentration dependence of the permeation coefficient P. An interesting example is a polymer foil produced with and without clusters of Pd atoms. For both samples the permeability of hydrogen was not changed very much, although the diffusivity decreased by more than two orders of magnitude within the polymer containing the palladium clusters as trapping centers.[132]

18. SMALL MOLECULES IN GLASSY POLYMERS

There are a variety of different applications where the solubility and diffusivity of small molecules play an important role such as penetration of CO_2 through PET beverage bottles, permeation of oxygen through plastic foils used to wrap food, drug release through polymer coatings, diffusion of dye molecules into fibers, separation of gases by polymeric membranes, and so on. If molecules are dissolved in the intermolecular space between the macromolecules of a polymer they interact mainly via Van der Waals or dipole-dipole forces with the matrix. Contrary to metals the matrix is in the majority of cases an amorphous one.

For polymers above the glass transition (rubbery or liquid state) the elastic constants are orders of magnitude smaller than in the glassy state and, therefore, any size misfit between the intermolecular site and the dissolved molecule is accommodated by a negligible amount of elastic stress, which most probably relaxes. As a consequence the volume change per dissolved molecule or its partial molar volume, respectively, is equal to the volume of the molecule (in its liquid state). Then the DOSE is represented by a Dirac delta function and for small

---

[132] D. Fritsch and K.-V. Peinemann, *Catalysis Today*, **25**, 277 (1995).

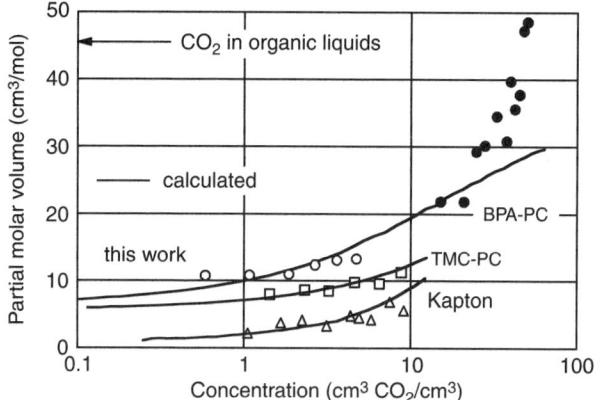

FIG. 27. Measured partial molar volumes of CO_2 in two glassy polycarbonates and in a glassy polyimide (Kapton) as a function of CO_2 concentration and calculated behavior (solid lines).[136] The partial molar volume increases because with increasing concentration smaller sites with a higher elastic energy have to be occupied. The partial molar volume in the liquid or rubbery state of the polymer as well as in many organic liquids is about 46 cm³/mol. This value is approached at high concentrations of CO_2 (closed circles) because the polycarbonate is swelling and finally transforms into the liquid state due to the strain induced by the dissolved CO_2.

concentrations Eq. (3.3) yields the well-known relation for an ideal dilute solution

$$\mu = E^o + k_B T \ln \frac{N}{N_o} = E^o + k_B T \ln c. \qquad (6.5)$$

where N_o is the total number of sites, which is usually not known because of an unknown structure (cf. the factor f in Eq. (5.5)). But N_o can be either estimated[133] or measured by positron annihilation spectroscopy.[134,135] Expressing μ by the partial pressure of the small molecules in the gas phase via Eq. (2.5) yields the proportionality between concentration and pressure. This is in agreement with experimental findings and it is called Henry's Law because the partial pressure is equivalent to the thermodynamic activity at low pressures.

Small molecules dissolved in polymers below the glass transition temperature behave very differently. The partial molar volume of the small molecules is much smaller than the volume of the molecule and it increases with increasing concentration (cf. Fig. 27 and Ref. [136]). This is interpreted via a distribution of site volumes belonging to the intermolecular volume. In accordance with the nomenclature used

---

[133] R. Kirchheim, *Macromolecules*, **25**, 6952 (1992).
[134] J. Bohlen, J. Wolff, and R. Kirchheim, *Macromolecules*, **32**, 3766 (1999).
[135] J. Bohlen and R. Kirchheim, *Macromolecules*, **34**, 4210 (2001).
[136] P. Gotthardt, A. Grüger, H. G. Brion, R. Plaetschke, and R. Kirchheim, *Macromolecules*, **30**, 8058 (1997).

in polymer science the sites are called holes in the following. Because the polymer structure is frozen in below the glass transition temperature, the free volume occupied by small molecules is not regenerated as in the case of rubbery or liquid polymers. For molecules having larger sizes than the holes, elastic energy has to be paid during dissolution and, therefore, larger holes are filled first. Increasing the number of dissolved molecules requires filling of the smaller holes as well, leading to an increasing volume change with increasing concentration. Assuming spherical shapes for both molecule and hole allows the calculation of the partial molar volume V_p in the framework of continuum mechanics[137]

$$V_p = \gamma (V_i - V_h) \tag{6.6}$$

where V_i is the volume of the interstially dissolved small molecule and V_h the one of the hole, γ is a factor of about unity taking into account the different elastic constants of the spherical molecule and the matrix.[136] Thus the smaller V_h, is the larger V_p.

The elastic energy E_{el} associated with the incorporation of the molecule is[137]

$$E_{el} = \frac{2\mu_s (V_i - V_h)^2}{3\gamma V_h}, \tag{6.7}$$

where μ_s is the shear modulus of the polymer. Thus a broad distribution of the volume of holes gives rise to a broad DOSE. The distribution of hole sizes is modeled according to Bueche[138] as a Gaussian stemming from volume fluctuations above the glass transition temperature

$$n(V_h) = \frac{1}{\sigma_V \sqrt{\pi}} \exp\left[-\frac{(V_h - V_h^o)^2}{\sigma_V^2}\right] \text{ with } \sigma_V = \sqrt{\frac{2kT_g V_h^o}{B}}, \tag{6.8}$$

where B is the bulk modulus at the glass transition temperature T_g. The original treatment by Ref. [138] is modified in as far as the temperature, T, was replaced by T_g, assuming that the volume fluctuation is quenched at the glass transition. For the simplified case of a linear expansion of the elastic energies around the average volume V_h^o, the Bueche distribution leads to a Gaussian DOSE[133] with an average energy E^o and a width σ_E given by

$$E^o = E_r + E_{el}^o = E_r + \frac{2\mu_s (V_i - V_h^0)^2}{3\gamma V_h^0} \quad \text{and}$$

$$\sigma_E = \frac{2\mu_s (V_i^2 - V_h^{0^2})}{3\gamma V_h^{0^2}} \sigma_V \quad \text{for } V_i \geq V_h^0, \tag{6.9}$$

---

[137] J. D. Eshelby, *Solid State Physics*, eds. F. Seitz and D. Turnbull, Academic, New York (1956).
[138] F. J. Bueche, *J. Chem. Phys.* **1953**, 21 (1850).

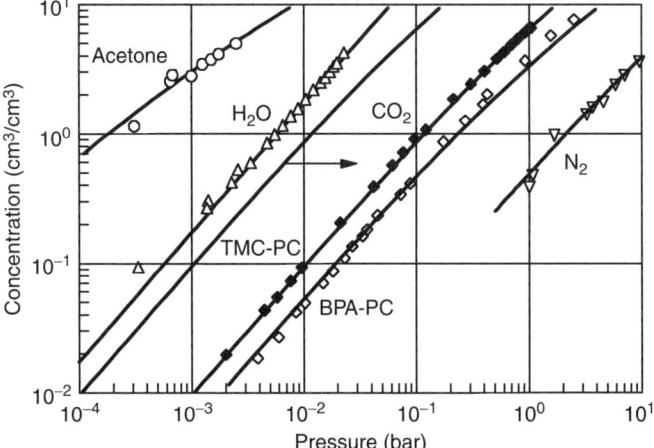

FIG. 28. Concentration pressure isotherms for various small molecules in polycarbonates.[139] The lines are calculated assuming a Gaussian DOSE and FD-Statistics. A slope of unity corresponds to ideal dilute behavior and the validity of Henry's Law, which is the case for water. The width of the Gaussian distribution determines slope and curvature of the isotherms and the smaller the slope the broader the width. Changes of the average energy E^o result in a parallel movement of the isotherms in the direction of the log c axis.

where E_r is that part of the dissolution energy stemming from Van der Waals interactions. Throughout this study site energies were considered, although the formula will not change if E is replaced by G (Gibbs free energies). Thus site entropies can be taken into account.

Experimental concentration-pressure isotherms are shown in Figure 28 and compared with the predictions of a Gaussian DOSE. It can be shown[139] that in a double logarithmic plot the shape of the calculated isotherms is solely determined by the width σ_E, whereas the second fitting parameter E^o moves the curves in the direction of the abscissa. Straight lines with a slope of unity correspond to Henry's Law, which is appropriate for water molecules in polycarbonate (see Fig. 28). This is a consequence of the size of H_2O molecules being smaller than the holes in polycarbonate. No elastic energy comes into play and the DOSE degenerates to a Dirac-delta function known to lead to Henry's Law.

---

[139] A. Grüger, P. Gotthardt, M. Pönitsch, H. G. Brion, and R. Kirchheim, *J. Polym. Sci.: Polym. Phys.* **36**, 483 (1998).

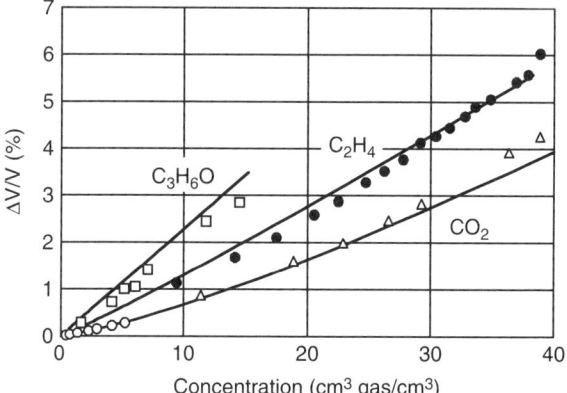

FIG. 29. Relative volume changes of polycarbonate for three different solute molecules (carbon dioxide, ethylene, and acetone). The changes scale with the size of the molecules and the lines are calculated with one fitting parameter for all three curves.

In textbooks on polymers[140,141] the solubility is described by the dual sorption model[142–144] where the DOSE is a two-level system. It allows fitting of c-p isotherms with three free parameters, yielding an agreement with experimental data as good as fitting with a Gaussian DOSE. However, the parameters of the dual sorption model do not have a rigorous physical meaning and they are not related analytically to other physical quantities like the two parameters of the Gaussian distribution as represented by Eqs. (6.8) and (6.9). For a more detailed comparison of the two approaches the reader is referred to Refs. [133, 136, 139, 145]. The quantities used in Eqs. (6.8) and (6.9) are compiled in Ref. [141] or obtained as follows. For the volume of the dissolved molecule, the partial molar volume in a rubbery polymer or in the liquid state of the small molecules was used. From a comparison of measured partial molar volumes with predicted ones, the average hole volume V_h^o is obtained. Once this value is known the volume change of other small molecules can be calculated without a fitting parameter (cf. Fig. 29).

By knowing values of V_h^o, values of the width of the Gaussian DOSE, σ_E, can be calculated via Eqs. (6.8) and (6.9). They are about 30–50 percent larger than the ones obtained from fitting c-p isotherms. This is considered to be good

---

[140] H. Batzer, *Polymere Werkstoffe*, Georg Thieme Verlag, Stuttgart (1985).
[141] D. W Van Krevelen and P. J. Hoftyzer, *Properties of Polymers*, 2nd ed., Elsevier, New York (1980), Chap. 4.
[142] R. M. Barrer, J. A. Barrie, and J. Slater, *J. Polym. Sci.*, **23**, 315 (1957).
[143] A. S. Michaels, W. R. Vieth, and W. R. Barrie, *J. Appl. Phys.*, **34**, 13 (1963).
[144] D. R. Paul, *Ber. Bunsen-Ges. Phys. Chem.*, **83**, 294 (1979).
[145] M. Pönitsch, P. Gotthardt, A. Grüger, H. G. Brion, and R. Kirchheim, *J. Polym. Sci.: Polym. Phys.* **35**, 2397 (1997).

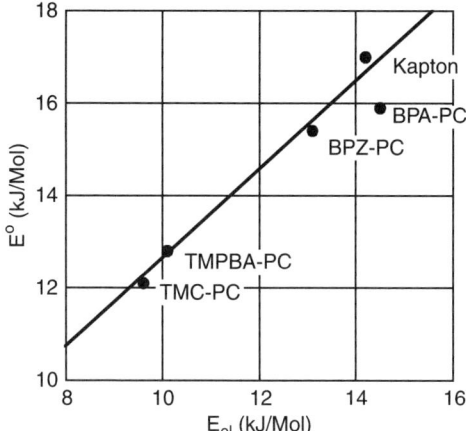

FIG. 30. Average energy of the DOSE of CO_2 for different glassy polymers obtained from concentration-pressure isotherms plotted versus the elastic energy associated with the incorporation of the molecule into a smaller hole of the polymer and as obtained from Eq. (6.7). The linear relation between the two quantities used as an assumption in Eq. (6.9) appeared to be fulfilled.

agreement in the light of the various crude assumptions made during the derivation of Eq. (6.9). The functional relations are fulfilled as well as shown in Figs. 30 and 31 (i.e., E^o plotted vs. E_{el} yields a straight line of slope 1).

Plotting σ_E vs. V_i squared should give a straight line with an intercept $(V_h^o)^2$ on the abscissa. For atoms or molecules such as He and H_2O the width is zero because $V_i < V_h^o$. Very large molecules such as ethene and acetone yield smaller values than predicted by the linear relation. This deviation at large V_i may arise because the assumption of an elastic incorporation of the small molecule is no longer valid. A calculation of the stresses within Eshelby's continuum approach[137] yields values that exceed the flow stress of the polymer considerably and, therefore, inelastic relaxation of macromolecules or plastic deformation, respectively, will occur. In other words, the larger the molecule gets the more it is incorporated substitutionally. The analogue in a crystalline lattice would be an atom being too large for an interstices, which kicks out a neighboring atom from its lattice site and becomes a substitutional solute. By generalizing these considerations a solute atom is incorporated in a material interstitially, if it is straining the matrix elastically only.

Diffusion of small molecules follows the predictions for a broad DOSE as discussed in Section III.6 and, therefore, the concentration dependence as obtained from measured c-p isotherms (c-μ isotherms, respectively) is in good agreement with experimental data as shown for a few examples in Figure 32. Again, only the reference diffusion coefficient D^o has been used as a fitting parameter, which in the presentation of Figure 32 moves the curves in the direction of the ordinate

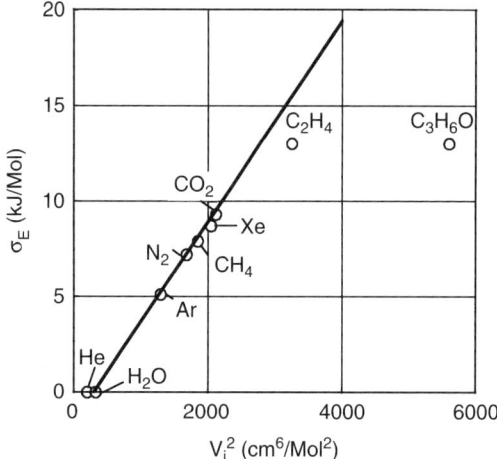

FIG. 31. Widths σ_E of a Gaussian DOSE for various small molecules in bisphenol-A polycarbonate (BPA-PC) plotted versus the squared molar volume of the small molecules. The width was obtained by fitting concentration-pressure isotherms as shown in Figure 28 and the linear relationship is predicted by Eq. (6.9). The line intercepts the abscissa at V_h^o, i.e. the average site volume in BPA-PC. For molecules being smaller than V_h^o, no elastic energy has to be provided during dissolution and the Gaussian degenerates to a Dirac-Delta Function.

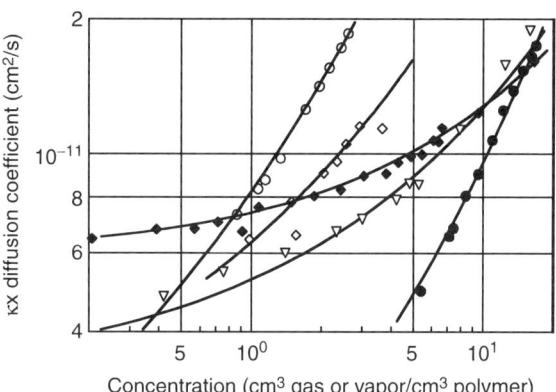

FIG. 32. Concentration dependence of the diffusion coefficient (scaled with a factor κ, in order to fit into the diagram) of different small molecules in different polymers.[145] The lines were calculated using measured concentration-pressure isotherms and Eqs. (2.13) and (3.17) and one fitting parameter D° (cf. Eq. (3.17)), which, like the scaling factor κ, moves the curves parallel in the direction of the ordinate only.

without changing their shape. The concentration dependence is larger the larger the size of the dissolved molecule. Figure 32 also shows that in cases where σ_E is smaller the diffusion coefficient becomes independent of c for $c \to 0$ in agreement with the expectation for a Gaussian DOSE.

19. Hydrogen in Amorphous Silicon and Germanium

Experiments on H and D solubility and interdiffusion of H and D in amorphous silicon and germanium were conducted by using Secondary Ion Mass Spectrometry (SIMS).[146–149] It was shown that the activation energy of diffusion decreased with increasing concentration. This was interpreted using the concept of the chemical potential and a site energy distribution as developed much earlier for metallic glasses.[14] While there are striking similarities, there are also remarkable differences. Most of the hydrogen atoms are bound to dangling bonds of the silicon because the amorphous semiconductor was prepared by CVD from SiH_4 at elevated temperature. H concentration was varied by varying the preparation temperature. This way the site energy distribution, or its fraction belonging to dangling bonds, depends on H concentration. This is schematically shown in Figure 33.

As the DOSE has small values around μ any remarkable increase of the H concentration requires a large change of μ. Thus the solubility seems to be constant or it is predetermined by the concentration of dangling bonds, respectively. Whereas in metallic glasses or glassy polymers the chemical potential increased because of a filling of a pre-existing distribution of site energies, the situation in amorphous semiconductors is different. In the latter case an increase in concentration is accompanied by changes of the distribution function leading to an increase of the chemical potential as well (cf. Fig. 33).

A peculiar behavior can be observed if a sample with a high content of H (or D), called sample B in the following, is brought in contact with a sample of low H (or D) content, called sample A. The total (H+D) concentration will not change in the two samples, although interdiffusion occurs as observed from the isotope redistribution measured by SIMS depth profiling. A step in the total concentration of H+D remains at the interface. In addition the diffusivity of H (or D) is now much faster in sample A with the low H content when compared with an experiment without a step in total concentration. This increase of H diffusivity in sample A stems from an increase of the chemical potential as imposed by sample B (cf. Fig. 33). In order to raise μ in sample A to about the same level as in B, only a negligible amount of H has to be transferred from B to A.

---

[146] R. A. Street, *Physica B* **170**, 69 (1991).
[147] W. Beyer and U. Zastrow, *J. Non-Cryst. Solids*, **230**, 880 Part B (1998).
[148] W. Beyer and H. Wagner, *Mat. Res. Soc. Symp. Proc.*, **336**, 323 (1994).
[149] W. Beyer, *Sol. Ener. Mat. Sol. C*, **78**, 235 (2003).

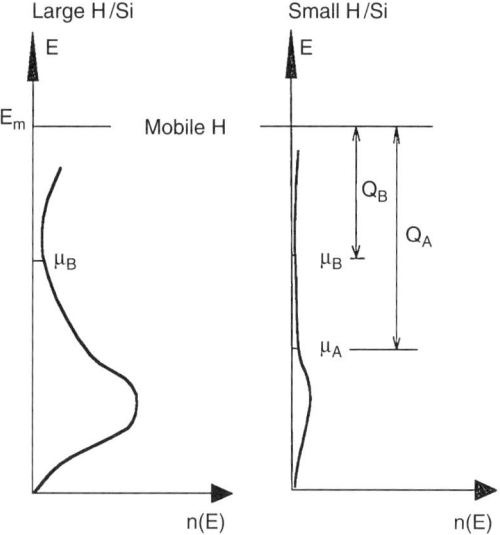

FIG. 33. Schematic presentation[146-149] of the DOSE for hydrogen in amorphous silicon (or germanium). The shown part of the distribution stems mostly from dangling bonds being saturated with H atoms. For high concentrations (sample B with the DOSE shown left) the distribution extends further to the energy E_m of "mobile" hydrogen and, therefore, the energy difference between μ_B and E_m is smaller when compared with the case of low H concentrations (sample A and distribution on the right). However, if two samples with the distributions shown are in contact, the chemical potential has to be equal. This is achieved by a few H atoms moving from B to A due to the low density $n(E)$ around μ_A. Thus the activation energy for diffusion in sample A is now much smaller and similar to B.

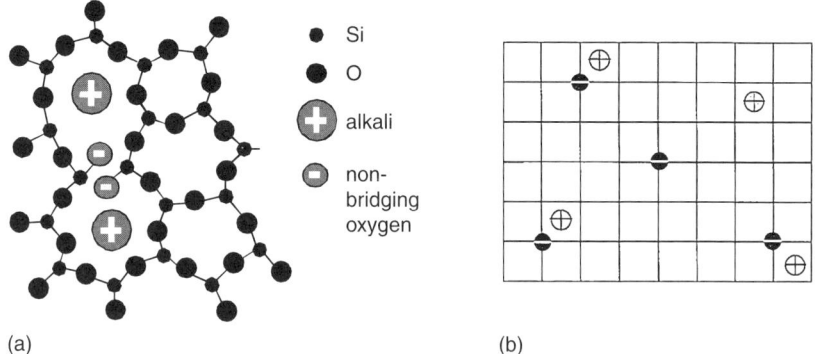

FIG. 34. (a) Schematic silicate network containing an alkali oxide. (b) Cation distribution in a regular lattice containing fixed anions as a model for cation diffusion in oxide glasses.

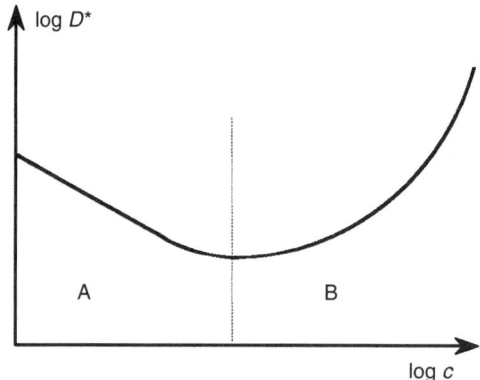

FIG. 35. Schematic presentation of the concentration dependence of the tracer diffusion coefficient of alkali ions in oxide glasses. Decreasing D^* in concentration range A and vice versa in B.

Then the diffusion coefficient is about the same in A and B in agreement with experiment.[146–149]

20. Ions in Oxidic Glasses

Mixing an alkali oxide with SiO_2 leads to non-bridging oxygen atoms in the network of SiO_4 tetrahedra.[150–152] These oxygen atoms are the centers of negative charge, which are immobile far below the glass transition temperature, whereas the alkali cations are still mobile. They migrate via the holes (interstices) of the silicate network (cf. Fig. 34). Due to the strong Coulomb interaction among cations and anions the sites next to the anions have the lowest energy. Because of charge neutrality the concentration of anions and cations has to be equal, in analogy to the equivalence of dangling bonds and H atoms in amorphous silicon. Therefore, the distribution of site energies is changed in silicate glasses by changing the alkali concentration.

Unlike materials previously discussed, a decrease of the diffusion coefficient has been determined for low alkali contents,[153] which was explained[69] by the "weak electrolyte" model (cf. Fig. 35). Assuming that sites next to the anions have a much lower energy compared to sites far away leads to a bimodal distribution of

---

[150] K. Hughes and J. O. Isard in *Physics of Electrolytes*, vol. 1, ed. J. Hladik, Academic Press, London (1972).
[151] G. H. Frischat, *Ionic Diffusion in Oxide Glasses*, TransTech, Aedermansdorf (1975), 161.
[152] M. D. Ingram, in *Materials Science and Technology*, vol. 9, eds. R. W. Cahn, P. Haasen, and E. J. Kramer, VCH-Verlagsgesellschaft, Weinheim (1991), 715.
[153] M. Tomazawa, in *Treatise on Materials Science and Technology*, vol. 12, eds. M. Tomazawa and R. H. Doremus, Academic Press, New York (1977), 335.

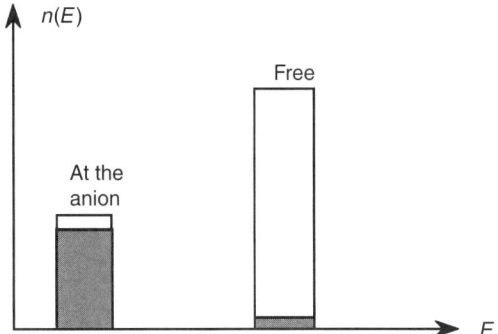

FIG. 36. Bimodal distribution of site energies and occupation according to the shaded area. The total area of the first peak increases proportionally with cation concentration. With this distribution the "weak electrolyte behavior" of cation mobility can be derived.[69]

site energies where both fractions vary with concentration, as shown in Figure 36. With this distribution the weak electrolyte behavior could be modeled for this solid material.[69] It is worth noting that a similar concentration dependence at low c is expected for H in amorphous silicon. The decrease of the diffusion coefficient with increasing concentration is in contradiction with the simple-minded interpretation of Eq. (3.17) used so far, where an increase of μ by increasing c lowers the activation energy of diffusion. This must not be used in this context, because the DOSE changes with concentration, whereas it has been tacitly assumed in the derivation of Eq. (3.17) that it will not change. The decrease of D for cations in glasses stems from an entropy effect.[69] As the number of anion and cation pairs increases, the number of "free" cations increases as well. However, the corresponding enhancement of the effective mobility is more than compensated by the increased trapping efficiency of the "naked" anions.

At high alkali contents the diffusivity increases as in the case of the other amorphous or glassy materials. Besides the interpretation given before either for amorphous metals or silicon another possible cause exists for a decrease of the activation energy and the concomitant increase of D^*. At high alkali concentrations a considerable modification of the network occurs accompanied by a decrease of the O atom density. Thus the network becomes more open, which could lower the activation energy for cation hopping. In addition it increases the mobility of neutral atoms such as He.[154] The mesh size of the silicate network can be changed with additions of alkali ions, or also by externally applied hydrostatic pressure. The effect of pressure on cation diffusivity can be easily measured by monitoring changes of the electrical conductivity yielding the activation volume of the cation

---

[154] R. Kirchheim, *Glass Sci. Technol.* **75**, 294 (2002).

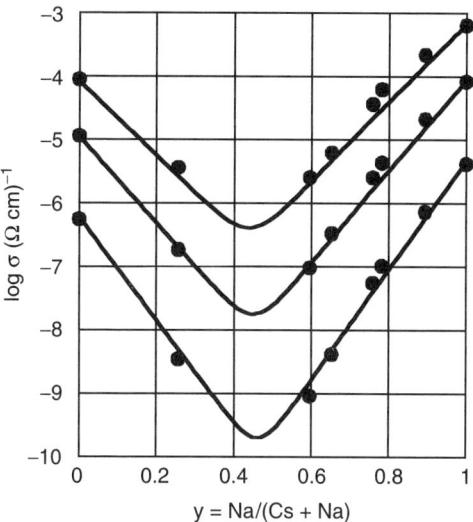

FIG. 37. dc conductivity of a mixed Na-Cs silicate glass[157] with 25 mol-% alkali oxide at 200, 300, and 400°C. The lines are calculated using a rectangular DOSE for the exchange energy of Cs and Na.[155,156]

diffusion coefficient. The average changes of the mesh size or strain, respectively, by external pressure can be calculated from continuum elastic theory and can be compared with the changes induced by the addition of alkali oxides.[154] Thus a quantitative treatment of mesh size effects is achieved. If we accept this mesh size effect and include the effect of a DOSE and FD-Statistics, a new and quantitative explanation of the mixed alkali effect can be offered.[155,156]

The mixed alkali effect[151–153,155] can be observed as a decrease of ion conductivity or mobility, respectively, by several orders of magnitude when one alkali ion is substituted by another one without changing the total alkali content (cf. Figs. 37 and 38). If we assume that all the alkali ions are distributed over a broad DOSE and that the smaller ones occupy the lower levels (because they come closer to the anions), the addition of smaller cations to a glass with larger ones reduces the mobility of the small ones as they are placed in the lowest energy levels. The mobility of the larger cations is reduced as well, because the mesh size of the network is reduced according to an average reduction of the mean cation size. This behavior can be treated quantitatively, too, yielding agreement with experimental data by fitting one free parameter, the width of the distribution, only.[155,156]

---

[155] R. Kirchheim, *J. Non-Cryst. Solids* **272**, 85 (2000).
[156] R. Kirchheim and D. Paulmann, *J. Non-Cryst. Solids*, **286**, 210 (2001).

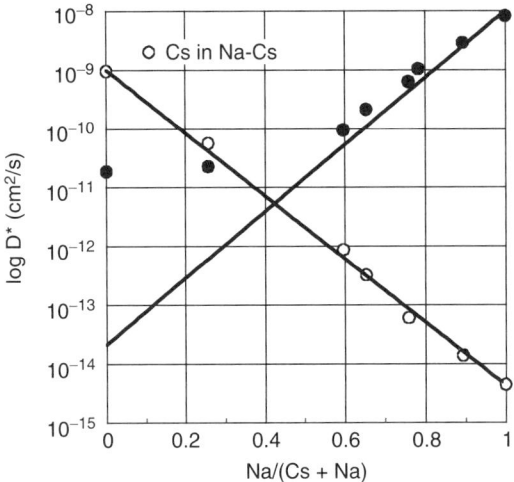

FIG. 38. Diffusion coefficient of Na$^+$ (solid circles) and Cs$^+$ (open circles) at 397°C in a 25 mol-% alkali oxide silicate glass.[157] The straight lines are fits to the linear behavior for A$^+$ ions at y → 1 and R$^+$ ions at y → 0, that is, in regions where the corresponding ions are the majority component.[155,156]

VII. Hydrogen in Systems with Reduced Dimensions

The behavior of hydrogen in samples with reduced dimensions is one of the more recent areas of research in metal–hydrogen systems.[20,158–170] The properties of these systems are affected by the proximity of surfaces or interfaces providing

---

[157] H. Jain, N. L. Peterson, and H. L. Downing, *J. Non-Cryst. Solids* **283–300**, 55 (1983).
[158] B. Hjörvarsson, J. Ryden, E. Karlsson, J. Birch, and J.-E. Sundgren, *Phys. Rev.* **B 43** 6440 (1991).
[159] G. Andersson, B. Hjörvarsson, and P. Isberg, *Phys. Rev.* **B 55**, 1774 (1997).
[160] G. Song, M. Geitz, A. Abromeit, and H. Zabel, *Phys. Rev.*, **B54**, 14093 (1996).
[161] G. Song, A. Remhof, D. Labergerie, and H. Zabel, *Phys. Rev. Lett.*, **79**, 5062 (1997).
[162] D. Nagengast, J. Erxmeyer, F. Klose, Ch. Rehm, P. Kuschnerus, G. Dortmann, and A. Weidinger, *J. Alloys Compd.* **231**, 307 (1995).
[163] B. Hjörvarsson, J. A. Dura, P. Isberg, T. Watanabe, T. J. Udovic, G. Anderson, and C. F. Majkrzak, *Phys. Rev Lett.*, **79**, 901 (1997).
[164] A. Züttel, Ch. Nützenagel, G. Schmid, Ch. Emmenegger, P. Sudan, and L. Schlapbach, *Appl. Surf. Sci.*, **162**, 571 (2000).
[165] Q. M. Yang, G. Schmitz, S. Fähler, H. U. Krebs, and R. Kirchheim, *Phys. Rev. B* **54**, 9131 (1996).
[166] U. Laudahn, A. Pundt, M. Bicker, U.v.Hülsen, U. Geyer, T. Wagner, and R. Kirchheim, *J. Alloys Compd.* **293–295**, 490 (1999).
[167] P. Kesten, A. Pundt, G. Schmitz, M. Weisheit, H. U. Krebs, and R. Kirchheim, *J. Alloys Compd.* **330–332**, 225 (2002).
[168] C. Sachs, A. Pundt, R. Kirchheim, M. Winter, M. T. Reetz, and D. Fritsch, "Solubility of hydrogen in single-sized palladium clusters," *Phys. Rev.* **B 64** 075408 (2001).
[169] M. Suleiman, N. M. Jisrawi, O. Dankert, M. T. Reetz, C. Bähtz, R. Kirchheim, and A. Pundt, *J. Alloys Compd.* **356–357**, 644 (2003).
[170] M. Suleiman, PhD thesis, University of Göttingen (2003).

additional sites to the DOSE. In this case the surface of the sample or its interface with a different material is just a source of new sites and reduction of dimensions leads to an increasing fraction of the new sites. Thus the interaction of hydrogen with free surfaces falls in the same category (cf. Section IV.9). In addition, stresses and strains play an increasing role. On a free surface the normal component of the stress is zero and this boundary condition is important for hydride formation in the elastic regime.[171,158] Inelastic processes such as dislocation emission are also affected by free surfaces as they act as sinks for defects. Reduced dimensions inhibit dislocation generation and motion[172] and, therefore, hinder that mechanism of hydride formation that requires dislocation emission. Samples adhering to a substrate or being embedded in a matrix are subject to boundary conditions with respect to their strain at the interface. Again the related stresses may not relax as easily as in the extended bulk sample. It has also been discussed (cf. Section VII.22) whether the proximity of a different material changes the electronic structure in the neighborhood of an interface.[159]

21. THIN FILMS

The first measurements of hydrogen dissolved in thin films were conducted by H. Zabel et al.[160] They determined the lattice expansion of niobium films in a hydrogen gas atmosphere of varying partial pressure. As the Nb film was adhering to a substrate, no expansion in plane could occur and compressive stresses developed. During the dissolution of hydrogen expansion out of plane takes place and it is accompanied by the Poisson effect, that is, additional expansion due to the increasing compressive stresses. By using H concentrations from bulk pc isotherms the lattice expansion per H atom was found to be much larger than in bulk niobium.[160] However, it was shown later on[166] that the lattice expansion per H atom was the same as in bulk niobium and the larger values obtained in Ref. [160] were most probably due to less reliable H concentration values obtained from the isotherms. Strains and stresses developing in thin Nb films as a function of H concentration are presented in Figures 39 and 40.

Within the α phase of Nb and Pd the out-of-plane expansion of thin films follows the laws of continuum mechanics. However, the associated compressive stresses are smaller by 10 to 30 percent when compared with calculated values.[166] This discrepancy may be attributed to some stress relaxation. A much more pronounced stress relaxation takes place when the terminal solubility of the α phase is reached and a hydride or a high-concentration phase is formed. Then the in-plane distance of lattice planes is changing, too. In some cases the pronounced stress relaxation occurs within the α phase, that is, before the terminal solubility is reached, because the compressive stresses simply exceed the yield strength of the thin metal film.

---

[171] H. Wagner in Ref. [30], p. 5.
[172] W. D. Nix, *Metall. Trans.*, **20**, 2217 (1989).

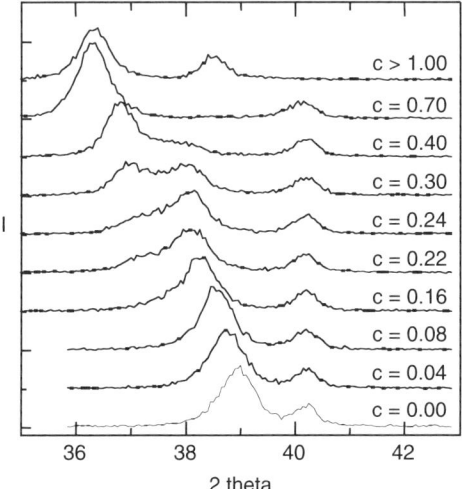

FIG. 39. Intensity, I, of X-rays in a θ–2θ scan of a 100-nm-thick Niobium film covered with a 10-nm-thick Palladium film. The latter protects Nb from oxidation and allows easy electrochemical H doping. With increasing H concentration (c = H/Nb) the Nb (110) peak moves to smaller angles, indicating the out-of-plane expansion of Nb. The Pd (111) peak at about 40.5 degrees does not move unless saturation of the Nb layer ($c > 1$) occurs. Splitting of the Nb peak into two peaks reveals the decomposition in a low-concentration phase of α-Nb and a hydride (β-phase). The formation of the high-concentration phase occurs at a higher terminal solubility when compared to bulk Nb.

The terminal solubility increases as the film thickness decreases, which may be caused by a lowering of the critical temperature in finite dimensions or a kinetic barrier for the necessary volume expansion.

The plot of stress vs. H concentration as shown in Figure 40 may be interpreted as a stress-strain curve, where the strain is internally imposed by H atoms instead of by an external machine. Therefore, similar features known from the deformation of bulk metals can be observed, such as single crystalline Nb-films have a much lower yield strength and a lower work hardening rate than polycrystalline films. After removal of the hydrogen a dislocation network remains in the film and at the subsequent second loading with hydrogen yield strength and work hardening are increased (cf. Figs. 41 to 43).

Recently, thin Y films have attracted considerable interest[173–176] because a metal/insulator transition occurs between H/Y = 2 and H/Y = 3, between the

---

[173] E. S. Kooij, A. T. M. van Gogh, and R. Griessen, *J. Electrochem. Soc.*, **146**, 2990 (1990).
[174] J. A. Alford, M. Y. Chou, E. K. Chang, and S. G. Louie, *J. Phys. Rev.* **B67**, 125110 (2003).
[175] S. J. van der Molen, J. W. J. Kerssemakers, J. H. Rector, N. J. Koeman, R. Dam, and R. Griessen, *J. Appl. Phys.*, **86**, 6107 (1999).
[176] P. van der Sluis, M. Ouwerkerk, and P. Duine, *Appl. Phys. Lett.*, **70**, 3356 (1997).

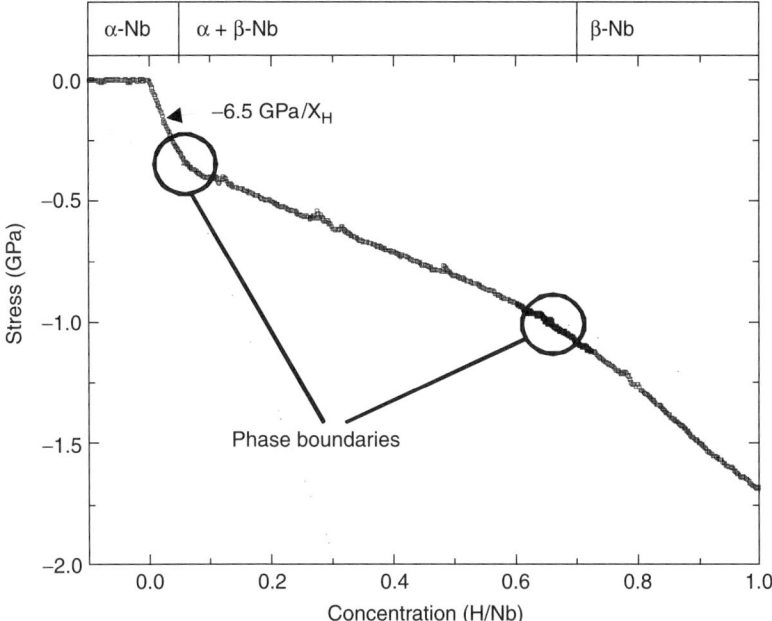

FIG. 40. Compressive stresses in a 190-nm-thick epitaxial Nb film on MgO as a function of H concentration. Stresses are determined from substrate curvature. At low H concentrations the compressive stresses increase linearly with c. Deviation from this steep increase, yielding, occurs at the decomposition in the two phases α and β. The stress of about 400 MPa where the deflection occurs is much larger than the flow stress of bulk Nb.

di- and trihydride. This transition can be used to build a switchable mirror, if the Y film is deposited on a transparent substrate. The Y-H system is also peculiar with respect to volume changes. Besides the usual lattice expansion observed for H/Y <2, a contraction occurs with increasing H concentration, followed by an expansion again. The corresponding tensional and compressive stresses cancel mostly and, therefore, the switching between di- and trihydride is not accompanied by large changes of film stresses.[177] Thus a large number of transitions between a reflecting and a transparent mirror are possible without failure of the device.

If the adhesion between film and substrate is weak, the films start to delaminate from the substrate during hydrogen loading. This has been observed for Nb films on mica[178] and Pd films on polycarbonate.[179]

---

[177] M. Dornheim, S. J. v.d. Molen, E. S. Kooij, J. Kerssemakers, H. Harms, U. Geyer, R. Griessen, R. Kirchheim, and A. Pundt, *J. Appl. Phys.* **93**, 8958 (2003).
[178] G. Song, A. Remhof, D. Labergerie, H. Zabel, private communication.
[179] A. Pundt, E. Nikitin, P. Pekarski, and R. Kirchheim, *Acta mater.* 52, 1579 (2004).

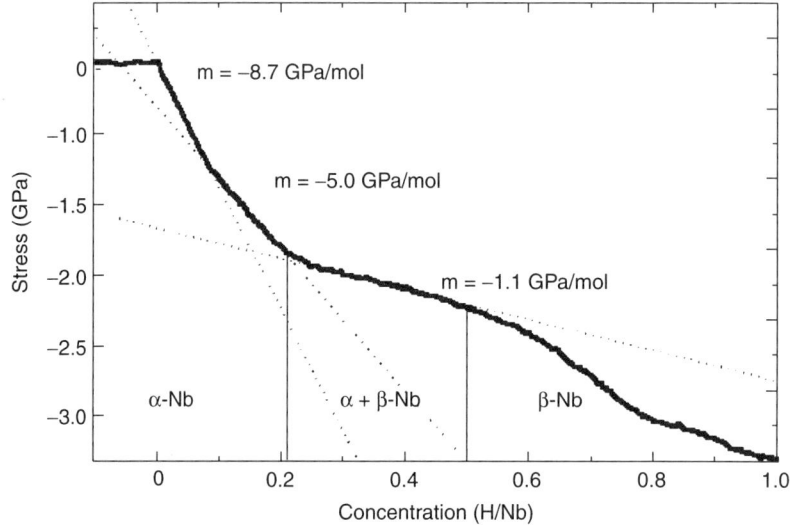

FIG. 41. Same as Figure 40 but for a 200-nm-thick nanocrystalline Nb film deposited on silicon by laser ablation. Here stress relaxation or yielding, respectively occurs within the α-phase. The relaxation is increased by entering the two-phase region.

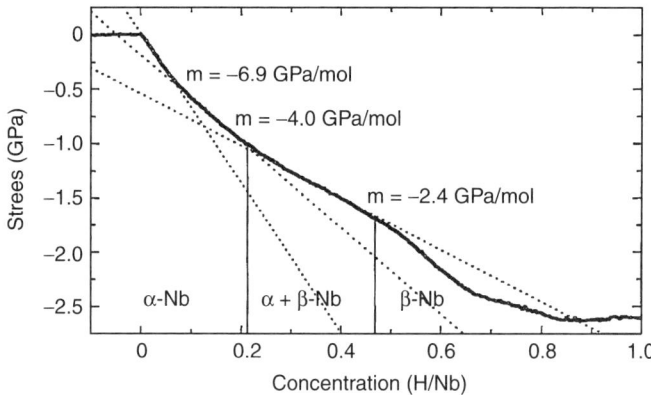

FIG. 42. Same as Figure 41 but for a 200-nm-thick nanocrystalline Nb film prepared by electron beam evaporation. Here yielding occurs at about 0.6 GPa within the α-phase. Compared to the laser-ablated film presented in the previous figure, the grain size is larger for the electron-deposited film.

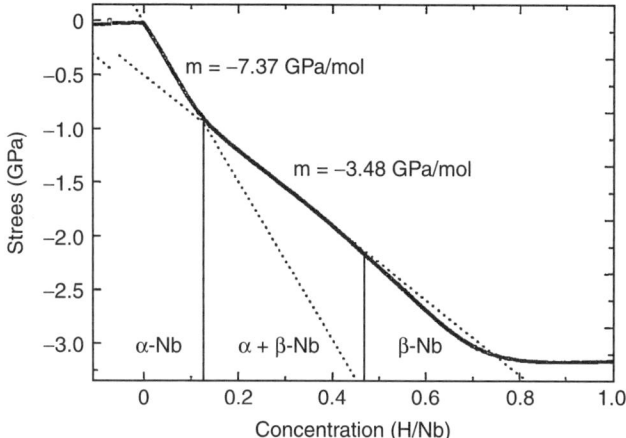

FIG. 43. Same sample as in Figure 42 after the removal of hydrogen and a second loading with hydrogen. Now stress relaxation starts at a higher stress of about 0.9 GPa.

22. MULTILAYERS

With the increasing interest in multilayers, the behavior of hydrogen in these systems has also attracted intensive research. In their pioneering work, Miceli, Zabel, and Cunningham[13,180] describe critical phenomena in Nb/Ta super-lattices. Hjörvarsson and co-workers[159] emphasized the effect of electron transfer at Mo/V interfaces on the hydrogen solubility in V layers. The influence of hydrogen on the magnetism of Fe/Nb and Fe/Ce multilayers was investigated by Weidinger et al.[162] and Felsch et al.[181] Actual measurements on strain relaxation and phase separation in Pd/Nb multilayers are presented in Ref. [165]. Transport of hydrogen through thin metallic films is essential for the development of coatings preventing bulk materials from hydrogen uptake. The influence of thin films of Pd, Ni, and Cu on the hydrogen permeation through Fe was investigated by Song and Pyun[182] and Takano et al.[183] Furthermore, to avoid oxidation and to enable hydrogen charging, thin films are often covered with an additional Pd surface, such that the specimens are multilayered structures (substrate, film, and Pd layer). The influence of Pd coatings on the hydrogen permeation was investigated by Züchner.[184] In the case

---

[180] P. F. Miceli and H. Zabel, *Phys. Rev. Lett.* **59**, 1224 (1987).
[181] F. Klose, J. Thiele, A. Schurian, O. Schulte, M. Steins, O. Bremert, and W. Felsch, *Z. Phys.* **B 90**, 82 (1993).
[182] R. H. Song and S. Pyun, *J. Electrochem. Soc.* **137**, 1051 (1990).
[183] N. Takano, Y. Murakami, and F. Terasaki, *Scr. Metall. Mater.* **32**, 401 (1995).
[184] H. Züchner, *Proceedings of 2. JIM International Symposium, Hydrogen in Metals JIM*, Minakami, Japan, (1979), 101.

of thin metallic films prepared on substrates, the hydrogen permeation through the substrate, as well as the influence of the interface and the property of the film, has to be taken into account. Assuming defect-free layers, far away from the interfaces, the hydrogen transport should be well described by the known diffusion behavior of the bulk elements. However, deviations from the ideal bulk diffusion can be expected at the interfaces, where possibly high densities of lattice defects, localized misfit strains, or different hydrogen solubility are present. Therefore, it is most important to understand whether the hydrogen transport through a layered specimen is correctly modeled by a layered structure with bulk materials of different hydrogen solubilities or by taking into account additional interface sites of different site energies.

The permeation method, that is, loading the sample at one side with hydrogen and determining the time dependence of the concentration change at the opposite side, can be used to measure hydrogen diffusivity through multilayers. Depending on the experimental conditions of hydrogen loading, the concentration change at the opposite side appears with a certain time lag.[184] In a single layer the time dependence of the concentration change can be solved analytically for various boundary conditions,[185] whereas in composite systems only the characteristic time lag to reach the steady state has been modeled so far.

Often four different boundary conditions are applied in steady-state permeation experiments: (A) constant current density at the input and output surface, (B) constant concentration at the input and output surface, (C) constant current density at the input and constant concentration at the output surface, and (D) constant concentration at the input and constant current density at the output surface. In their theoretical treatment of diffusion through multilayers, Ash and co-workers[186,187] derived the time lag for the permeation using boundary condition B with vanishing concentration at the output side. Their result was applied to interpret hydrogen diffusion in metallic bilayers and triple layers by Züchner[184] and Takano et al.[183] Song and Pyun[182] modified the hydrogen loading conditions by using a constant hydrogen current density at the input side (condition C) and derived the corresponding time lag for a bilayer system. In a recent paper by G. Schmitz et al.,[188] solutions for the conditions A and C were provided for the most general case of multilayers including the substrate as one of the layers. For a substrate of thickness s with $2N$ alternating layers of two metals of the thickness a, the time lag t_L is for $N \gg 1$

$$t_L = 3\frac{s^2}{6D_s} + \frac{1}{6}\frac{N^2 a^2}{kD_1} + \frac{N^2 a^2}{6D_1} + \frac{N^2 a^2}{6D_2}, \qquad (7.1)$$

---

[185] J. Crank, *The Mathematics of Diffusion*, Oxford University Press, Oxford (1975).
[186] R. Ash, R. M. Barrer, and J. H. Petropoulous, *Br. J. Appl. Phys.* **14**, 854 (1963).
[187] R. Ash, R. M. Barrer, and D. G. Palmer, *Br. J. Appl. Phys.* **16**, 873 (1965).
[188] G. Schmitz, P. Kesten, and Q. M. Young, *Phys. Rev.* **B 58**, 7333 (1998).

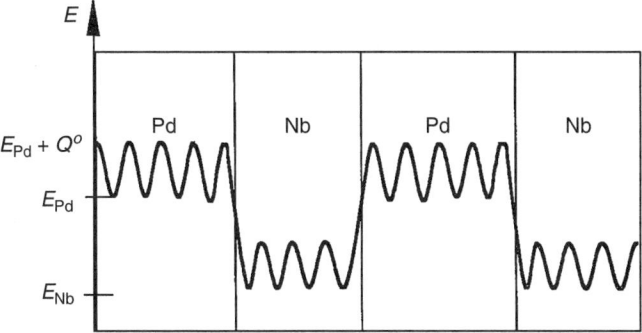

FIG. 44. Potential trace for hydrogen atoms in a Nb/Pd multilayer. Site energies E_{Pd} and E_{Nb} are known from solution energies in bulk metals. H diffusion through the layers is hindered by Nb sites acting as traps or by Pd layers acting as barriers, respectively.

where k is the ratio of H solubilities in both metals 1 and 2 for the same H_2 partial pressure chosen in a way that $k < 1$ and D are the corresponding diffusion coefficients. The first term on the righthand side of Eq. (7.1) is three times the time lag of the isolated substrate and the third and fourth terms are the time lags of the isolated layers of either metal 1 or 2. For nanostructured multilayers on a thick substrate usually $Na \ll s$ applies and the last equation simplifies to

$$t_L = 3\frac{s^2}{6D_s} + \frac{1}{6}\frac{N^2 a^2}{kD_1}. \tag{7.2}$$

The second term on the righthand side of the last equation cannot be neglected because k may be very small. The term can also be used to define an effective diffusion coefficient $D_{\it eff} = (2Na)^2/6t_L = 4kD_1$ for the multilayers. For a different thickness of the two metals the effective diffusion coefficient for a multilayer can be written independently of the boundary conditions as[186–189]

$$D_{\it eff} = k v_1 v_2 D_1 \tag{7.3}$$

where v_1 and v_2 are the volume fractions of metal 1 and metal 2.

In order to comprehend the physics behind Eq. (7.3), the potential trace a migrating H atom will experience is shown in Figure 44 for the case of Pd = metal 1 and Nb = metal 2. The site energies E_{Pd} and E_{Nb} correspond to the energies of hydrogen dissolution in Nb and Pd. In agreement with the definition of $k = \exp[(E_{Nb} - E_{Pd})/kT] < 1$ the experimental values correspond to $E_{Pd} > E_{Nb}$.

At room temperature k is as small as about 10^{-4} for Nb/Pd[188] and, therefore, the effective diffusion coefficient is decreased by four orders of magnitude, which is in

---
[189] P. Kesten, PhD thesis, University of Göttingen (2000).

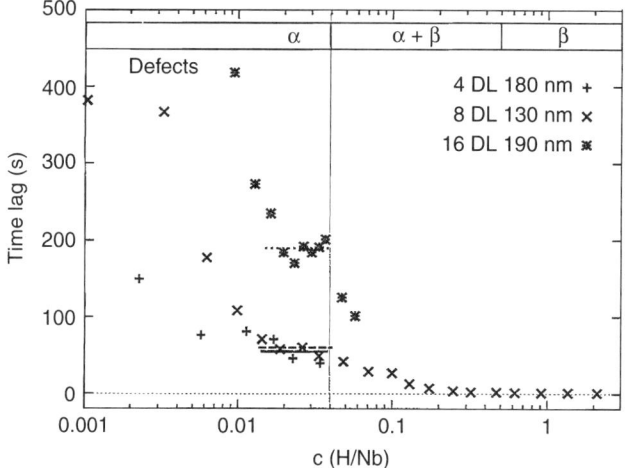

FIG. 45. Time lag versus H concentration for diffusion through multilayers on a Pd subtrate of 12.5 μm thickness. The multilayer consists of 4, 8, and 16 alternating double layers of Pd and Nb of the same thickness, and a total thickness as given in the inset.

agreement with experimental findings.[188] It is obvious from looking at Figure 44 that the Nb layers provide traps for the hydrogen diffusing in Pd, whereas Pd layers act as barriers for H in Nb.

Experimental results of time lag measurements of Pd/Nb multilayers[188] are shown in Figure 45 as a function of the partial hydrogen concentration in Nb or Pd, respectively. Due to the lower Gibbs free energy of H dissolution in Nb, nearly all of the hydrogen is dissolved in the Nb layers.[36] The total thickness of the layers was about 0.2 μm and they were produced on a Pd substrate of 12.5 μm thickness. With the solubility ratio k in Pd/Nb of about 10^{-4} and a room temperature diffusivity of $3.5 \cdot 10^{-7}$ cm^2/s in Pd the time-lag is calculated to be 0.75 s for the pure substrate and 14 s for the substrate plus multilayers. For H concentrations of $0.01 < c < 0.04$ experimental data are of the same order of magnitude as this estimated value. Unfortunately, the comparison is only approximate because scatter of data on Gibbs free energy of solution yields data on k which scatter by more than one order of magnitude.[188] It is interesting to note that the high diffusivity of H in Nb with $D = 8 \cdot 10^{-6}$ cm^2/s does not play a role and that despite that value being higher than in Pd the Nb layers retard the overall transport of H. At very low H concentrations of $c < 0.01$ or for the smallest double-layer thickness, the interaction with defects or with the interphase Pd/Nb boundaries may increase the time lag, that is, the effective diffusivity.

At a few at.-% of H in Nb the terminating solubility at room temperature is reached and the Nb layers contain two phases, a solid solution of H in Nb and a hydride phase. Then the time lag can no longer be calculated by Eq. (7.2), which was derived for the single-phase case. Experimental values in Figure 45 show that t_L is reduced in the two-phase region. Increasing the total hydrogen content further leads to a total conversion of the Nb layers into hydride layers, where the k values are higher because of the higher Gibbs free energies of solution in saturated Nb hydride. Then the second term on the righthand side of Eq. (7.2) becomes negligible compared to the first one and the time lag reduces to the value of the pure Pd substrate (0.75 s).

For the Ni/Pd system[190] the Gibbs free energies of H solution are such that H is enriched in the Pd layers according to k values of about 10^{-4} [36] and the diffusion coefficient of H in Ni is $5 \cdot 10^{-10}$ cm^2/s.[83] Inserting these values in Eq. (7.2) yields time lags of the order of thousands of seconds. However, experimental values[190] are about the same as the time lag of the Pd substrate; the Ni layers do not act as barriers for H diffusion. This discrepancy between experimental findings and theory is overcome by assuming that grain boundary diffusion in the Ni layers is relevant. The Ni layers of the samples used were polycrystalline and the concept described in Ref. [188] has to be modified by using an effective diffusion coefficient $D_{gb}\delta/d$, where D_{gb} is the diffusion coefficient of H in grain boundaries of Ni, δ is the thickness of these boundaries, and d is the diameter of the grains (about the same as the Ni-layer thickness). The ratio k has to be multiplied by a segregation factor S defined as the ratio of H concentrations in grain boundaries and grains. Experimental values of S and D_{gb}^{83} are so large for the H concentrations used in the experiment that the second term in Eq. (7.2) is smaller than the first one and in agreement with experimental findings the effective diffusion coefficient is not affected by the presence of the Ni/Pd layers.

Similar considerations should apply for the case of Pd/Nb layers where the Pd barrier layers were nanocrystalline, too. However, in this case the product $SD_{gb}\delta/d$ is smaller than the corresponding value of D for the grains. At the H concentrations expected in the experiments this is mainly caused by a lower diffusivity of H in grain boundaries of Pd when compared with the grains (see Section IV).

The interface boundaries between different metals produced by either sputtering or electron evaporation are not atomically sharp, but an intermixing zone of about 1 nm thickness has been detected by analyzing the multilayers with a tomographic atom probe.[191] This intermixing zone decreases the width of both adjacent layers and it does not absorb hydrogen at a given partial pressure,[192] that is, it acts as a dead layer. This way a different explanation is offered for the dead layers observed

---

[190] P. Kesten, diploma thesis, University of Göttingen (1996).
[191] T. Al-Kassab, H. Wollenberger, G. Schmitz, and R. Kirchheim, in *High-Resolution Imaging and Spectroscopy of Materials*, eds. F. Ernst and M. Rühle, Springer Verlag, Berlin (2002) 271–320.
[192] P. Kesten, A. Pundt, G. Schmitz, M. Weisheit, H. U. Krebs, and R. Kirchheim, *J. alloys compd.*, **330–332**, 225 (2002).

in Refs. [159, 193], where a transport of electrons across the interface was assumed to be responsible for the dead layer.

23. CLUSTERS

In order to study the solubility of hydrogen in metal clusters, a large number of clusters are necessary for the commonly applied techniques. However, clusters tend to agglomerate because of the large cohesion energy of metals. This tendency of forming a polycrystalline metal rather than remaining as isolated clusters is circumvented by stabilizing cluster with a surfactant shell, embedding them in a solid matrix, or depositing them on a substrate. Again the Pd-H system is the one first chosen by experimentalists as the nobility of the metal excludes the formation of an oxide layer, which otherwise would be an appreciable fraction of the sample. First measurements of pc isotherms were conducted by Flanagan et al.[194] with Pd-black having a surface area of about 40 m^2/g, which corresponds to a diameter of about 12 nm. Although changes are small when compared to bulk Pd, a narrowing of the miscibility gap was observed. Griessen et al.[195] observed a much more pronounced narrowing for Pd clusters deposited on an aluminum oxide substrate. However, in this case stresses will develop during hydrogen loading as in thin films and, therefore, the pc isotherms will be affected by these stresses. The interaction of hydrogen with free palladium clusters of 2 to 6 nm diameter which were stabilized by a surfactant shell has been described in detail in recent studies.[164,168–169] The results are summarized as follows.

The preparation of small Pd clusters is achieved by electrochemical dissolution of Pd and its reduction on a counter electrode to neutral Pd atoms. If a non-aqueous electrolyte such as tetrahydrofuron with surfactant molecules like octylammonium bromide is used, the Pd atoms agglomerate to clusters which are stabilized by the surfactant.[170,196] At a certain concentration of clusters they precipitate as a black powder. The diameter of the clusters and the width of their distribution can be varied by changing temperature, electrode distance, and current density. Thus clusters are produced with a rather narrow size distribution, with a variance of the diameter being 10 to 15 percent of the average diameter.

Detailed investigations of the atomic structure of these clusters with molecular dynamic simulations, high-resolution transmission electron microscopy, and X-ray diffraction revealed that they have an icosahedral structure for sizes below 4 nm whereas the larger ones have a fcc structure. All the clusters absorb hydrogen

---

[193] B. Hjörvarsson, J. Ryden, E. Karlsson, J. Birch, and J.-E. Sundgren, *Phys. Rev.* **B43** 6440 (1991). Also G. Andersson, B. Hjörvarsson, and P. Isberg, *Phys. Rev. B*, **55**, 1774 (1997).
[194] J. F. Lynch and T. B. Flanagan, *J. Phys. Chem.*, **77**, 2628 (1973).
[195] E. Salomons, R. Griessen, D. E. de Groot, J. H. Rector, and D. Fritsch, *Europhys. Lett.*, **5**, 449 (1988).
[196] M. T. Reetz and W. Helbig, *J. Am. Chem. Soc.* **116**, 7401 (1994).

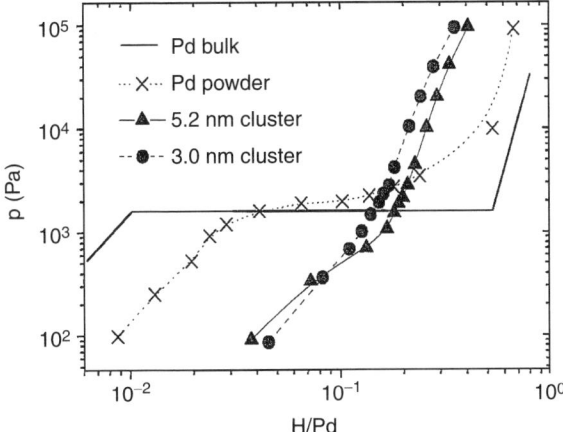

FIG. 46. Pressure composition isotherms of Pd clusters of 3 nm (circles) and 5.2 nm (triangles) in diameter compared with Pd powder (crosses) and Pd bulk (solid line).

from the gas phase, easily shown by monitoring the weight change in a hydrogen atmosphere. However, before hydrogen reacts with palladium it forms water, with oxygen being absorbed on the surface of the clusters. Pumping off the water and the residual hydrogen in the vacuum chamber leads to oxygen- and hydrogen-free Pd clusters. After this pre-treatment the clusters can be exposed to hydrogen with a given partial pressure and the absorbed amount is obtained from weight changes or from pressure drops. In the latter case the valves to the pumps and to the hydrogen inlet have to be closed. The absorption takes between a few minutes and a few hours to be complete. Pressure composition isotherms are obtained this way. Some of them are shown in Fig. 46.

At about the same pressure where bulk Pd decomposes into α- and β-Pd, the clusters exhibit a sloped plateau. However, the maximum concentration in the α phase is larger and the minimum one in the β phase is smaller or, in other words, the width of the miscibility gap is reduced and the reduction is larger the smaller the clusters are. The increased terminal solubility in the α phase can be explained by subsurface sites having lower site energies than bulk sites.[168] This is analogous to segregation in grain boundary sites. The surface sites were most probably occupied during the pre-treatment step, where the oxygen was removed with hydrogen. Because of the large binding energies of the order of 100 kJ/mol-H,[197] this hydrogen will not be removed at room temperature and high vacuum conditions.

Within the miscibility gap there is a hysteresis between absorption and desorption[168,170] in analogy with bulk behavior. However, the hysteresis cannot be

---

[197] R. J. Behm, V. Penka, M. G. Cattania, K. Christmann, and E. Ertl,, *J. Chem. Phys.* **78**, 7486 (1983).

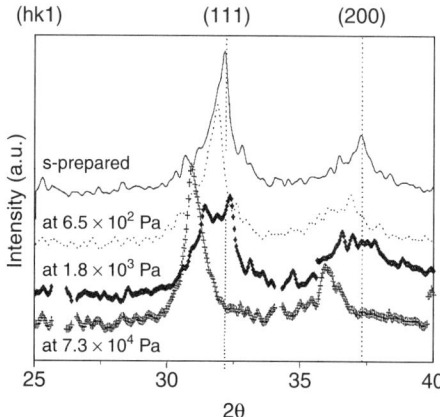

FIG. 47. X-ray diffraction peaks of Pd clusters of 6 nm in diameter at different partial pressures of hydrogen. At the pressure of $1.8 \cdot 10^3$ Pa, which corresponds to the two-phase field of α and β bulk Pd (cf. Fig. 46), peaks split into two or at least broaden.

explained by the work required to punch out dislocations around the new phases formed,[198] because the clusters are too small to allow the formation of dislocations. Thus a coherent phase transformation takes place, which leads to a hysteresis as well.[199,200]

Additional evidence for a phase separation in small Pd clusters is provided by X-ray diffraction, where at partial pressure of hydrogen belonging to the plateau region X-ray peaks split in two (cf. Fig. 47). With increasing H concentration the peaks shift to smaller angles, indicating a volume expansion as in bulk Pd. Calculating lattice parameters from the peak positions and plotting them versus the hydrogen pressure yields the results shown in Figure 48. Within the region of the sloped plateau there is a pronounced change of the lattice parameter according to the transition from the low to the high concentration phase of the Pd-H system. The overall change of the lattice parameter across the miscibility gap decreases with decreasing cluster size in accordance with the corresponding decrease of the width of the miscibility gap.

The change per H atom is about the same as in bulk Pd. As the changes of the lattice parameter scale with the H concentration there is also a hysteresis of the lattice parameter between absorption and desorption within the miscibility gap. The hysteresis is not a matter of sluggish kinetics because it is not observed within the one-phase region. Thus the hysteresis is an additional piece of evidence for a phase separation in the Pd clusters, besides the previously discussed sloped plateau

---

[198] T. B. Flanagan, B. S. Bowerman, and G. E. Biehl, *Scr. Metall.* **14**, 443 (1980).

[199] J. W. Cahn and F. Larche, *Acta Metall.*, **32**, 1915 (1984).

[200] R. B. Schwarz and A. G. Khachaturyan, *Phys. Rev. Lett.* **74**, 2523 (1995).

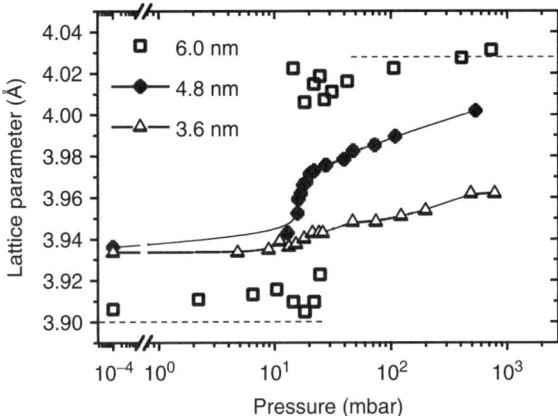

FIG. 48. Lattice parameter pressure isotherms for various Pd clusters. The change of the lattice parameter at about 20 mbar is decreasing with decreasing cluster size in accordance with the narrowing of the miscibility gap (dashed line = bulk behavior).

of the pc isotherms. Whether a cluster itself contains two phases in coexistence or is either α- or β-phase throughout its volume remains an open question. A piece of evidence for the first case is provided by the hysteresis. If it is caused by the coherent phase transformation as discussed in Refs. [199, 200], high and low concentration phase have to be present in the same cluster. These and other problems are the subjects of ongoing research, where MD simulation may become a suitable tool to answer part of the remaining questions. Some of the attraction of the cluster research actually stems from the possibility of conducting MD simulations with "samples" of the same size as one used by the experimentalist.

Acknowledgments

The author is thankful for the collaboration with many diploma students, PhD students, and post-docs working on metal hydrogen systems. They are listed here in alphabetical order: Denio Arantes, Jochen Bankmann, Martin Dornheim, Jürgen Gegner, Xianya Huang, Frank Jaggy, Philipp Kesten, Wolfgang Kieninger, Christian Kluthe, Uwe Laudhan, Ruth Lüke, Christof Marte, Michael Maxelon, Thomas Mütschele, Chenong Park, Astrid Pundt, Christian Sachs, Guido Schmitz, Ulrich Stolz, Mohamed Suleiman, Fang Wang, Quin-Min Yang. In addition, the cooperation with many colleagues, mostly leading to joint papers, is gratefully acknowledged. Most of the financial support stems from the Deutsche Forschungsgemeinschaft via several projects and via SFB 270, SFB 345, and SFB 602.

Author Index

Numbers in parentheses are reference numbers and indicate that an author's name is not cited in the text.

A

Aaronson, H.I., 2(1), 12(16), 186(146)
Abromeit, A., 278(160)
Ågren, J., 109(54), 151(112)
Ahrenkiel, S.P., 9(11), 9(12)
Akaiwa, N., 145(89), 145(90)
Alder, W.F., 62(32)
Alefeld, G., 205(11), 206(30), 206(31)
Alexander, J.I., 111(60)
Alfrord, J.A., 280(174)
Al-Jassim, M.M., 9(11)
Al-Kassab, T., 245(99), 245(100), 247(101), 287(191)
Alpay, S.P., 147(97)
Anderson, I.S., 253(116)
Andersson, G., 278(159), 278(163), 288(193)
Andren, H.O., 82(46)
Arantes, D.R., 239(83)
Ardell, A.J., 179(141)
Asaro, R.J., 121
Ash, R., 284(186), 284(187)
Ashby, M.F., 244(93)
Aziz, M.J., 125

B

Backhaus-Ricoult, M., 244(97)
Bähtz, C., 278(169)
Bai, Q., 87(49)
Balasubramaniam, R., 205(15)
Balluffi, R.W., 243(86)
Bankman, J., 259(128)
Baranowski, B., 213(52)
Barker, J., 234(78), 247(101)
Barnett, D.M., 2(1), 78(42), 186(146)
Barrer, R.M., 270(142), 284(186), 284(187)
Barrie, J.A., 270(142)
Barrie, W.R., 270(143)
Barvosa-Carter, W., 125(70)
Baskes, M.I., 206(29)
Batzer, H., 270(140)
Behm, R.J., 289(197)

Berkenpas, M.B., 179(140), 187(148)
Berry, B.S., 212(45), 212(46)
Besher, D.N., 218(63)
Beyer, W., 273(147), 273(148), 273(149)
Bicker, M., 278(166)
Biehl, G.E., 290(198)
Biesen, B.C., 253(114)
Bimberg, D., 6(6)
Birch, J., 278(158), 288(193)
Birnbaum, H.K., 205(12), 206(29), 215(61), 252
Blaurock, J., 231(75)
Blixt, A.M., 206(19)
Boes, H., 210(41)
Bohlen, J., 267(134), 267(135)
Boué, J.E., 237(79)
Bowerman, B.S., 290(198)
Bratkovski, A.M., 8(10)
Bremert, O., 283(181)
Brenner, S.S., 82(46)
Brion, H.G., 267(136), 269(139), 270(145)
Brower, R., 253(113)
Bueche, F.J., 268(138)

C

Cahn, J.W., 10, 13(18), 106, 109, 110(55), 129(78), 132(79), 142(83), 142(84), 142(85), 146(94), 162, 167(124), 174(132), 179(139)
Cannelli, A., 212(50)
Cantelli, R., 212(50)
Carstanjen, H.D., 207(34)
Cattania, M.G., 289(197)
Cermelli, P., 132(80)
Cerullo, M., 8(8)
Chan, J.W., 290(199)
Chang, E.K., 280(174)
Chen, L.-Q., 198(156)
Chiang, C.-S., 147(98), 155(115)
Chou, M.Y., 280(174)
Clewley, J.D., 206(25)
Cobet, J.W., 206(29)
Cookson, D.J., 214(58)
Copley, S.M., 187(147)

Cordero, F., 212(50)
Cottrell, A.H., 78(43)
Cotts, R.M., 263(130)
Crank, J., 284(185)
Cullis, A.G., 7(7)
Cunningham, J.E., 205(13), 283
Curtin, W.A., 258(121)

D

Dam, R., 280(175)
Dankert, O., 278(169)
Darken, L.S., 214(55)
Dauskardt, R.H., 259(126)
Davis, S.H., 121(64), 127(72)
Dean, J.A., 208(35)
de Groot, D.E., 288(195)
de Groot, D.G., 206(20)
Dekker, J.P., 206(20)
Deleo, G.G., 206(29)
Desai, R.C., 129(74)
Devanathan, M.A.V., 210(43)
Doi, M., 179(142)
Doremus. R.H., 275(153)
Dornheim, M., 281(177)
Dortmann, G., 278(162)
Downing, H.L., 278(157)
Driesen, G., 253(116)
Duine, P., 280(176)
Dura, J.A., 278(163)

E

Eaglesham, D.J., 8(8)
Easterling, K.E., 94(50)
Eastman, J.A., 243(87), 243(91)
Eckert, J., 259(127)
Eliaz, N., 206(28), 259(125)
Eliezer, D., 206(28), 259(125)
Emmenegger, Ch., 278(164)
Epperson, J.E., 237(79)
Erxmeyer, J., 278(162)
Eshelby, J.D., 62(32), 82(45), 268(137), 271
Estreicher, S.K., 206(29)
Evans, A.G., 244(93)

F

Fähler, S., 278(165)
Feenstra, R., 253(113)
Felsch, W., 283(181)

Filipek, S., 253(115)
Filipek, S.M., 239(84)
Finocchiaro, R.S., 253(114)
Flanagan, T.B., 205(15), 206(24), 206(25), 206(26), 213(52), 250(106), 288(194), 290(198)
Flynn, C.P., 205(10), 205(12)
Follstaedt, S.M., 9(11), 9(12)
Fried, E., 12
Frischat, G.H., 275(151)
Fritsch, D., 266(132), 278(168), 288(195)
Froes, F.H., 206(28)
Fromm, E., 209(36)
Fujii, H., 215(59), 259(122)
Fukai, Y., 206(21), 207(32)
Fukunaga, T., 259(122)
Fung, Y.C., 51(29)

G

Gahr, S., 215(61)
Gebert, A., 259(127)
Gebhardt, E., 209(36)
Gegner, J., 222(66), 243(89), 244(98)
Geitz, M., 278(160)
Geyer, U., 278(166), 281(177)
Gibbs, J.W., 10, 11, 14(20), 19, 99(51), 106, 109, 110, 247(102)
Glas, F., 161(120)
Gotthardt, P., 267(136), 269(139), 270(145)
Gray, L.J., 125(70)
Gray, E.MacA., 214(58)
Griessen, R., 206(20), 228(70), 253(113), 256, 280(173), 280(175), 281(177), 288, 288(195)
Griffiths, R.B., 147(108)
Grinfeld, M.A., 121
Gross, D., 193(152), 198(154), 198(155)
Grüger, A., 267(136), 269(139)
Guha, S., 8(9)
Gulff, G., 176(136)
Gurtin, M.E., 10(13), 12, 62(33), 100(52), 124(69), 132(80)
Gust, W., 240(85)
Guyer, J.E., 78(39), 128(73)

H

Haasen, P., 250(105)
Haller, E.E., 206(29)
Harms, H., 281(177)
Harris, J.H., 258(121)

Hasaki, M., 250(106)
Helbig, W., 288(196)
Hempelmann, R., 253(116)
Herring, C., 12
Heumann, Th., 209(38)
Heuser, B.J., 237(79)
Hill, T.L., 217(62)
Hillert, M., 109(54), 152(114)
Hilliard, J.E., 159(118)
Hirota, H., 214(57)
Hirscher, M., 256(118), 263(131)
Hirth, J.P., 81(44), 214(56), 229(72), 244(93), 251(107), 252 (108)
Hjörvarsson, B., 206(19), 278(158), 278(159), 278(163), 288(193)
Hoftyzer, P.J., 270(141)
Hörz, G., 222(66), 243(89)
Howe, J.M., 167(127)
Hsiao, C., 87(49)
Huang, X.Y., 239(83), 243(90), 243(91)
Hughes, K., 275(150)
Huh, J.-Y., 163(122), 167(127), 169(130)
Huiberts, J.N., 206(20)
Hülsen, U.V., 278(166)
Hut Mteall, J.Y., 84(47)

I

Ikeda, K., 215(59), 259(122)
Ingram, M.D., 275(152)
Inoue, A., 259(124)
Inui, H., 214(57)
Iooss, G., 188(149)
Isard, L.O., 275(150)
Isberg, P., 278(159), 278(163), 288(193)
Ismail, N., 259(127)

J

Jaffee, R.I., 62(32)
Jaggy, F., 258(120)
Jain, H., 278(157)
Jamienson, H.C., 206(23)
Jang, H., 244(94), 244(96)
Jaswon, M.A., 78(43)
Jena, P., 206(29)
Jisrawi, N.M., 278(169)
Johnson, H.H., 231(74)
Johnson, N.M., 206(29)
Johnson, W.C., 34(22), 57(30), 63(36), 78(41), 84(47), 110(57), 111(60), 142(81), 142(82), 145(87), 147(98), 147(104), 147(105), 147(107), 155(115), 163(122), 167(127), 169(130), 177(137), 179(139), 179(140), 180(144), 187(148), 200(157)
Johnson, W.L., 259(123)
Jones, E.D., 9(12)
Joseph, D.D., 188(149)
Jou, H.J., 193(151), 193(153)

K

Kalonji, G., 174(132)
Kamins, T.I., 8(10)
Kanninen, M.F., 62(32)
Kaplan, T., 125(70)
Karlsson, E., 278(158), 288(193)
Katoh, Y., 212(48)
Kaur, I., 240(85)
Kehr, K.W., 219(65)
Kerssemarkers, J.W.J., 280(175), 281(177)
Kestel, B.J., 243(87)
Kesten, P., 278(167), 284, 285(189), 287(190), 287(192)
Khachaturyan, A.G., 2(2), 290(200)
Kieninger, W., 258(120)
King, J.S., 237(79)
Kirchheim, R., 205(14), 205(15), 205(16), 206(26), 206(29), 210(42), 213(53), 214(54), 214(56), 222(66), 223(68), 227(69), 231(73), 232(76), 233(77), 234(78), 238(81), 239(83), 239(84), 243(88), 243(89), 243(90), 244(98), 245(99), 245(100), 247(101), 249(104), 251(107), 253(115), 258(120), 263(129), 267(133), 267(134), 267(135), 267(136), 269(139), 270(145), 276(154), 277(155), 277(156), 278(165), 278(166), 278(167), 278(168), 278(169), 281(177), 281(179), 287(191), 287(192)
Kitahara, K., 209(39)
Kittel, Ch., 222(67)
Klose, F., 206(18), 278(162), 283(181)
Kluthe, C., 245(99), 245(100), 247(101)
Kobayashi, R., 142(85)
Koeman, N.J., 206(20), 280(175)
Kooij, E.S., 280(173), 281(177)
Köster, U., 259(125)
Kostorz, G., 180(145)
Kownacka, I., 239(84)
Krebs, H.U., 278(165), 278(167), 287(192)

Kronmüller, H., 229(71), 256(118), 263(131)
Kuji, T., 250(106)
Kummnick, A.J., 231(74)
Kuschnerus, P., 278(162)

L

Labergerie, D., 278(161), 281(178)
Lacher, J.R., 205(9)
Larche, F., 290(199)
Larché, F.C., 10, 13(18), 132(79), 142(83), 142(84), 167(124)
Laudahn, U., 278(166)
Laughlin, D.E., 12(16), 174(132), 187(148)
Laurent, S., 244(97)
Lee, J.K., 2(1), 186(146)
Lee, S.R., 9(11), 9(12)
Lee, Y.S., 256(119)
Leiner, V., 206(19)
Lemier, C., 238(82)
Leo, P.H., 111(58), 131, 145(91), 145(92), 193(151), 193(153)
Leonard, F.C., 129(74)
Li, J.C.M., 214(55)
Liu, Z.-K., 151(112), 198(156)
Lothe, J., 81(44), 229(72)
Louat, N., 218(64)
Louie, S.G., 280(174)
Lowengrub, J.S., 193(151), 193(153)
Lüke, R., 206(26)
Lund, A.C., 3(4)
Lupis, C.H.P., 85(48)
Lynch, J.F., 206(24), 288(194)

M

Mader, W., 243(90), 243(91), 244(93), 244(95)
Madhukar, A., 8(9)
Majchrzak, S., 213(52)
Majkrzak, C.F., 278(163)
Malvern, L.E., 36(23)
Malyshkin, V.G., 78(40)
Manchester, F.D., 206(23)
Marsden, J.E., 62(33)
Marte, C., 239(83)
Mascarenhas, A., 9(12)
Mascrenhas, A., 9(11)
Masumot, T., 259(124)
Maxelon, M., 234(78)
Mbaye, A.A., 147(99)

McDowell, A.F., 263(130)
McFadden, G.B., 145(87)
McLelan, R.B., 210(42)
Mederios-Ribeiro, G., 8(10)
Merkle, K.L., 244(96)
Miceli, P.F., 205(13), 283
Michaels, A.S., 270(143)
Mirecki-Millunchick, J., 9(11), 9(12)
Mitzubayashi, H., 212(48), 212(49)
Miyazaki, T., 179(142)
Moody, N.R., 205(6), 205(7)
Mossinger, J., 263(131)
Moutinho, H., 9(11)
Mueller, R., 198(154), 198(155)
Mueller, W.H., 147(107)
Mullins, W.W., 12, 63(37), 123(67), 145(91)
Mura, T., 2(3)
Murakami, Y., 283(183)
Murayama, S., 212(49)
Murdoch, I., 100(52)
Mütschele, T., 243(88)
Mütschele, T., 238(81)
Myers, S.M., 206(29)

N

Nagengast, D., 206(18), 278(162)
Nagorny, U., 214(54)
Nakahigashi, J., 206(27)
Nakazato, K., 209(39)
Necker, G., 244(93), 244(94)
Neuberger, M., 161(121)
Nicholson, R.B., 179(141)
Nix, W.D., 78(42), 279(172)
Noh, H., 206(25)
Norden, H., 82(46)
Norman, A.G., 9(11)
Nowick, A.S., 212(45)
Noziéres, P., 62(34)
Nützenagel, Ch., 278(164)
Nye, J.F., 50(28)

O

Ohlberg, D.A.A., 8(10)
Okuda, S., 212(48)
Okuma, N., 206(21)
Onuki, A., 121(65)
Oriani, R.A., 214(55), 215(60)
Orimo, S., 215(59), 259(122)
Ouwerkerk, M., 280(176)

P

Palmer, D.G., 284(187)
Paul, D.R., 270(144)
Paulman, D., 277(156)
Pearton, S.J., 206(29)
Peinemann, K.-V., 266(132)
Peisl, H., 212(51)
Penka, V., 289(197)
Peterson, N.L., 278(157)
Petropoulous, J.H., 284(186)
Phillips. V.A., 180(143)
Pidduck, A.J., 7(7)
Piller, J., 82(46)
Pippel, E., 244(98)
Plaetschke, R., 267(136)
Pönitsch, M., 269(139), 270(145)
Porter, D.A, 94(50)
Pound, B.G, 244(92)
Pritchet, W.C., 212(46)
Pundt, A., 234(78), 245(100), 278(166), 278(167), 278(168), 278(169), 281(177), 287(192)
Pychho-Hintzen, W.,
Pyckhout-Hintzen, W., 234(78), 247(101)
Pyun, S., 283(182)

R

Rajkumar, K.C., 8(9)
Rector, J.H., 206(20), 280(175), 288(195)
Reetz, M.T., 278(169), 288(196)
Rehm, Ch., 206(18), 278(162)
Remhof, A., 278(161), 281(178)
Reno, J.L., 9(11)
Reetz, M.T., 278(168)
Rice, J.R., 252 (108)
Richards, P.M., 256(117)
Richter, D., 253(116)
Rissel, K., 186(146)
Robbins, A.J., 7(7)
Robertson, I.M., 215(61)
Rodriges, J.A., 232(76)
Rosenfeld, A.R., 62(32)
Ross, D.K., 206(17), 237(80)
Rottman, C., 110(56)
Roytburd, A.L., 147(95), 147(96), 147(97), 147(102), 147(103), 147(106), 155(116), 155(117)
Rühle, M., 244(93)
Ryden J., 278(158), 288(193)

S

Sachs, C., 278(168)
Sakamoto, Y., 250(106)
Salomons, E., 288(195)
Saruki, S., 259(122)
Satoh, S., 180(144)
Schlapbach, L., 204(5), 278(164)
Schmid, G., 278(164)
Schmidt, I., 193(152)
Schmitz, G., 206(26), 278(165), 278(167), 284, 287(191), 287(192)
Schober, T., 206(22)
Schulte, O., 283(181)
Schurian, A., 283(181)
Schwarz, R.B., 290(200)
Seidman, D.N., 244(94), 244(96)
Sekerka, R.F., 12(16), 111(58), 131, 145(91), 145(92), 174(132)
Senkov, O.N, 206(28)
Shaskolskaya, M.P., 176(135)
Shchukin, V.A., 6(6), 78(40)
Shewmon, P.G., 210(40)
Shih, D.S., 215(61)
Sirotin, Y.I, 176(135)
Sirovich, L., 62(33)
Slater, J., 270(142)
Slutsker, J., 147(106)
Smith, P.W., 7(7)
Sofronis, P., 252 (110)
Stolz, U., 233(77)
Song, G., 278(160), 278(161), 281(178), 283(182)
Spencer, B.J., 121(64), 127(72), 129(75), 129(76)
Spietling, A., 233(77)
Srolovitz, D.J., 121(63)
Stachurski, Z., 210(43)
Stavola, M.J., 206(29)
Stefanopoulus, K.L., 237(80)
Steins, M., 283(181)
Stevenson, D.A., 256(119)
Stolz, U., 212(47), 214(54), 227(69), 263(129)
Stoneham, A.M., 205(10)
Street, R.A., 273(146)
Stringfellow, G.B., 161(119)
Su, C.H., 5(5), 145(88)
Su, C.S., 193(150)
Sudan, P., 278(164)
Suh, D., 259(126)
Sulieman, M., 278(169), 278(170)

Summerfield, G.S., 237(79)
Sundgren, J.-E., 278(158), 288(193)
Sutton, A.P., 243(86)
Switendick, A.C., 209(37)
Szökefalvi-Nagy, A., 233(77), 253(115)

T

Takano, N., 283(183), 284
Tanimoto, H., 212(49)
Taylor, J.E., 177(138)
Tenhover, M.A., 258(121)
Terasaki, F., 283(183)
Tersoff, J., 129(75), 129(76)
Thiele, J., 283(181)
Thompson, A.W., 205(6), 205(7)
Thompson, L.J., 243(87)
Thompson, M.E., 146(93), 193(150)
Thornton, K., 145(89), 145(90)
Tien, J.K., 187(147)
Tiller, W.A., 121
Tomazawa, M., 275(153)
Tsai, C.L., 253(114)
Twesten, R.D., 9(11), 9(12)

U

Udovic, T.J., 278(163)
Uhlemann, M., 259(127)

V

Vaithyanathan, V., 198(156)
van der Molen, S.J., 280(175), 281(177)
van der Sluis, P., 280(176)
van Gogh, A.T.M., 280(173)
van Krevelen, D.W., 270(141)
Vargas, P., 229(71)
Vieth, W.R., 270(143)
Vinals, J., 145(91)
Völkl, J., 205(11), 206(30), 206(31)
Voorhees, P.W., 3(4), 5(5), 78(39), 78(41), 121(64), 124(69), 125, 127(72), 128(73), 129(75), 129(76), 142(81), 142(82), 145(87), 145(88), 145(89), 145(90), 147(104), 177(137), 193(150)

W

Wagner, C., 247(103), 253
Wagner, H., 273(148), 279(171)

Wagner, T., 278(166)
Wakatsuki, T., 179(142)
Wang, F., 245(100)
Wang, G., 78(42)
Wang, K.G., 219(65)
Watanabe, T., 278(163)
Wayman, C.M., 12(16), 174(132)
Weatherburn, 123(68)
Weatherly, G.C., 206(23)
Weidinger, A., 206(18), 278(162)
Weisheit, M., 278(167), 287(192)
Weismüller, J., 238(82)
Wenzl, H., 206(22)
Westerholt, K., 206(19)
Wheeler, J.W., 147(108)
Wichmann, T., 219(65)
Wicke, E., 231(75)
Wijnaarden, R.J., 206(20)
Williams, R.O., 147(100), 147(101)
Williams, R.S., 8(10)
Winter, M., 278(168)
Wipf, H., 207(33)
Wolff, J., 267(134)
Wollenberger, H., 287(191)
Woltersdorf, J., 244(98)
Wood, D.M., 147(99)
Wu, E., 214(58)

Y

Yamaguchi, M., 214(57)
Yamamoto, T., 214(57)
Yang, Q.M., 278(165)
Yoshimura, H., 206(27)
Young, D.J., 143(86)
Young, Q.M., 284
Yu, Y., 155(116)

Z

Zabel, H., 205(13), 206(19), 278(160), 278(161), 279, 281(178), 283
Zander, D., 259(125)
Zastrow, U., 273(147), 273(149)
Zhang, T., 259(124)
Zhang, Y., 9(12)
Zhu, J.Z., 198(156)
Züchner, H., 210(41), 210(44), 283(184), 284
Zunger, A., 147(99)
Züttel, A., 278(164)

Subject Index

A

Accretion, 114, 199
Actual state representation, 46–48
AFM (atomic force microscopy), 9
Ag/MgO interface, 244, 245
Aluminum-Ni system, 5–6
Amorphous alloys
  hydrogen in
    bulk metallic glasses, 259–262
    crystalline alloys, 252–253
    germanium, 273–274
    glassy polymers, 266–273
    metal/nonmetal glasses, 253–256
    modeling diffusion, 263–266
    oxidic glasses, 275–278
    silicon, 273–274
    transition metallic glasses, 257–259
Anisotropic system, 193–198
Arrhenius Law, 212
ATG instability, 121–122, 128
Atomic force microscopy (AFM), 9

B

Bifurcation breaking, 187–191
Binary substitution alloy
  crystal, 70–75
  no vacancies, 56–58
  vacancies, 58
Boundaries
  grain
    H-H interaction, 239–240
    nonocrystalline sample, 238–239
    perpendicular diffusion, 240–243
    solute segregation, 237–238
    transport determination, 241–242
    multilayer hydrogen systems, 284–285

C

Capillary properties, 94–95
Cartesian tensors, 34
Cauchy stress tensor, 44–45, 120
Chemical potential, 108–109
Common-tangent construction
  characterization, 152–155
  parallel plates, 155–157
Composition modulations, 9
Concentration gradient, 222–224
Configurational forces, 62
Copper-MgO interface, 244
Countour lines, 82–83
Crack tips, 251–252
Crystal-binary substitutional alloy, 14–19
Crystals
  accretion, 114
  BBC, 59
  disordered alloys, 252–253
  -fluid equilibrium
    characterization, 109–111
    conditions for, 109–111
    energy extremum, 112–120
    energy functional, 111–112
    interface condition, 128–132
    interface continuity, 201–202
    small-strain limit, 120–121
    solid-surface stability, 121–122
    surface diffusion, 122–128
    vacuum context, 122
  free energy densities, 26–29
  interface changes, 199
  single-phase systems
    diffusion potential, 70–75
    equilibrium, 67–69
    shape differentiation, 61–67
  substitutional binary, 70–75
  two-phase systems
    common-tangent, 152–157
    continuity, 200–201
    densities, 147–150
    displacement boundary, 157–162
    elastically anisotropic, 193–198
    elastic scaling, 176–179
    energy functional, 132–133
    external stress field, 187–191
    field diagrams, 150
    fields, 147–150
    Gibbs-Thompson, 141–146
    interface characterization, 131–132
    interfacial energies, 176–179

Crystals (*Continued*)
  isotropic ellipsoids, 186–187
  Neumann's Principle, 174–176
  parallel plates, 155–157
  particle morphology, 191
  particle shape, 173–174, 191
  phase diagrams, 150
  phase equilibria, 146–147
  phase rule, 150–152
  precipitate shape, 191–193
  shape transitions, 179–185
  small strain limit, 139–141
  symmetry, 174–176
  traction boundary, 163–173
Curie's principle, 187, 188

D

Deformation
  area changes, 41–44
  density on, 30–31
  elastic stress, 50–52
  lagrangian strain tensor, 34–38
  small strain tensor, 38–41
  volume changes, 41–44
Density, thermodynamic, 147–150
Density of site energies (DOSE)
  definition, 215–217
  delta functions, 263–264
  Gaussian functions, 264–266, 268–271
  glasses
    late/early transitions, 257–259
    metal/nonmetal, 253–256
  hydrogen/metal systems
    diffusivity, 220–227
    dislocation interaction, 229–237
    Henry's Law, 219–220
    H-H interaction, 227–228
    solubility, 218–219
    solute interaction, 228–229
    vacancy interaction, 228–229
  integration over, 217–218
Density of states (DOS), 216
Diffusion
  concentration gradient, 222–224
  concentration limit, 227
  general random walk, 224–227
  modeling, 263–266
  potential
    function, 106–107
    stress-dependence
      interstitial BCC solution, 84–85
      substitutional binary crystal, 70–75
  solid-vacuum surface, 122
  surface, 122–128
  tracer, 220–222
Dislocation interactions, 229–237
Distributions of saddle point energies (DOSPE)
  delta function, 263
  Gaussian function, 264, 265–266
Divergence theorem, 46

E

Eigenstrains
  compositional, 56–58
  definition, 52–53
  energy density, 60–61
  misfit, 53–55
  thermal, 55–56
  transformation, 53–55
Einstein summation convention, 34
Elastic stresses
  crystalline microstructure, 2–3
  deformation, 50–52
  eigenstrains
    compositional, 56–58
    energy density, 60–61
    misfit, 53–55
    thermal, 55–56
    transformation, 53–55
  film development, 6–7
  independent constants, 51
  spatial correlation, 3–4
Elastic work
  actual state representation, 46–48
  conditions, 45
  reference state representation, 48–50
Embrittlement
  concerns, 204–205
  context, 215
  crack tips, 251–252
  interaction, 251–252
Energy
  crystal-fluid equilibrium, 111–112
  defect formation, 247–251
  density, eigenstrains, 60–61
  DOSE (See Density of site energies (DOSE))
  extremum
    crystal-fluid system, 112–120
    two-phase crystal, 133–139
  functional, crystal-fluid system, 111–112
  functional, two-phase crystal, 132–133

SUBJECT INDEX

Gibbs free
  definition, 22
  derivatives, 130
  per-atom basis, 25–26
  interaction, 231
  interfacial, 176–179
  tetrakaidechedron, 183, 184
Equiaxed particles, 2–3
Equilibrium
  chemical, 104–105
  composition change, 75–78
  crystal-fluid
    characterization, 109–111
    conditions for, 19–21
    energy extremum, 112–120
    energy functional, 111–112
    interface condition, 128–132
    interface continuity, 201–202
    small-strain limit, 120–121
    surface diffusion, 122–128
    vacuum context, 122
  multi-phase, 10
  open-system elastic constants, 78–83
  single-phase systems, 67–69
  thermodynamic conditions, 101–106
  two-phase crystalline
    common-tangent, 152–157
    crystal symmetry, 174–176
    densities, 147–150
    diagrams, field/phase, 150
    displacement boundary, 157–162
    elastically anisotropic, 193–198
    elastic scaling, 176–179
    energy functional, 132–133
    field diagrams, 150
    fields, 147–150
    Gibbs-Thompson, 141–146
    interface characterization, 131–132
    interfacial energies, 176–179
    isotropic ellipsoids, 186–187
    Neumann's principle, 174–176
    parallel plates, 155–157
    particle morphology, 191
    particle shape, 173–174, 191
    phase diagrams, 150
    phase equilibria, 146–147
    phase rule, 150–152
    precipitate shape, 191–193
    shape transitions, 179–185
    small strain limit, 139–141
Eshelby's continuum approach, 271

Eulerian strain tensor, 37–38
External stress field, 187–191

F

Fermi Dirac (FD) statistics
  definition, 215–217
  hydrogen/metal systems
    diffusivity, 220–227
    Henry's Law, 219–220
    H-H interaction, 227–228
    solubility, 218–219
Fields
  densities and, 147–150
  diagrams, characterization, 150
Films
  composition modulations, 9
  morphological development, 6–7
  Nb, 279
  Si-Ge, 7–9
  thin, hydrogen in, 279–282
  Y-H system, 281
Fluids
  -crystal equilibrium
    characterization, 109–111
    conditions for, 19–21
    energy extremum, 112–120
    energy functional, 111–112
    interface condition, 128–132
    interface continuity, 201–202
    small-strain limit, 120–121
    solid-surface stability, 121–122
    surface diffusion, 122–128
    vacuum context, 122
  free energy densities, 23–26
  homogeneous, 13–14
  multi-phase, equilibrium, 10·
Free energy densities
  crystalline system, 26–29
  fluid system, 23–26
  Helmholtz, 90–94, 160–161
  interstitial solutions, 31–33
  pressure free crystal, 29–30
  thermodynamic potential, 94

G

General random walk, 224–227
Germanium alloys, 273–274
Germanium-Si films, 7–9
Gibbs-Duhem equation, 21–22

Gibbs equation for surface
  alternate formula, 100
  thermodynamic equilibrium, 112–113
  two-phase system, 111
Gibbs free energy
  definition, 22
  derivatives, 130
  per-atom basis, 25–26
Gibbs phase rule, 150–151
Gibbs-Thompson equation
  dimensionless form, 177–178
  phase at interface, 141–146
  solid immersed in fluid, 131
Gibbs variational approach, 62–63
Glasses
  bulk metallic, 259–262
  early transition, 257–259
  late transition, 257–259
  metal/nonmetal, 253–256
  oxidic, ions in, 275–278
Glassy polymers, 266–273
Gorsky effect, 211
Grain boundaries, 237–243

H

Helmholtz free energy
  composition space, 160–161
  crystal, 22
  density expression, 25
  as Legendre transform, 90–94
Helmholtz free energy density, 90–94
Henry's Law, 219–220, 227
HIF systems, 227–228
Homogeneous fluid, 13–14
Hydrogen
  in amorphous alloys
    bulk metallic glasses, 259–262
    crystalline alloys, 252–253
    germanium, 273–274
    glassy polymers, 266–273
    metal/nonmetal glasses, 253–256
    modeling diffusion, 263–266
    oxidic glasses, 275–278
    silicon, 273–274
    transition metallic glasses, 257–259
  characterization, 203–207
  in crystalline alloys, 252–253
  in defective metals
    defect formation energy, 247–251
    diffusivity, 220–227
    dislocation interaction, 229–237
    grain boundaries, 237–243
    Henry's Law, 219–220
    H-H interaction, 227–228
    history, 214–215
    metal/oxide boundaries, 243–247
    solubility, 218–219
    solute interaction, 228–229
    vacancy interaction, 228–229
  embrittlement
    concerns, 204–205
    context, 215
    crack tips, 251–252
    interaction, 251–252
  in glasses
    bulk metallic, 259–262
    early transition, 257–259
    late transition, 257–259
    metal/nonmetal, 253–256
  -metal systems
    diffusivity, 209–210
    electrochemical techniques, 210–211
    Gorsky effect, 211
    internal friction, 211–212
    mechanical spectroscopy, 211–212
    as models, 205–206
    partial molar volume, 212–214
    permeation, 211
    solubility, 207–209
    strain fields, 212–214
    stress fields, 212–214
  mobility, 204
  -Pd multilayers, 288–291
  in reduced dimension systems
    Nb/Pd multilayers, 283–288
    Pd-H clusters, 288–291
  properties, 279
  thin films, 279–282
Hydrostatically stressed crystals
  binary substitutional alloy, 14–19
  capillary properties
    applications, 106–109
    chemical potential, 108–109
    diffusion potential, 106–107
    equilibrium conditions, 101–106
    function, 94–95
    surface excess quantities, 95–101
  fluid equilibrium, 19–21, 33–34
  free energy densities
    crystalline system, 26–29
    deformation dependence, 30–31

fluid system, 23–26
  interstitial solutions, 31–33
  pressure free crystal, 29–30
homogeneous fluid, 13–14
interfacial properties
  applications, 106–109
  chemical potential, 108–109
  diffusion potential, 106–107
  equilibrium conditions, 101–106
  function, 94–95
  surface excess quantities, 95–101
thermodynamic relationships
  free energy densities, 23–31
  Gibbs-Duhem equation, 21–22
  interstitial solutions, 31–33

I

Instability, morphological, 122–128
Interfacial curvature, 127
Interfacial energies, 176–179
Interstital alloy, 58–60
Interstitial BBC solution
  characterization, 83–84
  compositional change, 87–90
  diffusion potentials, 84–85
  free energy densities, 90–94
  open-system constants, 85–87
  open-system elastic constants, 85–86
  tetragonal distortion, 59
Interstitial solutions, 31–33
Isolated system, 19–20
Isotropic ellipsoids, 186–187
Isotropic material, 51–52

K

Kronecker delta function, 37, 99

L

Lagrangian strain tensor, 34–38
Legendre transform, 90
Linearly stable site, 169

M

Magnesium oxide
  Ag/MgO interface, 245
  Cu/MgO interface, 244
Maxwell relations, 93–94
Mechanical equilibrium
  actual state representation, 46–48
  conditions, 45
  obtaining, 47
  reference state representation, 48–50
Mechanical spectroscopy, 211–212
Metal hydride batteries, 214
Metals
  defective, hydrogen in
    crack tips, 251–252
    defect formation energy, 247–251
    diffusivity, 220–227
    dislocation interaction, 229–237
    grain boundaries, 237–243
    Henry's Law, 219–220
    H-H interaction, 227–228
    history, 214–215
    metal/oxide boundaries, 243–247
    solubility, 218–219
    solute interaction, 228–229
    vacancy interaction, 228–229
  -hydrogen systems
    electrochemical techniques, 210–211
    Gorsky effect, 211
    internal friction, 211–212
    mechanical spectroscopy, 211–212
    as models, 205–206
    partial moral volume, 212–214
    permeation, 211
    solubility, 207–209
    strain fields, 212–214
    stress fields, 212–214
Misfit strain, 53–55
Monte-Carlo (MC) simulations, 263, 264–265
Multi-phase fluids, 10

N

Nanson's formula, 118–119
Nanson's relation, 45
Neumann's Principle, 174–176, 179
Nickel-Al system, 5–6
Niobium
  films, 279
  -Pd multilayers, 285–287
Nonhydrostatic stresses, 2

O

Open-system elastic constants
  characterization, 78–83
  interstitial BCC solution, 85–87

SUBJECT INDEX

P

Palladium
 -H multilayers, 288–291
 -Nb multilayers, 285–287
Parallel plates
 construction, 155–157
 displacement boundary, 157–162
Parcel, morphology i two-phase system, 191
Parcel, shape in two-phase system, 191
Phase
 diagrams, characterization, 150
 rule, characterization, 150–152
Piola-Kirchhoff stress tensor, 45, 95–96
Polymers, glassy, 266–273
Precipitates
 equilibrium shape, 2
 evolution, 2
 shape, two-phase system, 191–193
 solute redistribution, 82
 spherical, 82–83
Pressure-free crystal, 29–30

R

Reduced dimension systems
 Nb/Pd multilayers, 283–288
 properties, 279
 thin films, 279–282
Reference state representation, 48–50

S

Scaling of elastic stress, 176–179
Secondary Ion Mass Spectrometry (SIMS), 273
Self-consistent thermodynamic framework, 9
Shape transitions, symmetry-breaking, 183–185
Shape transitions, symmetry-conserving, 179–182
Silicon
 alloys, 273
 -Ge films, 7–9
Silver-MgO interface, 244
Single-phase systems
 diffusion potential, 70–75
 equilibrium, 67–69
 shape differentiation, 61–67
Small-angle neutron scattering (SANS), 234–235
Small molecules, 266–273
Small-strain limit

constitutive equations, 50–52
crystal-fluid system, 120–121
two-phase system, 139–141
Small-strain tensor, 38–41
Solids
 assumptions, 10–11
 -vacuum surface, 122
Solute redistribution
 hydrogen/defective metals, 228–229
 stress-induced
  composition change, 75–78
  interstitial BCC solution, 83–90
  open-system constants, 78–83
Spectroscopy, mechanical, 211–212
Strain
 fields, hydrogen-metal systems, 212–214
 reference state, 52–53
Stranski-Kranstanow growth, 7
Stresses
 Cauchy, 44–45
 diffusion potential dependence
  interstitial BCC solution, 84–85
  substitutional binary crystal, 70–75
 dislocations, 228–230
 external field, 187–191
 hydrogen-metal systems, 212–214
 Piola-Kirchhoff, 45
 solute redistribution
  composition change, 75–78
  interstitial BCC solution, 83–90
  open-system constants, 78–83
Symmetry-breaking shape transitions, 183–185
Symmetry-conserving shape transitions, 179–182

T

Tensors
 Cartesian, 34
 Cauchy stress, 44–45
 Lagrangian strain, 34–38
 Piola-Kirchhoff stress, 45
 small strain, 38–41
 transformation, 54–55
Tetrakaidechedron, 183
Thermal strain, 55–56
Thermodynamics
 equilibrium, multi-phase fluids, 10–11
 equilibrium conditions, 101–106
 framework, self-consistent, 9

free energy, 94
single-phase system
  diffusion potential, 70–75
  equilibrium, 67–69
  shape differentiation, 61–67
two-phase system
  common-tangent construction, 152–155
  densities, 147–150
  diagrams, 147–150
  displacement boundary conditions, 157–162
  field diagrams, 150
  fields, 147–150
  parallel plates, 155–157
  phase diagrams, 150
  phase rule, 150–152
  traction boundary, 163–173
Tracer diffusion, 220–222
Traction boundary conditions
  description, 163–165
  free energy, 170
  misfit strain, 167–169
  phase diagram, 165–167
  phase equilibria, 168–170
  temperature-composition, 170–173
Transformation strain, 54–55
Two-phase systems
  common-tangent, 152–157
  continuity, 200–201
  densities, 147–150
  displacement boundary, 157–162

elastically anisotropic, 193–198
elastic scaling, 176–179
energy functional, 132–133
external stress field, 187–191
field diagrams, 150
fields, 147–150
Gibbs-Thompson, 141–146
interface characterization, 131–132
interfacial energies, 176–179
isotropic ellipsoids, 186–187
Neumann's Principle, 174–176
parallel plates, 155–157
particle morphology, 191
particle shape, 173–174, 191
phase diagrams, 150
phase equilibria, 146–147
phase rule, 150–152
precipitate shape, 191–193
shape transitions, 179–185
small strain limit, 139–141
symmetry, 174–176
traction boundary, 163–173

V

Vegard's law, 57

Y

Y-H films, 281
Young's modulus, 51